Gmelin Handbook of Inorganic and Organometallic Chemistry

8th Edition

Gmelin Handbook of Inorganic and Organometallic Chemistry

8th Edition

Gmelin Handbuch der Anorganischen Chemie

Achte, völlig neu bearbeitete Auflage

PREPARED AND ISSUED BY

Gmelin-Institut für Anorganische Chemie
der Max-Planck-Gesellschaft
zur Förderung der Wissenschaften

Director: Ekkehard Fluck

FOUNDED BY Leopold Gmelin

8TH EDITION 8th Edition begun under the auspices of the Deutsche Chemische Gesellschaft by R. J. Meyer

CONTINUED BY E. H. E. Pietsch and A. Kotowski, and by Margot Becke-Goehring

Springer-Verlag
Berlin · Heidelberg · New York · London · Paris · Tokyo
1990

Gmelin-Institut für Anorganische Chemie
der Max-Planck-Gesellschaft zur Förderung der Wissenschaften

Volumes published on "Manganese" (Syst.-No. 56)

Manganese A 1 (in German)
History – 1980

Manganese B (in German)
The Element – 1973

Manganese C 1 (in German)
Compounds (Hydrides. Oxides. Oxide Hydrates. Hydroxides) – 1973

Manganese C 2 (in German)
Compounds (Oxomanganese Ions. Permanganic Acid. Compounds and Phases with Metals of the Main and Subgroups I and II) – 1975

Manganese C 3 (in German)
Compounds of Manganese with Oxygen and Metals of the Main and Subgroups III to VI. Compounds of Manganese with Nitrogen – 1975

Manganese C 4 (in German)
Compounds of Manganese with Fluorine – 1977

Manganese C 5 (in German)
Compounds of Manganese with Chlorine, Bromine, and Iodine – 1978

Manganese C 6 (in German)
Compounds of Manganese with Sulfur, Selenium, Tellurium, Polonium – 1976

Manganese C 7
Compounds of Manganese with Boron and Carbon – 1981

Manganese C 8
Compounds of Manganese with Silicon – 1982

Manganese C 9
Compounds with Phosphorus, Arsenic, Antimony – 1983

Manganese C 10
Electronic Spectra of Manganese Halides. Cumulative Substance Index of C 1 to C 10 – 1983

Manganese D 1 (in German)
Coordination Compounds 1 – 1979

Manganese D 2 (in German)
Coordination Compounds 2 – 1980

Manganese D 3
Coordination Compounds 3 – 1982

Manganese D 4
Coordination Compounds 4 – 1985

Manganese D 5
Coordination Compounds 5 – 1987

Manganese D 6
Coordination Compounds 6 – 1988

Manganese D 7
Coordination Compounds 7 – 1990

Manganese D 8
Coordination Compounds 8 – 1990 **(present volume)**

Gmelin Handbook of Inorganic and Organometallic Chemistry

8th Edition

Mn
Manganese

D 8

Coordination Compounds 8

With 41 illustrations

AUTHORS Helga Demmer, Karin Greiner, Karl Koeber, Mirjana Kotowski, Klaus-Dieter Scherfise, Edith Schleitzer-Rust, Dieter Tille

FORMULA INDEX Ursula Hettwer

EDITORS Helga Demmer, Mirjana Kotowski, Edith Schleitzer-Rust, Dieter Tille

CHIEF EDITOR Edith Schleitzer-Rust

System Number 56

Springer-Verlag
Berlin · Heidelberg · New York · London · Paris · Tokyo
1990

LITERATURE CLOSING DATE: 1988
IN SOME CASES MORE RECENT DATA HAVE BEEN CONSIDERED

Library of Congress Catalog Card Number: Agr 25-1383

ISBN 3-540-93618-1 Springer-Verlag, Berlin · Heidelberg · New York · London · Paris · Tokyo
ISBN 0-387-93618-1 Springer-Verlag, New York · Heidelberg · Berlin · London · Paris · Tokyo

Printed in Germany

Typesetting, printing, and bookbinding: LN-Druck Lübeck

Preface

The present volume "Manganese" D 8, completes the description of manganese complexes. The introduction on p. 1 shows the classes of complexes that have already been described in Chapters 1 through 36 in Volumes D 1 to D 7. Complexes with ligands containing boron, silicon, phosphorus, arsenic, antimony, and tin are described in Chapters 37 to 41 of this volume.

Silylamide ligands have been used to stabilize low coordination numbers. The $Mn^{II}L_2$ complex formed with bis(trimethylsilyl)amine (HL) is dimeric in the solid state and monomeric in the gas phase and organic solvents. The corresponding $Mn^{III}L_3$ complex is the first example of a well characterized, three-coordinate manganese(III) complex. The triorganylboranamine complex reported on p. 5 also exhibits a low coordination number. Due to steric requirements and a weak tendency toward bridging, the ligand forms a two-coordinate complex with manganese.

Complexes with phosphorus-containing ligands make up the majority of the present volume. The characteristic features of complexes with monophosphanes (L), predominantly with trialkyl-, triphenyl-, or mixed alkylphenylphosphanes, are summarized on p. 36. The $MnLX_2$ complexes, which represent the most important group, are doubly halide-bridged oligomeric or polymeric compounds. These complexes and their THF solvates are able to bind small molecules, especially O_2, NO, SO_2, CO, or CS_2.

Manganese(II) bis(phosphinates) are interesting because of their polymeric structure. A number of single-crystal X-ray studies show different coordination modes and polymer configurations of the complexes. The conformations of the polymers are determined by the size and the nature of R in the $R_2PO_2^-$ ligands, see Chapter 39.7.

Increasing attention has been paid to the study of the complex-forming properties of phosphonic acids and their esters as well as esters of phosphoric acid. They are used as extractants for Mn^{2+} ions from aqueous solutions. Alkylenepolyaminepoly(phosphonic acids) have been shown to have an enhanced coordination capacity compared to their carboxylic acid analogs. An increased stability of the complexes with organophosphorus complexones has been observed.

A formula index at the end of this volume listing the empirical formulas and the linearized structural formulas of the ligands is intended to expedite locating specific compounds.

Frankfurt/Main
October 1990

Edith Schleitzer-Rust

Table of Contents

Page

Coordination Compounds of Manganese

(Continued)

Introduction

Arrangement. In Series D, coordination compounds of manganese, with the exception of the organometallic compounds, are described. The volumes "Mangan" D 1, D 2 and "Manganese" D 3 to D 7 contain the following chapters:

"Mangan" D 1, 1979

1) Review
2) Complexes with H_2O
3) Complexes with Alcohols
4) Complexes and Salts with Phenols and Other Aromatic Hydroxy Compounds
5) Complexes with Aldehydes
6) Complexes with Ketones
7) Complexes with Quinones
8) Complexes with Ethers and O-Heterocycles

"Mangan" D 2, 1980

9) Complexes and Salts of Carboxylic Acids and Their Derivatives
10) Cyanomanganate Complexes
11) Cyanato, Thiocyanato, and Selenocyanato Complexes

"Manganese" D 3, 1982

12) Complexes with Ammonia
13) Complexes with Amines
14) Complexes with Hydrazine and Its Derivatives
15) Complexes with Hydroxylamine
16) Complexes with N-Heterocycles

"Manganese" D 4, 1985

16) Complexes with N-Heterocycles (Continued)
17) Complexes with Aminoalcohols, -phenols, and -naphthols
18) Complexes with Aminoethers and Aminooxo Compounds
19) Complexes with Amino Acids
20) Complexes with Peptides
21) Complexes with Proteins

"Manganese" D 5, 1987

22) Complexes with Amine-N-polycarboxylic Acids
23) Complexes with Hydrazinecarboxylic Acid and Derivatives
24) Complexes with Amides and Related Compounds
25) Complexes with Hydrazides
26) Complexes with Derivatives of Hydroxylamine
27) Complexes with Oximes and Nitroso Compounds
28) Complexes with Azo Compounds
29) Complexes with Triazenes

"Manganese" D 6, 1988

30) Complexes with Schiff Bases
31) Complexes with Hydrazones or Related Compounds
32) Complexes with Carbazones, Thiocarbazones, and Formazans

"Manganese" D 7, 1990

33) Complexes with Nitriles or Related Compounds
34) Complexes with Nitro Hydrocarbons
35) Complexes with Ligands Containing Sulfur
36) Complexes with Ligands Containing Selenium or Tellurium

The present volume deals with manganese complexes with ligands containing boron (Chapter 37), silicon (Chapter 38), phosphorus (Chapter 39), arsenic (Chapter 40), antimony and tin (Chapter 41).

Rules and Definitions. Generally, the names of the ligands correspond to IUPAC nomenclature; trivial names are also used.

The stepwise stability (formation) constants (K_n) for the formation of the complexes in solution from a central atom (M) and ligands (L) and the cumulative constants (β_n) are defined as follows:

$K_n = [ML_n]/[MnL_{n-1}]\cdot[L]$ in L/mol for the equilibria $ML_{n-1} + L \rightleftharpoons ML_n$ (n = 1, 2, 3, ...)

$\beta_n = [ML_n]/[M]\cdot[L]^n$ in L^n/mol^n for the equilibria $M + nL \rightleftharpoons ML_n$ (n = 1, 2, 3, ...)

The formation of complexes with protonated ligands is described by:

$K^M_{MH_pL} = [MH_pL]/[M]\cdot[H_pL]$ for the equilibria $M + H_pL \rightleftharpoons MH_pL$ (p = 1, 2, 3, ...)

Enthalpy (ΔH), Gibbs free energy (ΔG), or entropy changes (ΔS) are given the same subscript as the corresponding K: e.g., ΔH_1 for constant K_1. For reactions represented by cumulative constants (β_n), the notation $\Delta H_{\beta n}$ is used. Ionic strengths are given in mol/L.

With respect to magnetic properties, the conventions of Carlin, R. L., Magnetochemistry, Berlin – Heidelberg – New York 1986 are followed. Magnetic measurements were generally performed by the Gouy method. If another technique was used, that method is reported.

Definitions of ESR or NMR parameters: A is the isotropic hyperfine coupling constant, D the isotropic zero-field splitting parameter, and g the isotropic g-factors.

Abbreviations and Dimensions. Temperatures are normally given in °C; K stands for Kelvin. Abbreviations used with temperatures are m.p. for melting point and dec. for decomposition. With thermodynamic data, (s) is used to label solids, (l) is used for liquids, and (g) is used to designate the gaseous state.

The vibrational spectra are labeled as IR (infrared) or R (Raman). The symbol ν is used for stretching vibrations and δ for deformation vibrations; wavenumbers are given in cm^{-1}. The intensities are placed in parentheses (w = weak, m = medium, s = strong, vs = very strong, etc.); sh means shoulder; br means broad. The UV-visible absorption maxima of the electronic spectra are given in nm (λ_{max}) or cm^{-1} (ν_{max}), the extinction coefficient ε is given in $L\cdot mol^{-1}\cdot cm^{-1}$.

Abbreviations for ligands are listed on p. 221, the first page of the ligand formula index.

37 Complexes with Ligands Containing Boron

Remark. In this chapter complexes with organic ligands containing boron are described. For reasons of congruity some of these compounds have already been treated in the Gmelin volumes of the boron series together with other metal complexes. Therefore, information is given at the beginning of each chapter with references to corresponding sections of the boron series and additional, more recent data for the manganese complexes are added.

Tetrahydroborate complexes of manganese are reported in the volume "Borverbindungen" 8, 1976, p. 37, and "Manganese" C 7, 1981, pp. 61 and 92. Complexes of the $[MnL_n]X_2$ or $[MnL_n]X_3$ type with $X = BF_4^-$ or $B(C_6H_5)_4^-$ anions are described in the "Manganese" D series in the sections of the special ligand (L) type.

Metal complexes with carborane anions (metallocarboranes) containing metal–carbon bonds will be described in the series of organometallic compounds. Manganese carborane complexes containing additional carbonyl groups have been reported together with carborane complexes of other transition metals in "Borverbindungen" 6, 1980, pp. 63/5, and "Boron Compounds" 3rd Suppl. Vol. 4, 1988, p. 184.

37.1 Complexes with Hydrogen Alkanediolatoborates

Coordination compounds of hydrogen alkanediolatoborates have been described in "Borverbindungen" 8, 1976, pp. 118/20. In connection with compounds of other metals the preparation and properties of a manganese complex with hydrogen bis[2,2-bis(hydroxymethyl)-1,3-propanediolato(4-)-O^1,O^3]borate(1-), $H_n[B(OCH_2)_2C(CH_2O)_2]_n$, of composition $[Mn(C_5H_8BO_4)_2 \cdot NH_3 \cdot 9H_2O]_n$ were reported. For more recent data on this complex, see [1 to 4].

References:

[1] Shvarts, E. M.; Belousova, R. G.; Ievin'sh, A. F. (Latvijas PSR Zinatnu Akad. Vestis Kim. Ser. **1974** No. 2, pp. 149/55; C.A. **81** [1974] No. 32735).

[2] Shvarts, E. M.; Balode, M. M.; Ievin'sh, A. F. (Latvijas PSR Zinatnu Akad. Vestis Kim. Ser. **1975** No. 2, pp. 131/5; C.A. **83** [1975] No. 87561).

[3] Shvarts, E. M.; Bambergs, K.; Belousova, R. G.; Ievin'sh, A. F. (Ref. Dokl. Soobshch. 11th Mendeleevsk. S'ezd Obshch. Prikl. Khim., Alma-Ata 1975, Vol. 7, pp. 64/5; C.A. **88** [1978] No. 189069).

[4] Belousova, R. G.; Shvarts, E.; Ievin'sh, A. F. (Latvijas PSR Zinatnu Akad. Vestis Kim. Ser. **1976** No. 5, pp. 507/13; C.A. **86** [1977] No. 25394).

37.2 With Carboxylato- and Hydroxy Carboxylatoborates

A number of hydrogen hydroxy carboxylatoborates have already been described in "Borverbindungen" 8, 1976, and "Boron Compounds" 3rd Suppl. Vol. 2, 1987. These ligands with additional references are listed in the table:

name of hydrogen hydroxy carboxylatoborates	formula	boron series	Ref.
bis(malato-O^1,O^2)borate(1-)	$H[B(C_4H_4O_5)_2]$ ($=C_8H_9BO_{10}$)	"Boron Compounds" 3rd Suppl. Vol. 2, 1987, p. 119	—
dihydroxo(tartrato(4-)-O^2,O^3)borate(3-)	$H_3[B(OH)_2(C_4H_2O_6)]$ ($=C_4H_7BO_8$)	"Borverbindungen" 8, 1976, pp. 130/2	—

name of hydrogen hydroxy carboxylatoborates	formula	boron series	Ref.
bis(tartrato(4-)-O^2,O^3)borate(5-)	$H_5[B(C_4H_2O_6)_2]$ (= $C_8H_9BO_{12}$)	"Borverbindungen" 8, 1976, pp. 130/2	—
dihydroxo(2,3,4-trihydroxy-glutarato(3-)-O^1,O^2)borate(2-)	$H_2[B(OH)_2(C_5H_5O_7)]$ (= $C_5H_9BO_9$)	—	[1]
bis(citrato(2-)-O^2,O^3)borate(1-)	$H[B(C_6H_6O_7)_2]$ (= $C_{12}H_{13}BO_{14}$)	"Boron Compounds" 3rd Suppl. Vol. 2, 1987, p. 119	[2]
bis(mandelato(2-)-O^1,O^2)borate(1-)	$H[B(C_8H_6O_3)_2]$ (= $C_{16}H_{13}BO_6$)	—	[3]
bis(salicylato(2-)-O^1,O^2)borate(1-)	$H[B(C_7H_4O_3)_2]$ (= $C_{14}H_9BO_6$)	"Borverbindungen" 8, 1976, p. 138	—
bis(4-aminosalicylato(2-)-O^1,O^2)borate(1-)	$H[B(C_7H_5NO_3)_2]$ (= $C_{14}H_{11}BN_2O_6$)	"Boron Compounds" 3rd Suppl. Vol. 2, 1987, p. 110	—

The pentaacetato-μ-oxodiborate complex, **$Mn^{II}[B_2O(CH_3COO)_5]_2$**, was prepared by refluxing anhydrous $Mn(CH_3COO)_2$ with stoichiometric amounts of $B_2O(CH_3COO)_4$ in a mixture of equal volumes of acetic anhydride and anhydrous acetic acid for 10 min. The crystals obtained on cooling were washed with acetic anhydride and anhydrous ether, excluding all moisture, and dried in a vacuum desiccator. The colorless orthorhombic needles grow in characteristic star-shaped crystal forms. The complex can be recrystallized from acetic acid and is stable in a desiccator for several days, but releases acetic anhydride on prolonged storing [4].

The X-ray powder diagram is almost identical with those of the related complexes of Mg, Cr^{II}, Fe^{II}, Co^{II}, Ni^{II}, Cu^{II}, and Zn. Susceptibility measurements (Gouy balance) at 293 K and different field strengths yield the molar susceptibility, $\chi_{mol} = 14116.9 \times 10^{-6}$ cm^3/mol (corrected for diamagnetism), and the magnetic moment, $\mu_{eff} = 5.78$ μ_B. The IR spectrum recorded as Nujol or Tripen mulls between KBr disks in the 1800 to 1500 cm^{-1} region shows characteristic absorption bands at 1720, 1671, 1599, 1565, and 1517 cm^{-1}. The strong band at 1720 cm^{-1} is assignable to the $\nu_{as}(CO)$ vibrations of free carbonyl groups whereas the other bands may correspond to $\nu_{as}(CO)$ and $\nu_s(CO)$ vibrations of associated carbonyl groups, or vibration modes of acetato moieties in variable bonding. The IR spectrum may be interpreted by polymeric ring- or chain-shaped anions consisting of pentaacetatodiborate units, $[B_2O(CH_3COO)_5]_n^{n-}$, with the B–O–B groups connected by acetate moieties. These polymeric anions may form structures with cavities enabling octahedral coordination of suitably-sized cations such as Mn^{2+}. A monomeric pentaacetatoborate anion with an intramolecular ring including boron and one of the five acetate groups was also discussed. Thermal analysis (DTG, TG, heating rate 2.5 to 3°C/min) reveals decomposition of $Mn[B_2O(CH_3COO)_5]_2$ in four steps with the release of $(CH_3CO)_2O$ at 95, 125, 185, and 200°C to give $Mn[B_2O_{1.5}(CH_3COO)_4]_2$, $Mn[B_2O_2(CH_3COO)_3]_2$, $Mn[B_2O_{2.5}(CH_3COO)_2]_2$, and $Mn(CH_3COO)_2 + 2B_2O_3$, respectively. Further heating yields $MnO \cdot 2B_2O_3$, $(CH_3CO)_2O$, and its decomposition products such as ketones.

$Mn[B_2O(CH_3COO)_5]_2$ dissolves in acetic acid, is almost insoluble in diethyl ether, petroleum ether, and cold $(CH_3CO)_2O$, and is slightly soluble in boiling $(CH_3CO)_2O$. When heated in other anhydrous organic solvents, such as tetrahydrofuran, dioxane, benzene, $CHCl_3$, and CH_2Cl_2, it dissolves with decomposition. The complex is soluble in water, methanol, and ethanol, undergoing complete solvolysis [4].

References:

[1] Shvarts, E. M.; Ievin'sh, A. F. (U.S.S.R. 170055 [1964/65]; C.A. **63** [1965] 6640).
[2] Shvarts, E. M.; Vitol, I. M.; Sergeeva, G. S.; Piloyan, G. O.; Drozdova, O. V. (J. Therm. Anal. **31** [1986] 351/9).
[3] Berniyazova, D. G.; Kalacheva, V. G.; Shvarts, E. M.; Vitol, I. M.; Skorikov, S. N.; Leonov, I. D. (Zh. Neorgan. Khim. **32** [1987] 1543/6; Russ. J. Inorg. Chem. **32** [1987] 924/6).
[4] Kibbel, H. U. (Z. Anorg. Allgem. Chem. **359** [1968] 272/83; Z. Chem. [Leipzig] **4** [1964] 104/5).

37.3 Complex with a Triorganylboranamine

$(= C_{27}H_{34}BN = HL)$

$Mn^{II}(C_{27}H_{33}BN)_2 \cdot 3C_7H_8$ (C_7H_8 = toluene). The colorless complex was prepared by dropwise addition of two equivalents of $Li(C_{27}H_{33}BN) \cdot 2(C_2H_5)_2O$ (obtained "in situ" from *tert*-butyllithium and the ligand in a mixture of diethyl ether and tetrahydrofuran) to the tetrahydrofuran suspension of $MnBr_2 \cdot 2thf$ at 0°C. The orange solution was stirred at room temperature for 16 h, the volatiles were removed under reduced pressure, and the residue was dissolved in toluene. After filtration, the solution was concentrated and cooled at −20°C, then allowed to crystallize. For the single-crystal X-ray structure determination at 130 K the complex was coated with a layer of hydrocarbon oil upon removal from the Schlenk tube. A suitable crystal was selected, attached to a glass fiber using silicon grease, and immediately placed in the low-temperature nitrogen stream. The complex crystallizes in the monoclinic system, space group $C2/c-C^6_{2h}$ (No. 15) with a = 21.964(6), b = 13.048(4), c = 21.675 Å, β = 93.49(2)°; Z = 4. The structure was solved to R = 0.077. Selected interatomic distances and angles are tabulated below:

distances	in Å	angles	in °	angles	in °
Mn–N	2.046(4)	N–Mn–N′	160.4(2)	N–B–C(1)	110.5(5)
N–B	1.402(7)	Mn–N–B	104.6(3)	N–B–C(10)	127.5(5)
B–C(1)	1.618(8)	Mn–N–C(19)	128.2(3)	C(1)–B–C(10)	122.0(5)
B–C(10)	1.603(8)	C(1)–Mn–C(1′)	151.9(3)		
N–C(19)	1.428(7)				
Mn···C(1)	2.536(5)				

The molecular structure of the complex is shown in **Fig. 1**, p. 6 (description of the disorder of the toluene of solvation is not given in the paper). It is a rare example of a two-coordinate transition metal complex, due to the low bridging tendency and steric versatility of the ligand. The nitrogen centers are terminal and not bridging. A crystallographic twofold axis passes through manganese and bisects the line connecting N and N'. However, the geometry is not linear, but is distorted to an angle of 160.4°. This is presumably due to an interaction between manganese and the C(1) atom of one of the boron mesityl groups. Further evidence for this interaction comes from the large asymmetry (17°) in the N–B–C(1) and N–B–C(10) angles and the fact that the MnC(1) vector deviates only 12.5° from the normal to the C(1) ring plane. The distortion is probably due to the extreme electron deficiency of the manganese center caused by the low coordination number and the poor π-donor ability of the ligand. The Mn–N bond length of 2.046 Å is slightly longer than the terminal bonds in the three-coordinate trimethylsilylamide complexes, which are bridged products in the solid (see pp. 27 and 29). This longer distance is further evidence for the strong interaction with the mesityl ring and the complex can be regarded as possessing pseudo-four-coordination. The structure reveals that all the nitrogen and boron centers are planar with a twist angle of only 7.5° between B and N planes. The B–N bond distance of 1.402 Å is very close to that of the free ligand (1.407 Å), indicating that the multiplicity of the B–N bonds is unchanged in the complex. As a result, the ligand is expected to have little σ-bridging character.

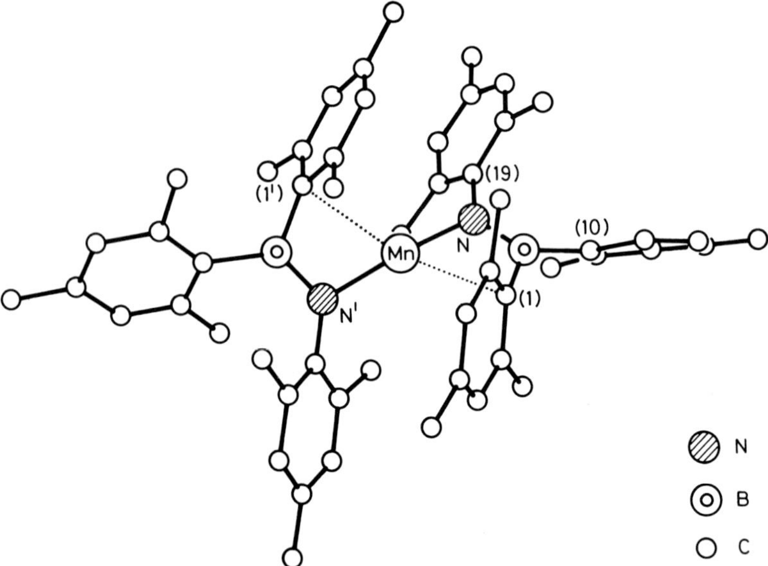

Fig. 1. Molecular structure of $Mn^{II}(C_{27}H_{33}BN)_2$ (hydrogen atoms are omitted).

Reference:

Bartlett, R. A.; Feng, X.; Olmstead, M. M.; Power, P. P.; Weese, K. J. (J. Am. Chem. Soc. **109** [1987] 4851/4).

37.4 Complexes with Borates $[H_{4-n}BR_n]^-$ (R = N-Heterocycle)

Survey. During the last decade, syntheses of a novel class of ligands, derived from tetrahydroborate anions by substituting nitrogen-containing moieties for the hydrogen atoms, have been successfully achieved. Such borate anions, $[H_{4-n}BR_n]^-$, where n = 2, 3, or 4, and R = N-heterocyclic rings, act as chelating agents for the transition metals. Generally, $Mn^{II}L_2$ and $[Mn^{II}LCl]$ complexes are formed by the reaction of a manganese(II) salt with the potassium salt of the ligand, KL, in aqueous or organic solution. The type of coordination to the metal depends on steric and electronic effects, determined primarily by the number of heterocyclic rings attached to boron. The (pyrazolato)borates were the first and most studied complexes. With this class of ligands manganese(III), manganese(IV), and mixed-valence Mn^{III}, Mn^{IV} complexes have also been described in addition to the manganese(II) compounds. In the binuclear manganese(III) complexes with the hydrotris(1H-pyrazolato-N[1])borate ligand, $[Mn^{III}_2O(RCOO)_2(C_9H_{10}BN_6)_2]$ where R = H, CH_3, and C_2H_5, the Mn^{3+} ions are bridged by one μ-oxo and two μ-carboxylato groups. Two acetonitrile solvates of the acetato complex, $[Mn^{III}_2O(CH_3COO)_2(C_9H_{10}BN_6)_2]\cdot 4CH_3CN$ and $[Mn^{III}_2O(CH_3COO)_2(C_9H_{10}BN_6)_2]\cdot CH_3CN$ with different structures have been described.

37.4.1 With (2,5-Pyrrolidinedionato-N[1])- or (1H-Isoindole-1,3(2H)-dionato-N[2])borates

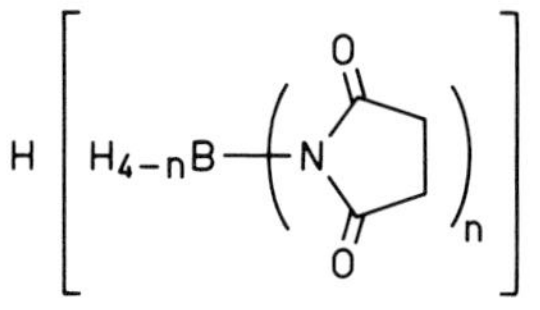

ligands 1 to 3 (=HL)

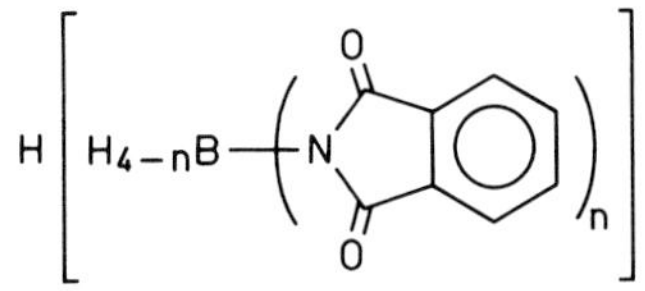

ligands 4 to 6 (=HL)

ligand	n	formula
1	2	$C_8H_{11}BN_2O_4$
2	3	$C_{12}H_{14}BN_3O_6$
3	4	$C_{16}H_{17}BN_4O_8$
4	2	$C_{16}H_{11}BN_2O_4$
5	3	$C_{24}H_{14}BN_3O_6$
6	4	$C_{32}H_{17}BN_4O_8$

$Mn^{II}L_2$. The complexes with ligand 2, 3 [1], 5, or 6 [2] were prepared by adding a hot aqueous solution of $MnCl_2$ to a hot aqueous solution of the potassium salt of the ligand (1:2 mole ratio). The precipitate that appeared after heating the reaction mixture on a water bath for several hours was filtered, washed repeatedly with water and ethanol, and finally dried in vacuum. Color, magnetic moments at room temperature, and characteristic IR bands of the complexes in KBr or Nujol (KL bands in parentheses) are listed below:

HL No.	complex	color	μ_{eff} in μ_B	IR data in cm^{-1} ν(CO)	ν(Mn–O)	Ref.
2	$Mn(C_{12}H_{13}BN_3O_6)_2$	black	5.23	1680, 1660 (1760, 1675)	330	[1]
3	$Mn(C_{16}H_{16}BN_4O_8)_2$	black	4.69	1640, 1590 (1760, 1685)	370	[1]
5	$Mn(C_{24}H_{13}BN_3O_6)_2$	yellow	5.86	1780, 1685 (1780, 1740)	365	[2]
6	$Mn(C_{32}H_{16}BN_4O_8)_2$	yellow	4.86	1730, 1680 (1770, 1750)	375	[2]

The magnetic moments are considerably less than expected for an octahedral high-spin Mn^{II} ion, indicating the presence of antiferromagnetic interactions [1, 2]. A tetrahedral geometry was assumed for the complex with ligand 5, $Mn(C_{24}H_{13}BN_3O_6)_2$ [2]. The IR spectra of the complexes show a considerable negative shift of the ligand ν(CO) stretching frequency, indicating that the coordination of ligand to metal occurs through the oxygen atoms of the carbonyl groups. This observation is also supported by the presence of a weak band in the 375 to 330 cm^{-1} region, assignable to the ν(Mn–O) vibration mode. There is no significant change in the ν(BH), ν(CN), and ν(BN) stretching frequencies of the ligand, which were observed in the ranges 2355 to 2320, 1455 to 1420, and 1380 to 1340 cm^{-1}, respectively, excluding the possibility of coordination through the nitrogen atoms [1, 2].

Bands observed in the solid state diffuse reflectance spectra are tabulated below:

HL No.	complex	electronic spectra, transitions from $^6A_{1g}$ in cm^{-1} $\rightarrow ^4T_{2g}(D)$	$\rightarrow ^4A_{1g}(G)$	$\rightarrow ^4T_{2g}(G)$	$\rightarrow ^4T_{1g}(G)$	Ref.
2	$Mn(C_{12}H_{13}BN_3O_6)_2$	—	—	22700	19200	[1]
3	$Mn(C_{16}H_{16}BN_4O_8)_2$	27700	25300	21700	—	[1]
5	$Mn(C_{24}H_{13}BN_3O_6)_2$	—	24635	21739	—	[2]
6	$Mn(C_{32}H_{16}BN_4O_8)_2$	—	—	22727	18867	[2]

The electronic spectra of the complexes with ligands 2, 3, and 6 are consistent with an octahedral geometry around the manganese atom. Ligand field parameters 10 Dq = 9672 cm^{-1}, B = 806 cm^{-1}, and $\beta = 0.83$ were calculated for $Mn(C_{12}H_{13}BN_3O_6)_2$; 10 Dq = 9252 cm^{-1}, B = 771 cm^{-1}, $\beta = 0.80$ for $Mn(C_{16}H_{16}BN_4O_8)_2$ [1]. These complexes also exhibit charge-transfer bands at 40000 and 41600 cm^{-1}, respectively [1]. An unassignable band at 11627 cm^{-1} in the electronic spectrum of $Mn(C_{24}H_{13}BN_3O_6)_2$ and $Mn(C_{32}H_{16}BN_4O_8)_2$ can probably be attributed to the ligand transitions [2].

$Mn(C_{24}H_{13}BN_3O_6)_2$ and $Mn(C_{32}H_{16}BN_4O_8)_2$ are stable toward air and moisture [2]. The high decomposition temperature of the $Mn^{II}L_2$ compounds in the 335 to 360°C range suggests a polymeric nature [1, 2]. The complexes $Mn(C_{12}H_{13}BN_3O_6)_2$ and $Mn(C_{16}H_{16}BN_4O_8)_2$ are insoluble in the common aprotic and protic organic solvents [1].

Mn^{II}LCl. The black complex with ligand 1, $Mn(C_8H_{10}BN_2O_4)Cl$, was obtained on addition of a hot solution of $K(C_8H_{10}BN_2O_4)$ to a hot aqueous solution of $MnCl_2$ (2:1 mole ratio) [3]. The yellowish brown complex with ligand 4, $Mn(C_{16}H_{10}BN_2O_4)Cl$, was prepared by pouring a DMF solution of $K(C_{16}H_{10}BN_2O_4)$, obtained in situ, into an ice-cold aqueous solution of $MnCl_2$ with continuous stirring (2:1 mole ratio). The precipitate, which appeared after keeping the solution in an ice bath for about 8 h, was filtered, washed repeatedly with water and alcohol, and finally dried in vacuum [4]. The complex with ligand 1 melts at 360°C [3], the complex with ligand 4 decomposes at 260°C [4].

The IR spectra of the complexes in Nujol recorded in the 4000 to 600 cm^{-1} range exhibit the characteristic ν(CO) bands (bands of the potassium salt of the respective ligand are given in parentheses) at 1650 (1665) and 1550 (1560) cm^{-1} for $Mn(C_8H_{10}BN_2O_4)Cl$ and at 1650 (1740) and 1530 (1710) cm^{-1} for $Mn(C_{16}H_{10}BN_2O_4)Cl$. As with the $Mn^{II}L_2$ complexes, the IR data support the involvement of carbonyl oxygen in coordination and suggest that the imide nitrogens are coordination-inactive. The diffuse reflectance spectra show maxima at 22700 and 19200 cm^{-1} for $Mn(C_8H_{10}BN_2O_4)Cl$ and 22800 and 17900 cm^{-1} for $Mn(C_{16}H_{10}BN_2O_4)Cl$; these were assigned to the electronic transitions $^6A_{1g} \rightarrow ^4T_{2g}(G)$ and $^6A_{1g} \rightarrow ^4T_{1g}(G)$ of octahedrally surrounded Mn^{II}. The magnetic moments of the complexes between 4.44 and 4.56 μ_B indicate an antiferromagnetic exchange. The complexes were assumed to be polymeric: Two carbonyl oxygen atoms from each ligand molecule are coordinated to one metal, whereas the other two are coordinated to an adjacent metal atom. The bridging chlorine atoms lie above and below the plane [3, 4].

$Mn(C_8H_{10}BN_2O_4)Cl$ is insoluble in water and common organic solvents [3]. $Mn(C_{16}H_{10}BN_2O_4)Cl$ is also insoluble in the usual organic solvents [4].

References:

[1] Zaidi, S. A. A.; Jaria, M.; Siddiqi, Z. A. (Bull. Soc. Chim. France **1987** 599/603).
[2] Zaidi, S. A. A.; Jaria, M.; Siddiqi, Z. A. (Syn. React. Inorg. Metal-Org. Chem. **16** [1986] 1067/87).
[3] Zaidi, S. A. A.; Jaria, M.; Khan, S. (Indian J. Chem. **24** [1985] 314/7).

[4] Zaidi, S. A. A.; Jaria, M.; Kureshy, R.; Yamin, M.; Siddiqi, Z. A. (Bull. Soc. Chim. France **1985** 177/9).

37.4.2 With (1H-Pyrazolato-N[1])borates

$$H\left[R_{4-n}B\left(\text{pyrazol-1-yl with } R' \text{ at 3,5 and } R'' \text{ at 4}\right)_n\right] \quad (=HL)$$

ligand	n	R	R′	R″	formula
1	2	H	H	H	$C_6H_9BN_4$
2	2	H	CH_3	H	$C_{10}H_{17}BN_4$
3	3	H	H	H	$C_9H_{11}BN_6$
4	3	H	CH_3	H	$C_{15}H_{23}BN_6$
5	3	H	CH_3	CH_3	$C_{18}H_{29}BN_6$
6	3	H	H	Cl	$C_9H_8BCl_3N_6$
7	3	n-C_4H_9	H	H	$C_{13}H_{19}BN_6$
8	3	C_6H_5	H	H	$C_{15}H_{15}BN_6$
9	4	—	H	H	$C_{12}H_{13}BN_8$

Remark. The syntheses and coordination chemistry of (pyrazolato)borates were described in detail in "Borverbindungen" 5, 1975, pp. 8/38. In this connection MnL_2 complexes with HL = ligands 1 to 9 and numerous carbonyl compounds containing pyrazolatoborate anions are reported together with the corresponding complexes of other transition metals.

37.4.2.1 Manganese(II) Compounds

MnL_2 complexes with ligand 1, 3 [1], 4, 5 [2], or 9 [1, 3] were prepared by adding an equimolar amount of manganese(II) salt to a stirred solution of the potassium salt of ligand 1 in H_2O, of ligand 4 or 5 in H_2O/DMF, and ligand 3 or 9 in DMF. In the case of the complex with ligand 4 or 5, a slight excess of the metal salt was used [2]. For the preparation of $Mn(C_{10}H_{16}BN_4)_2$, a DMF solution of $K(C_{10}H_{16}BN_4)$, obtained in situ by refluxing KBH_4 and ligand 2 (1:3 mole ratio), was treated with a slight excess of aqueous $MnSO_4$ solution. $Mn(C_9H_7BCl_3N_6)_2$ was prepared as follows: The potassium salt of ligand 6 was obtained in situ by melting together KBH_4 and the ligand in a 1:4 mole ratio. The melt was cooled, dissolved in water, and then treated with aqueous $MnSO_4$. The complex with ligand 7, $Mn(C_{13}H_{18}BN_6)_2$, was obtained by adding pyrazole (4 mol) and n-$C_4H_9B(OH)_2$ (1 mol) to a THF solution of the sodium salt of pyrazole (1 mol in 800 mL). The reaction mixture was stirred and heated until THF, water, and pyrazole had distilled. The melt was cooled and dissolved in 3 L water. The suspension was stirred and filtered, and the clear solution treated with a manganese(II) salt solution. The complex with ligand 8, $Mn(C_{15}H_{14}BN_6)_2$, was prepared by adding over 2 h a toluene solution of dichlorophenylborane (0.12 mol in 35 mL) to a toluene solution of pyrazole (1.46 mol in 240 mL). The reaction was exothermic with a temperature rise to 45 to 55°C. The reaction mixture was cooled to room temperature and stirred with 170 mL of a 0.5 M Mn^{II} salt solution [2]. The products were isolated either by filtration (complexes with ligands 2, 4 to 6, 8, and 9) or by extraction with dichloromethane (complexes with ligands 1, 3, and 7). The filtrates were washed with water and dried; the extracts were dried, filtered, and stripped. The white compounds were further purified by recrystallization from toluene, cyclohexane, xylene, or 1,2-dichlorobenzene [1 to 3]. In the case of the air-sensitive $Mn(C_6H_8BN_4)_2$ complex with ligand 1, the preparation procedure was done in an inert atmosphere [1].

The complex with ligand 9 sublimes at 280°C/1 mm [3]. The melting points of the complexes with ligands 1 to 9 are shown below:

HL	complex	m.p. in °C	Ref.
1	$Mn(C_6H_8BN_4)_2$	155 [a], 162 [b]	[1]
2	$Mn(C_{10}H_{16}BN_4)_2$	190 to 192	[2]
3	$Mn(C_9H_{10}BN_6)_2$	283 to 284	[1]
4	$Mn(C_{15}H_{22}BN_6)_2$	366 to 368	[2]
5	$Mn(C_{18}H_{28}BN_6)_2$	369 to 370	[2]
6	$Mn(C_9H_7BCl_3N_6)_2$	248 to 249	[2]
7	$Mn(C_{13}H_{18}BN_6)_2$	362 to 367	[2]
8	$Mn(C_{15}H_{14}BN_6)_2$	>430 (dec.)	[2]
9	$Mn(C_{12}H_{12}BN_8)_2$	342 to 343	[1, 3]

[a] Sublimed. – [b] Recrystallized from heptane.

The IR spectra of the $Mn(C_6H_8BN_4)_2$, $Mn(C_9H_{10}BN_6)_2$, and $Mn(C_{12}H_{12}BN_8)_2$ complexes with ligands 1, 3, and 9, respectively, in the 4000 to 650 cm^{-1} range are illustrated in the paper [1]. The low-frequency IR spectra in the 650 to 150 cm^{-1} range (Nujol mulls on CsI and polyethylene) have been measured for the complexes with ligand 3, 5, or 9. Based on the metal-isotope substitution method for the corresponding complexes with Fe^{II}, Ni^{II}, Cu^{II}, and Zn, bands at 236 and 213 cm^{-1} for $Mn(C_9H_{10}BN_6)_2$, at 231 cm^{-1} for $Mn(C_{18}H_{28}BN_6)_2$, and at 232 cm^{-1} for $Mn(C_{12}H_{12}BN_8)_2$ are assigned to ν(Mn–N) vibrations [5]. Susceptibility measurements in CH_2Cl_2 or $CHCl_3$ solutions at room temperature by an NMR method yielded magnetic moments of 5.99 ± 0.02, 6.11 ± 0.05, or 5.98 ± 0.05 μ_B for the complex with ligand 2, 3, or 9, respectively. The NMR spectra of the complex with ligand 1, 3, or 9 show ^{11}B shifts of −2590, +1880, or +1690 cps, respectively, relative to trimethyl borate as an external standard (14.2 Mcps) [4]. The change in the sign of the ^{11}B shift indicates a change in geometry determined by the number of pyrazolyl groups attached to boron: A tetrahedral geometry was proposed for the complex with the bidentate bis(pyrazolato)borate (ligand 1) and octahedral geometry for the complexes with the tridentate tris(pyrazolato)borate (ligand 3) and tetrakis(pyrazolato)borate (ligand 9) [1, 2, 4].

$Mn(C_9H_{10}BN_6)_2$ is sparingly soluble in polar solvents such as alcohols or acetone but is readily soluble in halocarbons and aromatic hydrocarbons [1]. $Mn(C_{12}H_{12}BN_8)_2$ is soluble in CH_2Cl_2, hot aromatic hydrocarbons, and dilute aqueous mineral acids. From the acid solution it may be recovered by treatment with base [1, 4]. Solvent extraction of Mn^{2+} into $CHCl_3$ using ligands 3 and 9 was investigated as a function of pH [6].

References:

[1] Trofimenko, S. (J. Am. Chem. Soc. **89** [1967] 3170/7).
[2] Trofimenko, S. (J. Am. Chem. Soc. **89** [1967] 6288/94).
[3] Trofimenko, S. (Inorg. Syn. **12** [1970] 99/108).
[4] Jesson, J. P.; Trofimenko, S.; Eaton, D. R. (J. Am. Chem. Soc. **89** [1967] 3148/58).
[5] Hutchinson, B.; Hoffbauer, M. (Spectrochim. Acta A **32** [1976] 1785/92).
[6] Yasuda, N.; Nakao, N.; Sohrin, Y.; Kihara, S.; Matsui, M. (Proc. Symp. Solvent Extr., Osaka 1987, pp. 203/8).

37.4.2.2 Manganese(III) Compounds

$[Mn_2O(RCOO)_2(C_9H_{10}BN_6)_2]$ (R = H, CH_3, C_2H_5). The complex with R = H, $[Mn_2O(HCOO)_2(C_9H_{10}BN_6)_2]$, was prepared (similarly to the isomorphous iron(III) analog) by exchange of bridging carboxylate ligands. A solution of 0.5 mmol $[Mn_2O(CH_3COO)_2(C_9H_{10}BN_6)_2]$ (see below) in 25 mL dichloromethane was stirred for 1.5 h with 25 mL of an aqueous solution containing 1M HCOOH and 1M $NaHCO_3$ at room temperature. The dichloromethane layer was separated, washed with distilled water, and filtered. The solvent was evaporated and the residue recrystallized from dichloromethane/acetonitrile to yield purple needles [1].

$[Mn_2O(CH_3COO)_2(C_9H_{10}BN_6)_2]$ can be obtained by generating manganese(III) acetate from excess $KMnO_4$ and $Mn(CH_3COO)_2 \cdot 4H_2O$ in glacial acetic acid, followed by the addition of the potassium salt of ligand 3, $K(C_9H_{10}BN_6)$, (method A). Excess $KMnO_4$ was found to be necessary in order to remove Mn^{II} traces and $Mn(C_9H_{10}BN_6)_2$, which is difficult to separate from the final product. Attempts to use commercial manganese(III) acetate in glacial acetic acid failed to isolate the desired complex. For this reason 10.2 mmol of $Mn(CH_3COO)_2 \cdot 4H_2O$ and 5.1 mmol of $KMnO_4$ were added to 50 mL of glacial acetic acid. The solution was heated to 80°C for 45 min with stirring, filtered and cooled to room temperature (color change from purple to chocolate brown). After the addition of 9.92 mmol of $K(C_9H_{10}BN_6)$ the solution was stirred for 15 min, then filtered and evaporated to dryness in vacuum with heating up to 60°C. The glassy brown residue was dissolved in dichloromethane. Vacuum removal of the solvent from the filtered solution gave a purple solid. For purification it was dissolved in dichloromethane, filtered, diluted with 5 times its volume of acetonitrile, and allowed to stand 3 h at room temperature. Subsequent cooling to −5°C for 24 to 48 h yielded deep purple prisms. The complex was also obtained (method B) by adding a solution of 4 mmol of $K(C_9H_{10}BN_6)$ in dry acetonitrile to a solution of manganese(III) acetate dihydrate (3.8 mmol) in degassed acetonitrile. The reaction mixture was heated to 65°C under nitrogen for 75 min, cooled to room temperature, and filtered to remove a brown solid. After evaporating the solvent under vacuum, a purple-black precipitate formed on triturating with cold ether. Filtration and drying in air yielded a purple solid. $[Mn_2O(C_2H_5COO)_2(C_9H_{10}BN_6)_2]$ was obtained in excellent yield (91%) by generating manganese(III) propionate in propionic acid. Shiny purple crystals were isolated [1].

Selected IR and Resonance Raman (RR) spectral bands (ν in cm^{-1}) for $[Mn_2O(CH_3COO)_2$-$(C_9H_{10}BN_6)_2]$ with proposed assignments are given in the following table:

ν	$\nu_{as}(OCO)$	$2\nu_{as}$(Mn–O–Mn)*)	ν_{as}(Mn–O–Mn)	ν_s(Mn–O–Mn)	ν(Mn–$OOCCH_3$)	ν(Mn–N)
IR	1580	—	712	559	340	272
RR ...	—	1432	717	558	343	271

*) Overtone.

Other notable IR absorption bands are ν(BH) at 2479 cm^{-1}, the broad intense peaks of the carboxylate bridging vibrations at 1580 cm^{-1}, and a series of sharp, intense peaks for the pyrazole rings in the 1500 to 600 cm^{-1} range. A Raman spectrum of the complex with $R = C_2H_5$ and its ^{18}O derivative is given in the paper [1].

Magnetic susceptibility measurements of powdered acetonitrile adducts (see below) indicate a weak antiferromagnetic exchange coupling of approximately 0.5 cm^{-1}, resulting in an S=0 ground state with thermally accessible excited spin states at liquid helium temperatures [1]. X-band ESR spectra of $[Mn_2O(CH_3COO)_2(C_9H_{10}BN_6)_2]$ in butyronitrile/propionitrile were obtained with parallel-polarized and conventional perpendicular-polarized microwaves at 4.2 K. The observed signals are consistent with a formal S = 2 state. In the parallel polarization a trough with $\mu_{eff} = 8.7\ \mu_B$ appears, and a weaker signal appears at higher field in the perpendicular polarization. Partially-resolved hyperfine structure with a spacing of 52 G is visible in the parallel polarization spectrum and the presence of at least 10 lines is consistent with an exchange-coupled dimer as the origin of the signal [2]. The greater paramagnetism and rapid spin relaxation of the complexes lead to large isotropic shifts and narrow lines in the proton NMR spectra. All protons were observed in the +67 to −56 ppm region and most could be assigned on the basis of deuterium substitution. The methyl protons of the bridging acetate groups of the complex with R = CH_3 occur at 65.6 and 10.5 ppm for the analogous iron compound. The resonance at −1.2 ppm was assigned to the B–H proton. Chemical shifts and relative intensities are listed in the paper [1]. The results suggest that substitution of Mn^{III} for Fe^{III} (if it could be experimentally achieved) in proteins containing the $[Fe_2O]^{4+}$ core, would be a powerful way to probe the nature of these centers in iron-oxo proteins. The electronic spectrum of the complex with R = CH_3, measured in CH_2Cl_2, is shown in the paper [1].

A cyclic voltammogram of $[Mn_2O(CH_3COO)_2(C_9H_{10}BN_6)_2]$ in acetonitrile (with 0.1M $(n\text{-}C_4H_9)_4NClO_4$ as supporting electrolyte) displays a quasi-reversible one-electron oxidation with $E_{1/2}$ = 0.51 V versus internal ferrocenium⁺/ferrocene, corresponding to a mixed-valence $Mn^{III}Mn^{IV}$ complex and an irreversible reduction at −1.20 V [1]. Chemical exchange of both the oxo bridge (Mn–O–Mn) and carboxylate bridges (Mn–$(RCOO)_2$–Mn) occurs readily in contact with buffered aqueous solution and is catalyzed by H^+ ion. Below pH 1 rapid loss of the carboxylate bridges occurs. A mixed-valence $Mn^{III}Mn^{IV}$ complex (see p. 13) was also generated by electrochemical oxidation of the complex $[Mn_2O(CH_3COO)_2(C_9H_{10}BN_6)_2]$ with aqueous $KMnO_4$, or by contact with air [1, 3].

$[Mn_2O(CH_3COO)_2(C_9H_{10}BN_6)_2]\cdot 4CH_3CN$ (I) and **$[Mn_2O(CH_3COO)_2(C_9H_{10}BN_6)_2]\cdot CH_3CN$ (II).** Deep red-purple truncated square-pyramidal crystals of complex I were obtained by cooling to 0°C a saturated solution of the unsolvated complex in dry CH_3CN. Purplish brown rectangular plates of complex II were isolated directly from the 4:1 acetonitrile/dichloromethane mother liquor (see method A for preparation of the unsolvated compounds above) [1].

Structural studies by X-ray diffraction at −65°C (complex I) or room temperature (complex II) reveal two six-coordinate Mn atoms bridged by one μ-oxo and two μ-acetato groups and capped by two tridentate hydrotris(1H-pyrazolato-N¹)borate ligands. Crystal parameters (a, b, and c in Å, β in °) are given in the following table together with the calculated densities (in g/cm^3) [1]:

complex	a	b	c	β	space group	Z	D
I	13.115(3)	15.097(6)	21.662(5)	107.26(2)	$P2_1/n\text{-}C^5_{2h}$ (No. 14)*)	4	1.355
II	19.162(2)	19.162(2)	53.547(8)	120	$R\bar{3}c\text{-}D^6_{3d}$ (No. 167)	18	1.248

*) Standard setting $P2_1/c$.

Crystals of I were found to be isomorphous with its iron analog. The structure of I was refined to R = 0.041, that of II to R = 0.071. Final positional parameters are given in the paper [1]. The molecular structure of the $[Mn_2O(CH_3COO)_2(C_9H_{10}BN_6)_2]$ complex in I is shown in **Fig. 2**. A similar structure of the complex in II from a different perspective is given in the paper [1].

Complex I has no crystallographically-required symmetry, whereas in the crystals of II a twofold symmetry axis passes through the μ-oxo atom, relating the two halves of the molecule. The angle Mn(1)–O(1)–Mn(2) in I was found to be 125.1(1)°, in II 125.0(3)°. Selected interatomic distances (in Å) are shown below:

distance	complex I	complex II	distance	complex I	complex II
Mn(1)–O(1)	1.773(2)	1.790(3)	Mn(1)–N(3)	2.196(2)	2.106(6)
Mn(1)–O(2)	2.044(2)	2.001(6)	Mn(1)–N(5)	2.166(3)	2.232(7)
Mn(1)–O(3)	2.083(2)	2.133(6)	Mn(1)–N(1)	2.067(2)	2.051(6)

Notable is that the average distances of the Mn–N bond *trans* to the bridging oxo atoms are shorter than the *cis* Mn–N bonds, due to the short Mn–O(oxo) bond raising the energy of the

dz^2 orbital directed along the Mn–O(oxo) bond vector. This highest lying d orbital is therefore empty in high-spin (d^4) Mn^{III}, resulting in a shortened *trans* Mn–N bond. The rhombic distortions in the two crystalline forms are not equivalent. The nonequivalence in the length of the N–Mn–O bond vectors is significantly larger in II than in I [1].

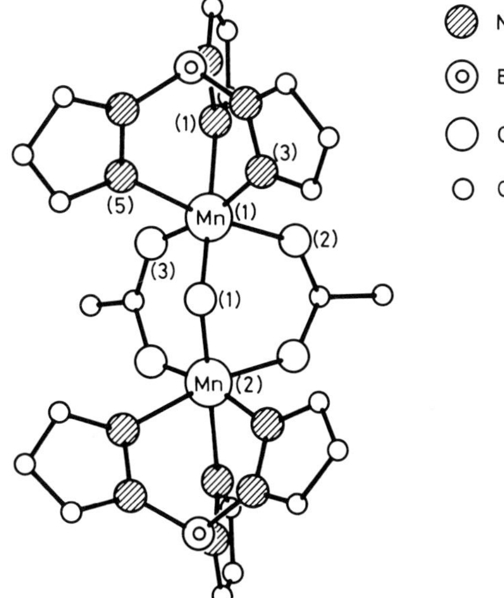

Fig. 2. Molecular structure of the complex $[Mn_2O(CH_3COO)_2(C_9H_{10}BN_6)_2]$ in the tetraacetonitrile solvate (I). Hydrogen atoms are omitted [1].

Temperature-dependent molar susceptibilities and effective magnetic moments per Mn for both complexes (determined by the Evans technique) are consistent with two isolated high-spin Mn^{III} centers with weak antiferromagnetic exchange coupling between them. The magnetic moment of complex I increases nearly linearly with temperature (μ_{eff} = 4.89, 4.77, 4.43, 4.03 μ_B at 278, 100, 20, or 6.01 K, respectively); that of complex II decreases slightly from μ_{eff} = 4.96 to 4.88 μ_B per Mn between 300 and 15.7 K and decreases further to 4.61 μ_B at 5.4 K [1].

References:

[1] Sheats, J. E.; Czernuszewicz, R. S.; Dismukes, G. C.; Rheingold, A. L.; Petrouleas, V.; Stubbe, J. A.; Armstrong, W. H.; Beer, R. H.; Lippard, S. J. (J. Am. Chem. Soc. **109** [1987] 1435/44).

[2] Dexheimer, S. L.; Gohdes, J. W.; Chan, M. K.; Hagen, K. S.; Armstrong, W. H.; Klein, M. P. (J. Am. Chem. Soc. **111** [1989] 8923/5).

[3] Sheats, J. E.; UnniNair, B. C.; Petrouleas, V.; Artandi, S.; Czernuszewicz, R. S.; Dismukes, G. C. (Progr. Photosynth. Res. Proc. 7th Intern. Congr. Photosynth., Providence 1986 [1987], Vol. 1, pp. 721/4).

37.4.2.3 Manganese(III, IV) Compound

A mixed-valence $Mn^{III}Mn^{IV}$ compound is formed by electrochemical oxidation of $[Mn_2O(CH_3COO)_2(C_9H_{10}BN_6)_2]$. Oxidation in $KMnO_4$ solution leads to a complex formulated as $[(C_9H_{10}BN_6)_2Mn(O)_2(CH_3COO)Mn(C_9H_{10}BN_6)_2]$. The Raman spectrum reveals four peaks in the region characteristic of a four-membered $Mn^{III}(O_2)Mn^{IV}$ ring. The expected isotopic shifts were observed on substitution with ^{18}O. The ESR spectra of the neutral di-μ-oxo complex at 6 to 200 K yield a 16-line multiplet with a width of 1180 G. Good agreement with simulated spectra is obtained using hyperfine parameters of 77 and 159 G for the two manganese ions. The temperature dependence of the EPR signal is consistent with that of a simple dimer, having a ground state spin of S = ½ [1, 2].

References:

[1] Sheats, J. E.; Czernuszewicz, R. S.; Dismukes, G. C.; Rheingold, A. L.; Petrouleas, V.; Stubbe, J. A.; Armstrong, W. H.; Beer, R. H.; Lippard, S. J. (J. Am. Chem. Soc. **109** [1987] 1435/44).

[2] Sheats, J. E.; UnniNair, B. C.; Petrouleas, V.; Artandi, S.; Czernuszewicz, R. S.; Dismukes, G. C. (Progr. Photosynth. Res. Proc. 7th Intern. Congr. Photosynth., Providence 1986 [1987], Vol. 1, pp. 721/4).

37.4.2.4 Manganese(IV) Compounds

$[Mn^{IV}(C_{15}H_{22}BN_6)_2](ClO_4)_2$ and **$[Mn^{IV}(C_9H_{10}BN_6)_2][Mn^{II}(H_2O)_6](ClO_4)_4$**. The deep blue complex, $[Mn(C_{15}H_{22}BN_6)_2](ClO_4)_2$, was obtained by adding 2 mmol of the potassium salt of ligand 4 to a solution of $Mn(ClO_4)_2 \cdot 6H_2O$ (2.45 mmol) in a 3:2 acetone-H_2O mixture. Addition of $NaMnO_4$ (1.97 mmol) resulted in precipitation of a brown solid. Treatment with 60% aqueous $HClO_4$, followed by extraction with acetonitrile, gave a deep blue solution. After evaporating a large amount of solvent and filtering, crystals separated from the solution. Attempts to prepare an analogous Mn^{IV} species with ligand 3 by the same procedure leads to formation of $[Mn(C_9H_{10}BN_6)_2][Mn(H_2O)_6](ClO_4)_4$. Both compounds crystallize in the triclinic system, space group $P\bar{1}-C_i^1$ (No. 2). The crystal parameters (a, b, and c in Å; α, β, and γ in °) are shown below:

$[Mn(C_{15}H_{22}BN_6)_2](ClO_4)_2$				$[Mn(C_9H_{10}BN_6)_2][Mn(H_2O)_6](ClO_4)_4$			
a	10.694(2)	α	78.66(2)	a	9.328(2)	α	105.20(2)
b	11.535(2)	β	81.45(2)	b	9.425(3)	β	86.63(2)
c	16.231(5)	γ	73.16(2)	c	13.227(3)	γ	119.21(2)
	Z = 2				Z = 1		

The structure of $[Mn(C_{15}H_{22}BN_6)_2](ClO_4)_2$ was refined to R = 3.84. It consists of two independent centrosymmetric $[Mn(C_{15}H_{22}BN_6)_2]^{2+}$ complex ions and well-separated perchlorate anions in general positions. A representation of the cation $[Mn^{IV}(C_{15}H_{22}BN_6)_2]^{2+}$ is shown in **Fig. 3**.

Selected interatomic distances (in Å) and angles (in °) are as follows:

	$[Mn(C_{15}H_{22}BN_6)_2](ClO_4)_2$	$[Mn(C_9H_{10}BN_6)_2][Mn(H_2O)_6](ClO_4)_4$
Mn–N(1)	1.976(3)	1.960(2)
Mn–N(3)	1.967(3)	1.957(2)
Mn–N(5)	1.971(2)	1.948(2)
N(1)–Mn–N(3)	89.8(1)	88.9(1)
N(1)–Mn–N(5)	90.7(1)	88.7(1)
N(3)–Mn–N(5)	90.3(1)	88.4(1)

The manganese coordination environment is nearly octahedral with Mn–N bond distances in the range 1.966(3) to 1.982(3) Å and *cis* N–Mn–N angles between 89.0(1)° and 90.7(1)° for both of the cations. The average Mn–N bond distances are somewhat shorter than the average Mn^{IV}–N separations in comparable compounds.

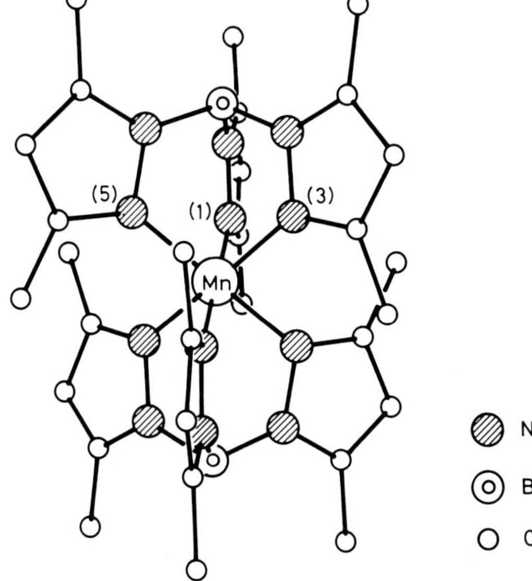

Fig. 3. Molecular structure of the $[Mn^{IV}(C_{15}H_{22}BN_6)_2]^{2+}$ ion. Hydrogen atoms are omitted.

Calculated densities are $D = 1.51\,g/cm^3$ for $[Mn(C_{15}H_{22}BN_6)_2](ClO_4)_2$ and 1.77 g/cm^3 for $[Mn(C_9H_{10}BN_6)_2][Mn(H_2O)_6](ClO_4)_4$.

The electronic absorption spectra of both complexes are dominated by intense charge-transfer transitions at 599 nm ($\varepsilon = 12000\ L \cdot mol^{-1} \cdot cm^{-1}$) and 545 nm ($\varepsilon = 12000\ L \cdot mol^{-1} \cdot cm^{-1}$), respectively. Magnetic susceptibility measurements on the solid state at room temperature ($\mu_{eff} = 3.81\,\mu_B$) confirm a d^3 electronic configuration. The X-band ESR spectrum of $[Mn(C_{15}H_{22}BN_6)_2]$-$(ClO_4)_2$ in acetonitrile at 77 K is in contrast to the expectation that nearly octahedral coordination around Mn may give rise to a small axial zero-field splitting parameter ($2D \gg h \cdot \nu \approx 0.31\,cm^{-1}$). The asymmetric appearance of the low-field signal and the fact that the crossing point of the same signal is at $g < 4$ (3.73) reveal a noticeable rhombic distortion.

The cyclic voltammogram of $[Mn(C_{15}H_{22}BN_6)_2](ClO_4)_2$ in acetonitrile (with 0.1 M $(C_2H_5)_4NClO_4$ as supporting electrolyte) reveals a quasi-reversible wave corresponding to the Mn^{IV}/Mn^{III} couple (1.35 V versus sodium-saturated calomel electrode) and an irreversible Mn^{III}/Mn^{II} couple at 0.02 V.

Reference:

Chan, M. K.; Armstrong, W. H. (Inorg. Chem. **28** [1989] 3777/9).

37.4.3 With (1H-Indazolato-N[1])borates

(= HL)

ligand	n	R	formula
1	2	H	$C_{14}H_{13}BN_4$
2	2	5-NO_2	$C_{14}H_{11}BN_6O_4$
3	2	6-NO_2	$C_{14}H_{11}BN_6O_4$
4	3	H	$C_{21}H_{17}BN_6$
5	3	6-NO_2	$C_{21}H_{14}BN_9O_6$
6	4	H	$C_{28}H_{21}BN_8$
7	4	6-NO_2	$C_{28}H_{17}BN_{12}O_8$

$Mn^{II}(C_{14}H_{10}BN_6O_4)_2$. A solution of the potassium salt of ligand 2 (formed by refluxing KBH_4 with 5-nitroindazole in DMF for 8 h) was added to an aqueous solution of $MnCl_2$ in a 3:1 mole ratio. The precipitate, which formed immediately, was warmed on a water bath for about 30 min, filtered, washed repeatedly with hot distilled water, and dried in a vacuum. The dark brown complex is fairly stable to air and moisture, and melts at 248°C with decomposition [1].

The IR spectrum recorded in Nujol in the 4000 to 200 cm^{-1} range exhibits characteristic absorption bands (in cm^{-1}, $K(C_{14}H_{10}BN_6O_4)$ bands in parentheses) assigned as follows: 2380, 2320 (2380, 2280) to ν(BH), 1615 (1620) to ν(ring), 1575 (1595) to ν(CN), 1380 (1395) to ν(BN), 1340 (1340) to $\nu(NO_2)$, and 520 (br) to ν(Mn–N). The shifts of ν(BH), ν(BN), and ν(CN) in the spectrum of the complex indicate that the ligand coordinates via both nitrogen atoms of the heterocyclic ring, analogously to the $Mn^{II}L_2$ complexes with (1H-imidazolato-N[1])borates (see p. 19). The nitro groups are not involved in coordination. The magnetic moment, $\mu_{eff} = 5.89\ \mu_B$, calculated from susceptibility data obtained at 311 K by the NMR method and corrected for diamagnetism is very close to that required for an octahedral environment of the metal atom. The electronic diffuse reflectance spectrum shows a charge-transfer band at 41662 cm^{-1}, a ligand transition band at 38461 cm^{-1}, and three bands at 29412, 25316, and 18518 cm^{-1} ascribed to the Mn^{II} transitions ${}^6A_{1g} \rightarrow {}^4E_g(D)$, ${}^6A_{1g} \rightarrow {}^4A_{1g}(G)$, and ${}^6A_{1g} \rightarrow {}^4T_{1g}(G)$, respectively. The ligand field parameters derived from the data are: 10 Dq = 8437 cm^{-1}, B = 767 cm^{-1}, β = 0.80. The nephelauxetic parameter, β, indicates a low degree of covalency and is in agreement with the data reported for the complexes with (1H-pyrazolato-N[1])borates or (1H-imidazolato-N[1])borates. Each of the tridentate ligand molecules is assumed to be bonded to manganese by two N(1) and one N(2) nitrogen atoms. The complex is soluble only in dimethylformamide. The molar electrical conductivity of a 10^{-3}M solution in DMF at 25°C, $\Lambda = 4.82\ cm^2 \cdot \Omega^{-1} \cdot mol^{-1}$, demonstrates that the complex is a nonelectrolyte. The electronic spectrum of the solid complex and its solution spectrum in DMF are identical, suggesting that the complex is not altered in such solution [1].

$Mn^{II}LCl$ complexes with ligand 1 [2], 4 [3], or 6 [4] were prepared by the reaction of the potassium salt of the appropriate ligand with manganese(II) chloride (1:1 mole ratio) in ethanol. After keeping the reaction mixture at room temperature for a few hours, the solid product was washed with ethanol and dried at 150°C [3, 4] or 180°C [2]. The complexes with ligand 3, 5, or 7 were prepared by using the DMF solution of the ligand obtained in situ, since the nitro-substituted (1H-indazolato-N[1])borates can be isolated only in poor yield. This solution was poured into an ice-cold aqueous solution of manganese(II) chloride. The solid precipitate was washed with cold, warm, and hot water repeatedly, and finally dried in vacuum over P_4O_{10} [5]. The complexes with ligand 1, 4, or 6 are white, those with ligand 3, 5, or 7 are brown. Color, melting points, magnetic moments, and characteristic IR bands of the complexes in Nujol or KBr are tabulated below; ligand bands (potassium salt) are given in parentheses:

No.	complex	m.p. in °C	μ_{eff} in μ_B	IR data in cm^{-1} ν(BH)	ν(CN)	ν(NO_2)	ν(BN)	ν(Mn–N)	Ref.
1	$Mn(C_{14}H_{12}BN_4)Cl$	>360	paramagnetic	2320, 2340	1510	—	—	455	[2]
				(2420, 2370)	(1500)	—	—	—	
3	$Mn(C_{14}H_{10}BN_6O_4)Cl$	>258(dec.)	5.90 [a]	2465, 2360	1650	1530	1390	415	[5]
				(2460, 2380)	(1680)	(1520)	(1390)	—	
4	$Mn(C_{21}H_{16}BN_6)Cl$	356 to 357	paramagnetic	—	—	—	—	—	[3]
5	$Mn(C_{21}H_{13}BN_9O_6)Cl$	>255(dec.)	5.58 [a]	2440	1640	1530	1390	420	[5]
				(2440)	(1670)	(1520)	(1380)	—	
6	$Mn(C_{28}H_{20}BN_8)Cl$	360	5.90 [b]	—	1505	—	—	390	[4]
				—	(1480)	—	—	—	
7	$Mn(C_{28}H_{16}BN_{12}O_8)Cl$	>250(dec.)	5.89 [a]	—	1650	1520	1380	420	[5]
				—	(1680)	(1520)	(1380)	—	

[a] At 23°C. – [b] At 30°C.

Bands observed at 230 or 280 cm^{-1} in the spectra of the complexes with ligands 4 and 7, respectively, were assigned to ν(Mn–Cl) vibration modes [3, 5]. Due to electronic and steric effects differences between the nitro-substituted and the unsubstituted species were observed. Formation of the MnL_2 complex with the 5-nitro derivative (see above) and the MnLCl complexes with the 6-nitro derivatives show that the position of the nitro group in the indazole ring also affects the chelating pattern of the ligand. The IR spectra of $Mn(C_{14}H_{12}BN_4)Cl$ and $Mn(C_{21}H_{16}BN_6)Cl$ complexes indicate coordination of ligand 1 or 4 through both N(1) and N(2) nitrogen atoms [2, 3], whereas the bulky ligand 6 is assumed to be coordinated only through the N(2) atom [4]. The IR spectra of complexes with the nitro-substituted ligands 3, 5, and 7 are very similar to those of the potassium salts of the same ligand. Only the ν(CN) vibration mode is shifted to lower wavenumbers. Therefore, coordination through the N(2) nitrogen atom was suggested and coordination via the N(1) atom or nitro group ruled out. The ν(Mn–Cl) bands lie in the region for bridging chloride ion, being lower than those reported for terminal chloride ions [5]. The observed magnetic moments are indicative of an octahedral geometry [2 to 5]. The somewhat lower (than that expected for an octahedral arrangement) value of the complex with ligand 5 has been attributed to the magnetically nondilute nature of the complex. The magnetic moment of 5.89 μ_B for $Mn(C_{28}H_{16}BN_{12}O_8)Cl$ indicates that the bulky ligand 7 causes a greater separation of the paramagnetic sites [5]. Bands observed in the diffuse reflectance spectra of the solid MnLCl complexes, transitions from $^6A_{1g}$ (in cm^{-1}), and ligand field parameters are shown below:

HL No.	complex	$\rightarrow {}^4A_{1g}(G)$	$\rightarrow {}^4E_g(G)$	$\rightarrow {}^4T_2(G)$	$\rightarrow {}^4T_1(G)$	10 Dq	B	β	Ref.
1	$Mn(C_{14}H_{12}BN_4)Cl$	25160	24750	—	18650	—	—	—	[2]
3	$Mn(C_{14}H_{10}BN_6O_4)Cl$	—	—	22200	21500	5880	745	0.78	[5]
4	$Mn(C_{21}H_{16}BN_6)Cl$	25270	24520	—	18640	—	—	—	[3]

HL No.	complex	$\rightarrow{}^4A_{1g}(G)$	$\rightarrow{}^4E_g(G)$	$\rightarrow{}^4T_2(G)$	$\rightarrow{}^4T_1(G)$	10 Dq	B	β	Ref.
5	$Mn(C_{21}H_{13}BN_9O_6)Cl$	—	25000	22200	18900	8449	768	0.80	[5]
6	$Mn(C_{28}H_{20}BN_8)Cl$	25230	24510	—	18670	—	—	—	[4]
7	$Mn(C_{28}H_{16}BN_{12}O_8)Cl$	—	25000	22200	18900	8371	761	0.79	[5]

$Mn(C_{14}H_{10}BN_6O_4)Cl$, $Mn(C_{21}H_{13}BN_9O_6)Cl$, and $Mn(C_{28}H_{16}BN_{12}O_8)Cl$ complexes also exhibit a charge-transfer band at 41700, 40300, and 41700 cm^{-1}, respectively, and a ligand transition band at 28600 cm^{-1}. The bands at 34440 and 36530 in the spectrum of $Mn(C_{28}H_{20}BN_8)Cl$ were also attributed to ligand transitions. The values of the nephelauxetic parameter, β, suggest a low degree of covalency [5]. Most of the spectral data are consistent with an octahedral geometry. A distorted tetrahedral ligand arrangement was assumed for the complex with ligand 3, $Mn(C_{14}H_{10}BN_6O_4)Cl$, which exhibits an additional band at 24100 cm^{-1}, assigned to the transition ${}^6A_{1g}\rightarrow{}^4A_1(G)$, ${}^4E(G)$ [5]. A polymeric structure with bidentate bridging ligands is tentatively proposed for all complexes [2, 4]. The complex with ligand 4 is soluble in DMF, DMSO, and nitrobenzene [3], the complexes with ligand 3, 5, or 7 [5] are soluble only in DMF. The insolubility in common organic solvents together with high decomposition temperatures underline the polymeric character [5]. The low molar conductivities of the complexes with ligands 1 [2], 4 [3], and 6 [4] in DMSO (Λ_m = 43.7, 4.2, and 42.8 $cm^2 \cdot \Omega^{-1} \cdot mol^{-1}$) and those of the complexes with ligands 3, 5, and 7 [5] in DMF (Λ_m = 19.9, 16.3, and 31.8 $cm^2 \cdot \Omega^{-1} \cdot mol^{-1}$) are consistent with their nonionic nature.

References:

[1] Siddiqi, Z. A.; Khan, S.; Zaidi, S. A. A. (Syn. React. Inorg. Metal-Org. Chem. **12** [1982] 433/53).

[2] Zaidi, S. A. A.; Neyazi, M. A. (Transition Metal Chem. [Weinheim] **4** [1979] 164/7).

[3] Siddiqi, K. S.; Neyazi, M. A.; Siddiqi, Z. A.; Majid, S. J.; Zaidi, S. A. A. (Indian J. Chem. A **21** [1982] 932/3).

[4] Siddiqi, K. S. S.; Neyazi, M. A.; Zaidi, S. A. A. (Syn. React. Inorg. Metal-Org. Chem. **11** [1981] 253/65).

[5] Siddiqi, Z. A.; Khan, S.; Siddiqi, K. S.; Zaidi, S. A. A. (Syn. React. Inorg. Metal-Org. Chem. **14** [1984] 303/24).

37.4.4 With (1H-Imidazolato-N[1])borates

$H[H_{4-n}B(C_3H_2RN_2)_n]$

(= HL)

ligand	n	R	formula
1	2	H	$C_6H_9BN_4$
2	2	CH_3	$C_8H_{13}BN_4$
3	3	H	$C_9H_{11}BN_6$
4	3	CH_3	$C_{12}H_{17}BN_6$
5	4	H	$C_{12}H_{13}BN_8$
6	4	CH_3	$C_{16}H_{21}BN_8$

$\mathbf{Mn^{II}L_2}$ complexes with ligands 1 and 3 were prepared by mixing aqueous solutions of $MnCl_2$ and the stoichiometric amount of the appropriate ligand and refluxing the mixture for 8 h [1] or 2 h [2], respectively. The complex with ligand 4, 6 [3], or 5 [4] was obtained similarly by using a suspension of the respective ligand in DMF. The complexes separated out on cooling, were filtered, washed with water and ethanol, and dried. Color, magnetic moments at room temperature, and characteristic IR bands of the complexes in Nujol with ligand bands (potassium salt) in parentheses are shown below:

No.	complex	color	μ_{eff} in μ_B	ν(BH)	ν(BN)	Ref.
1	$Mn(C_6H_8BN_4)_2$	green	5.80	2340 (2380) 2420 (2420)	1400 (1390)	[1]
3	$Mn(C_9H_{10}BN_6)_2$	gray	5.83	2410 (2440)	1385 (1400)	[2]
4	$Mn(C_{12}H_{16}BN_6)_2$	brown	5.71	2380 (2420)	1400 (1410)	[3]
5	$Mn(C_{12}H_{12}BN_8)_2$	gray	5.81	—	1390	[4]
6	$Mn(C_{16}H_{20}BN_8)_2$	brown	5.84	—	1395	[3]

The magnetic moments of all the complexes are in agreement with the presence of a high-spin state for the metal ions. The shifts of ν(BH) and ν(BN) observed upon complexation indicate that the uninegative ligands are coordinated to manganese through the (N1) nitrogen atom. The ring-stretching vibrations of the imidazolyl moiety in the 1660 to 1330 cm^{-1} range are scarcely different from those in the free ligand, as is usually observed for other metal imidazolato complexes. Nevertheless, bonding of the N(3) atoms is discussed. The electronic spectra indicate an octahedral arrangement around Mn^{II} [1 to 4]. Bands observed in the diffuse reflectance spectra of the solid compounds, assigned to transitions from $^6A_{1g}$ (in cm^{-1}), and ligand field parameters are shown below:

No.	$\rightarrow {}^4A_{1g}(G)$	$\rightarrow {}^4E_g(G)$	$\rightarrow {}^4T_{2g}(G)$	$\rightarrow {}^4T_{1g}(G)$	10 Dq	B	β	Ref.
1[a]	25641	25000	23809	19230	13740	916	0.80	[1]
3	25625	24775	—	18822	10400	800	0.83	[2]
4	—	26315	23934	19780	10260	855	0.89	[3]
5	25000	—	23752	19047	10777	829	0.86	[4]
6	26500	—	24000	19900	9936	828	0.86	[3]

[a] The complex with ligand 1, $Mn(C_6H_8BN_4)_2$, exhibits an additional band at 32258 cm^{-1}, assigned to the transition $^6A_{1g} \rightarrow {}^4T_{1g}(P)$ and a charge-transfer band at 43478 cm^{-1} [1].

For the $Mn(C_{12}H_{16}BN_6)_2$ and $Mn(C_{16}H_{20}BN_8)_2$ complexes it was suggested that one ligand molecule coordinates through two N(1) and one N(3) atoms, and the second ligand molecule through one N(1) and two N(3) atoms. The polymeric structure of these complexes where ligand 4 or 6 acts as a bridge bonded via N(1) and N(3) atoms, was discussed in [3]. The nephelauxetic parameter, β, in the 0.80 to 0.86 range indicates a low covalency in each complex [1].

The complexes with ligands 4 and 6 are insoluble in inert organic solvents [3].

$Mn^{II}(C_8H_{12}BN_4)Cl$ was obtained by the reaction of manganese(II) salt with the freshly prepared DMF solution of potassium salt of ligand 2.

Characteristic bands (in cm^{-1}) in the IR spectrum of the dark brown complexes were assigned as follows (bands of $K(C_8H_{12}BN_4)$ in parentheses): $\nu(CH_3)$ at 2860, 2940 (2850, 2950), ν(BH) at 2340, 2410 (2400, 2480), ν(BN) at 1380 (1395), ν_{ring} at 1590, 1600, 1540, 1520, 1340 (1610, 1620, 1530, 1430, 1340), and ν(Mn–N) at 300. As with the $Mn^{II}L_2$ complexes described above, the IR spectrum suggests that the coordination occurs through the N(1) and N(3) atoms of the ligand. The splitting of ν(BH) indicates that the two hydrogen atoms on the boron do not lie in the same plane. The diffuse reflectance electronic spectrum exhibits two bands at 23255 and 17857 cm^{-1}, ascribed to the electronic transitions $^6A_{1g} \rightarrow ^4T_{2g}(F)$ and $^6A_{1g} \rightarrow ^4T_{1g}(G)$, respectively, in agreement with an octahedral geometry at the manganese atom. The ligand field parameters are: 10 Dq = 13356 cm^{-1}, B = 891 cm^{-1}, and β = 0.92. A comparison with the ligand field parameters for $Mn(C_6H_8BN_4)_2$ complex with ligand 1 (see table above) shows that dihydrobis(1H-imidazolato-N^1)borate behaves as a stronger coordinating ligand than the methyl-substituted ligand 2. An octahedral polymeric structure was proposed for the complex: For each ligand one N(1) and one N(2) nitrogen atom are coordinated to one manganese ion and the other N(1) and N(2) nitrogen atom to an adjacent manganese ion. The chlorine atom coordinates as a bridging ligand in an array above and below the plane. The observed magnetic moment, $\mu_{eff} = 4.88\ \mu_B$, is considerably lower than expected for high-spin Mn^{II} complexes, and might be due to antiferromagnetic exchange through the bridging ligand. The high decomposition temperature (>300°C) and the insolubility of the complex in common organic solvents also indicate its polymeric nature [5].

References:

[1] Zaidi, S. A. A.; Khan, T. A. (Syn. React. Inorg. Metal-Org. Chem. **14** [1984] 717/29).
[2] Zaidi, S. A. A.; Khan, T. A.; Zaidi, S. R. A.; Siddiqi, Z. A. (Polyhedron **4** [1985] 1163/6).
[3] Zaidi, S. A. A.; Khan, T. A.; Shaheer, S. A.; Zaidi, S. R. A. (Acta Chim. Hung. **125** [1988] 229/37).
[4] Zaidi, S. A. A.; Khan, T. A.; Shaheer, S. A.; Zaidi, S. R. A.; Siddiqi, Z. A. (Bull. Soc. Chim. France **1986** 536/9).
[5] Zaidi, S. A. A.; Khan, T. A.; Siddiqi, Z. A. (Bull. Soc. Chim. France **1984** I 149/52).

37.4.5 With a (Benzotriazolato-N)borate

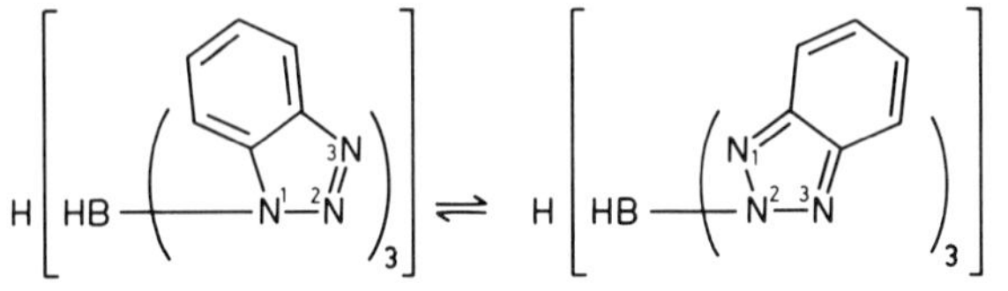

H[HB(N-N=N benzotriazolyl)$_3$] ⇌ H[HB(N benzotriazolyl)$_3$] (= $C_{18}H_{14}BN_9$ = HL)

A B

$Mn^{II}(C_{18}H_{13}BN_9)_2$ was prepared by the reaction of the potassium salt of the ligand (more crowded isomer A is more favored thermodynamically than isomer B) with the manganese(II) salt in aqueous solution. It is sparingly soluble in organic solvents.

Reference:

Lalor, F. J.; Miller, S.; Garvey, N. (J. Organometal. Chem. **356** [1988] C57/C60).

37.4.6 With (2(3H)-Benzothiazolethionato-N[3])borates

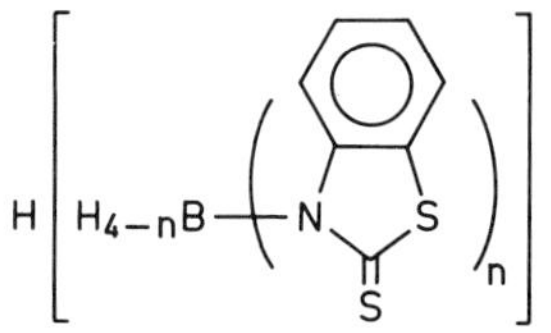

$H\left[H_{4-n}B\left(\text{benzothiazole-2-thione-N}\right)_n\right]$ (=HL)

ligand	n	formula
1	2	$C_{14}H_{11}BN_2S_4$
2	3	$C_{21}H_{14}BN_3S_6$
3	4	$C_{28}H_{17}BN_4S_8$

$Mn^{II}L_2$ complexes with ligands 1, 2, and 3 were prepared by adding stoichiometric amounts of $MnCl_2$ to a DMF solution of the potassium salt of the respective ligand (obtained in situ, procedure given in the paper). An immediate precipitation occurred with the evolution of heat. The precipitate was digested on a water bath, filtered, and dried in a vacuum desiccator. Color, magnetic moments (μ_{eff} in μ_B) at room temperature, characteristic IR bands, and the bands observed in the diffuse reflectance spectra (in cm^{-1}) of the complexes are summarized below:

ligand	1	2	3
complex	$Mn(C_{14}H_{10}BN_2S_4)_2$	$Mn(C_{21}H_{13}BN_3S_6)_2$	$Mn(C_{28}H_{16}BN_4S_8)_2$
color	dirty white	white	yellow
μ_{eff}	4.25	5.72	5.80
ν(BH)	2450	2430	—
ν(BN)	1090	1070	1070
ν(Mn–N)	380	—	380
ν_{max}	40816, 31250, 22727	38461, 31250, 14285, 11904	31250, 26021, 16129

The magnetic moment of 4.25 μ_B for the $Mn(C_{14}H_{10}BN_2S_4)_2$ complex is discussed as intermediate between spin-free and spin-paired Mn^{II}, i.e., partial spin pairing was suggested. The magnetic moment of the complexes with ligands 2 and 3 is consistent with an octahedral configuration. The absorption band at 380 cm^{-1}, assigned to the ν(Mn–N) vibration mode, supports the coordination of the metal through the nitrogen atom. The electronic spectra of the complexes suggest square-planar coordination for the complex with ligand 1, and octahedral geometry for complexes with ligands 2 and 3. The band maxima in the 31250 to 40816 cm^{-1} range were assigned to charge transfer, and the band at 26021 cm^{-1} in the spectrum of $Mn(C_{21}H_{13}BN_3S_8)_2$ to the $^6A_{1g}\rightarrow{}^4A_{1g}(G)$ octahedral transition.

The complexes are fairly stable towards air and moisture. They are sparingly soluble in chloroform. $Mn(C_{28}H_{16}BN_4S_8)_2$ is also soluble in acetone.

Reference:

Siddiqi, K. S.; Shah, M. A. A.; Kureshy, R. I.; Jaria, M.; Zaidi, S. A. A. (Acta Chim. Hung. **117** [1984] 57/65).

38 Complexes with Ligands Containing Silicon

38.1 Complexes with Silanes

ligand 1 $(CH_3)_3SiH$ ($=C_3H_{10}Si=HL$)
ligand 2 $(CH_3)_2C_6H_5SiH$ ($=C_8H_{12}Si=HL$)
ligand 3 $[(CH_3)_3Si]_3SiH$ ($=C_9H_{28}Si_4=HL$)

General Reference:

Tilley, T. D.; Transition-Metal Silyl Derivatives in: Patai, S.; Rappoport, Z.; The Chemistry of Organic Silicon Compounds, Wiley, New York 1989, pp. 1415/77.

Remark. In many respects the structure and reactivity of transition metal complexes with silyl ligands parallel those of the analogous alkyl derivatives, but there are some important differences. Many complexes with silyl ligands are quite inert toward insertion reactions, consistent with the observation that metal–silicon bonds are often stronger than expected for single covalent bonds. It has been suggested that the R_3Si anion can stabilize low coordination numbers and oxidation states of a metal center, but the large majority of these complexes are coordinatively saturated, containing additional ligands such as carbon monoxide, cyclopentadienyl, or phosphane. Carbonyl complexes of manganese containing additional silyl ligands will be described in the series of organometallic compounds.

$Mn(C_3H_9Si)_2Cl_2$(?) and **$Mn(C_3H_9Si)_2(O_2)_2Cl_2$.** A chloro complex of probable composition $Mn(C_3H_9Si)_2Cl_2$ was prepared under anhydrous conditions as follows: The solution of hexamethyldisilane (5 mmol) in 3 mL THF was added with stirring to a mixture of methylpotassium (5 mmol), crown ether (5 mmol), and $MnCl_2$ (4 mmol) in 25 mL THF. The reaction mixture was stirred for 40 h at ambient temperature, the separated white product filtered, washed several times with THF, and dried over P_4O_{10}. Upon exposure to dioxygen, this white compound undergoes a marked color change to give a brown oxygen carrier complex, postulated as $Mn(C_3H_9Si)_2(O_2)_2Cl_2$. The compound was characterized by IR and ^{1}H NMR spectra. The reversible elimination of O_2 and the thermal stability of the complex were studied [1].

$[Mn^{II}(C_9H_{27}Si_4)_2(C_4H_{10}O_2)]$ ($C_4H_{10}O_2$ = 1,2-dimethoxyethane). Reaction of $H_3CSiOSO_2CF_3$ with $Li(C_4H_{10}O_2)_2[Mn^{II}(C_9H_{27}Si_4)_2Cl]$ (see below) in toluene probably yields $Li(C_4H_{10}O_2)_2[Mn^{II}(C_9H_{27}Si_4)_2(OSO_2CF_3)]$. The triflate ligand is weakly coordinated and can be readily displaced by 1,2-dimethoxyethane to give the neutral $[Mn(C_9H_{27}Si_4)_2(C_4H_{10}O_2)]$ complex. Its magnetic moment in benzene is $\mu_{eff}=5.4\ \mu_B$. A four-coordinate structure is assumed for the complex: Manganese is bonded to two silicon atoms of ligand 3 and both oxygen atoms of 1,2-dimethoxyethane. The mass spectrum of the adduct contains no parent ion, but features a strong $Mn(C_9H_{27}Si_4)_2^+$ peak at 549.18 m/e. The complex starts decomposing at 161°C [2].

$Li(C_4H_{10}O_2)_2[Mn^{II}(C_9H_{27}Si_4)_2Cl]$ ($C_4H_{10}O_2$ = 1,2-dimethoxyethane). This yellow compound with an anionic silyl complex was prepared from $MnCl_2$ and $Li(C_9H_{27}Si_2)\cdot 3THF$ (1:2 mole ratio) in 1,2-dimethoxyethane. It was isolated by crystallization from a diethyl ether-pentane mixture. The magnetic moment, $\mu_{eff}=5.1\ \mu_B$, was determined by the Evans method in benzene.

The ESR spectrum in toluene at ambient temperature exhibits a broad signal with g = 2.003 and $A=1.87\times10^{-3}\ cm^{-1}$. An unusual three-coordinate geometry was assumed for the complex anion with the bulky ligand 3. A similar structure has been determined for the analogous Fe^{II} complex. The compound is air- and moisture-sensitive, but is stable at room temperature under nitrogen. On heating, it begins to decompose at 143°C. The coordinative unsaturation of the compound provides a high reactivity. The chloride ligand may be abstracted by $CH_3SiOSO_2CF_3$, giving CH_3SiCl and $[Mn(C_9H_{27}Si_4)_2(OSO_2CF_3)]^-$ (see above) [2].

Other Mn^{II} Compounds. $(R_3Si)_3MnMgCH_3$ complexes with ligand 1 or 2 were prepared by the reaction of a cold THF solution of $R_3SiMgCH_3$ (obtained from stoichiometric amounts of CH_3Li, the appropriate ligand, and CH_3MgI) with $MnCl_2$ in a 3:1 mole ratio. The compounds were not isolated, but the reaction solutions have been successfully used as agents for hydrosilation of olefins [3 to 6] or homoconjugate addition to dienyl- or vinylcyclopropanes [7].

References:

[1] Li, Guangnian; Li, Guangquing (Huaxue Xuebao **43** [1985] 1053/5).
[2] Roddick, D. M.; Tilley, T. D.; Rheingold, A. L.; Geib, S. J. (J. Am. Chem. Soc. **109** [1987] 945/6).
[3] Hibino, J.; Nakatsukasa, S.; Fugami, K.; Matsubara, S.; Oshima, K.; Nozaki, H. (J. Am. Chem. Soc. **107** [1985] 6416/7).
[4] Morimoto, Y.; Higuchi, Y.; Wakamatsu, K.; Oshima, K.; Utimoto, K.; Yasouka, N. (Bull. Chem. Soc. Japan **62** [1989] 639/41).
[5] Miyasaka, T.; Ishisu, H.; Sawada, A.; Fujimoto, A.; Noguchi, S. (Chem. Letters **1986** 871/4).
[6] Fugami, K.; Oshima, K.; Utimoto, K.; Nozaki, H. (Tetrahedron Letters **27** [1986] 2161/4).
[7] Fugami, K.; Oshima, K.; Utimoto, K.; Nozaki, H. (Bull. Chem. Soc. Japan **60** [1987] 2509/15).

38.2 Complexes with Silanols or Their Derivatives

ligand 1	$[(CH_3)_3Si]_3CSi(CH_3)_2OH$	$(=C_{12}H_{34}OSi_4=HL)$
ligand 2	$C_2H_5Si(OH)_3$	$(=C_2H_8O_3Si)$
ligand 3	$C_6H_5Si(OH)_3$	$(=C_6H_8O_3Si=HL)$

General. Silanols, $R_nSi(OH)_{4-n}$, the alcohols of organosilicon chemistry, differ from their carbon analogs by an enhanced acidity. They form numerous classes of derivatives by spontaneous or forced condensation of the hydroxy groups. Metal compounds with silanolates and their derivatives containing an Si–O–M linkage are named metallosiloxanes. These compounds appear to be oligomers of various sizes in organic solvents, and probably also in the solid state, though no structural details are available.

38.2.1 Isolated Monomeric or Dimeric Compounds

$Mn^{II}(C_{12}H_{33}OSi_4)_2 \cdot LiCl$. The complex with ligand 1 was prepared by dropwise addition of $Li(C_{12}H_{33}OSi_4)$ (7.98 mmol in 20 mL THF) to a suspension of $MnCl_2$ (3.99 mmol in 20 mL THF) at −78°C. On warming to room temperature, an orange solution was obtained. After stirring for 1 h, the solution was evaporated to dryness and the residue extracted with heptane. A white microcrystalline product precipitated from the concentrated extract during 2 d at ambient temperature. It melts at 230 to 233°C. The magnetic moment, $\mu_{eff}=5.9\ \mu_B$, was determined at room temperature by the Evans' method. The highest peak in the mass spectrum (m/e = 665) corresponds to the $[Mn(C_{12}H_{33}OSi_4)_2]^+$ ion. The air-sensitive complex is insoluble in nonpolar solvents [1].

$Mn^{II}[C_6H_5Si(OH)_2O]_2$. The amorphous complex with ligand 3 was prepared by reaction of the anhydrous monosodium salt of silanetriol, $C_6H_5Si(OH)_2ONa$, with anhydrous manganese chloride in absolute ethanol. It can be isolated only when the reaction is carried out under mild conditions (dilute solutions, low temperatures). Cryoscopic measurements in benzene also reveal the formation of a dimeric compound, **$\{Mn^{II}[C_2H_5Si(OH)O_{0.5}O]_2\}_2$**, formed by intermolecular condensation of the OH groups [2]. If the reaction takes place in an acid medium (to prevent the formation of manganese hydroxides), poly(manganophenylsiloxane) with a low metal: silicon ratio is obtained [3].

The IR spectra of the complex with ligand 3, $Mn[C_6H_5Si(OH)_2O]_2$, and of corresponding metallophenylsiloxanes containing Al^{III}, Zn^{II}, and Zr^{IV} have been investigated in 5×10^{-3}M CCl_4 solutions in the 3700 to 3300 cm^{-1} range. A broad band was found in each of the 3550 to 3600 and 3350 to 3400 cm^{-1} regions. Since molecular association of the hydroxy groups is practically excluded at low concentration, these bands are probably due to hydroxyl groups joined by an intramolecular hydrogen bond. The presence of two bands indicates that the OH groups in the compounds are nonequivalent. A tetrahedral structure in which the manganese atom is coordinated by four oxygen atoms was suggested. Two hydroxyl groups are assumed to be joined by an intramolecular hydrogen bond and two hydrogen atoms are displaced into the outer coordination sphere [2]. Low-temperature condensation in boiling benzene formed polymeric species [4], see below.

$Mn[C_6H_5Si(OH)_2O]_2$ is soluble in common organic solvents. $\{Mn[C_2H_5Si(OH)O_{0.5}O]_2\}_2$ is soluble only in alcohol and benzene [2]. It is a satisfactory inhibitor of the thermal oxidation of poly(dimethylsiloxanes) [5].

38.2.2 Oligomeric Manganosiloxanes

Several polysiloxanes with low Mn: silicon ratios were prepared and their properties determined. Polyphenyl- or polyvinylmanganosiloxane were prepared by direct interaction of manganese acetate with $C_6H_5SiCl_3$ [6 to 8] or $H_2C{=}CHSiCl_3$ [9], respectively. Oligomers containing methylphenylsiloxane [10] and ethylphenylsiloxane [11] were prepared by polycondensation of equimolar ratios of alkylphenyldiethoxysilanes, $R(C_6H_5)Si(OC_2H_5)_2$, and manganese acetate. Oligoethylethoxysiloxane [12] and oligophenylethoxysiloxane [13] were obtained analogously by using the respective triethoxysilane, $RSi(OC_2H_5)_3$. Polymeric manganosiloxanes can also be prepared by heating the monomeric or dimeric compound. Low-temperature condensation of $Mn[C_6H_5Si(OH)_2O]_2$ and $\{Mn[C_2H_5(OH)O_{0.5}O]_2\}_2$ in boiling benzene formed oligomers with molecular weights of 3600 and 2700, indicating degrees of polymerization of 10.1 and 6.8, respectively [4].

Polymeric metallosiloxanes are brittle, glass-like substances. X-ray structural analysis of several polyphenylmetallosiloxanes showed that they are amorphous, but electron microscopy indicated some degree of crystallinity that depended on the metal content [6, 12]. The structure of the oligomers was studied by IR [6, 9, 10, 13] and ESR spectroscopy [14, 15]. Cyclolinear structures, consisting of SiOSi groups with the metal incorporated into the main chain, were suggested [4, 9, 13]. Coordination interaction between the metal atom of one molecule and oxygen atoms of another molecule, in addition to the normal intermolecular interaction, is discussed in [8].

The polymeric compounds have attracted considerable attention because of their thermostability, heat resistance, and high dielectric constants over wide temperature intervals. They have therefore been used as electroisolation compounds and modifiers of semiconductors. In addition they are adhesives for glass, bactericides, and catalysts. The effect of the metal on the thermal stability and thermal degradation [6, 7, 9, 11, 13, 16, 17] as well as on the electrical properties (dielectric constant, dielectric loss tangent, dielectric permeability, dielectric strength, volume resistivity) [10, 18 to 21] was studied for the oligomeric metallosiloxanes of several transition and rare earth metals. The heat resistance and high dielectric indices of the polymers depend on the polarizing effect of the metal, indicating an ionized state of the metal inside the polymer structure [6, 11, 18].

The reaction of poly(manganophenylsiloxane) with acetic acid has been investigated by NMR techniques [21]. The polymer exhibits catalytic activity in cracking reactions [22]. It is soluble in benzene, toluene, acetone, dichloroethane, and methyl ethyl ketone, but is insoluble in alcohols, unsaturated hydrocarbons, and water [6, 7]. Heating the polymers to 350°C resulted in complete loss of solubility [6]. It is stable to water, concentrated acids (except HNO_3), and weak alkalies [7].

References:

[1] Blanchard, H.; Hursthouse, M. B.; Sullivan, A. C. (J. Organometal. Chem. **341** [1988] 367/71).
[2] Zhdanov, A. A.; Andrianov, K. A.; Levitskii, M. M. (Izv. Akad. Nauk SSSR Ser. Khim. **1976** 395/9; Bull. Acad. Sci. USSR Div. Chem. Sci. **25** [1976] 376/9).
[3] Bartholin, M.; Guyot, A. (Compt. Rend. C **264** [1967] 1694/6).
[4] Zhdanov, A. A.; Andrianov, K. A.; Levitskii, M. M. (Vysokomol. Soedin. A **18** [1976] 2509/14; Polym. Sci. [USSR] **18** [1976] 2864/71).
[5] Papkov, V. S.; Bulkin, A. F.; Zhdanov, A. A.; Slonimskii, G. L.; Andrianov, K. A. (Vysokomol. Soedin. A **19** [1977] 830/41; Polym. Sci. [USSR] **19** [1977] 962/76).
[6] Kirichenko, E. A.; Damaeva, A. D.; Markov, B. A.; Ivanova, S. M.; Yermakov, A. I. (Vysokomol. Soedin. A **17** [1975] 1482/5; Polym. Sci. [USSR] **17** [1975] 1703/9).
[7] Kreshkov, A. P.; Kirichenko, E. A.; Damaeva, A. D.; Kostylev, I. M. (Tr. Mosk. Khim. Tekhnol. Inst. No. 70 [1972] 191/3; C.A. **79** [1973] No. 19483).
[8] Damaeva, A. D.; Alekseev, A. A.; Akutin, M. S.; Kirichenko, E. A. (Izv. Vysshikh Uchebn. Zavedenii Khim. Khim. Tekhnol. **23** [1980] 482/5; C.A. **93** [1980] No. 150964).
[9] Damaeva, A. D.; Mashutina, G. G.; Kirichenko, E. A. (Deposited Doc. VINITI-5072-80 [1980] 1/15; C.A. **96** [1982] No. 104885).
[10] Aref'eva, L. G.; Golubkova, G. L.; Doronina, T. V. (Deposited Doc. VINITI-395-84 [1983] 56/8).

[11] Damaeva, A. D.; Mashutina, G. G.; Kirichenko, E. A. (Vysokomol. Soedin. A **24** [1982] 1684/9; Polym. Sci. [USSR] **24** [1982] 1920/6).
[12] Damaeva, A. D.; Ponomarev, V. S.; Kirichenko, E. A.; Buinitskaya, T. A. (Deposited Doc. VINITI-3630 [1979] 1/10; C.A. **94** [1981] No. 85041).
[13] Popov, V. M. (Deposited Doc. VINITI-575-82 [1981] 104/8; C.A. **98** [1983] No. 198846).
[14] Saratov, I. E.; Khankhodzhaeva, D. A.; Reikhsfel'd, V. O.; Lazarev, S. Ya.; Borgul, E. A. (Kremniiorg. Soedin. Mater. Ikh Osn. Tr. 5th Soveshch. Khim. Prakt. Primen. Kremniiorg. Soedin., Leningrad 1981 [1984], pp. 111/3; C.A. **102** [1985] No. 25344).
[15] Saratov, I. E.; Khankhodzhaeva, D. A.; Reikhsfel'd, V. O. (Zh. Obshch. Khim. **51** [1981] 111/5; Russ. J. Inorg. Chem. **51** [1981] 99/102).
[16] Damaeva, A. D.; Kirichenko, E. A.; Ponomarev, V. S. (Vysokomol. Soedin. B **23** [1981] 494/7; C.A. **95** [1981] No. 151507).
[17] Kirichenko, E. A.; Markov, B. A.; Damaeva, A. D.; Grigor'ev, V. V. (Vysokomol. Soedin. A **18** [1976] 1508/14; Polym. Sci. [USSR] **18** [1976] 1726/34).
[18] Kreshkov, A. P.; Kirichenko, E. A.; Damaeva, A. D.; Kostylev, I. M. (Plasticheskie Massy **1980** No. 6, pp. 30/1; C.A. **93** [1980] No. 169028).
[19] Kostylev, I. M.; Shmanev, S. V.; Kirichenko, E. A.; Kopylov, V. M. (Plasticheskie Massy **1988** No. 5, pp. 18/9; C.A. **109** [1988] No. 38821).
[20] Shmanev, S. V.; Lukina, S. P.; Kostylev, I. M.; Kirichenko, E. A.; Kopylov, V. M. (Lakokras. Mater. Ikh Primen. **1988** No. 6, pp. 30/1; C.A. **110** [1989] No. 137025).

[21] Kirichenko, E. A.; Damaeva, A. D.; Kostylev, I. M. (Zh. Fiz. Khim. **52** [1978] 733/4; Russ. J. Phys. Chem. **52** [1978] 413/4).
[22] Panchenkov, G. M.; Kolesnikov, I. M.; Andrianov, K. A.; Zhdanov, A. K.; Belov, N. N.; Levitskii, M. M. (Izv. Akad. Nauk SSSR Ser. Khim. **1974** 2652; Bull. Acad. Sci. USSR Div. Chem. Sci. **1974** 2567).

38.3 Complexes with Bis(triorganylsilyl)amines

38.3.1 General

Silylamides have been used intensively to stabilize low coordination numbers and oxidation states of several transition metal complexes. Stable, two-coordinate, open shell compounds are currently known only in the cases of Mn^{II}, Fe^{II}, and Co^{II}. Characteristic features of these complexes are the small number of electrons in their valence shells and a high reactivity toward protic reagents and nucleophiles. A survey of this complex type is given in [1].

The $Mn^{II}L_2$ complex formed with bis(trimethylsilyl)amine (= hexamethyldisilazane) is dimeric in the solid state and monomeric in the gas phase and organic solvents. Mixed ligand complexes of composition $Mn^{II}L_2 \cdot L'_2$ are formed with additional ligands, L′ (thf, py, C_6H_6). The $[MnL_2(thf)]$ complex can be distilled from the reaction mixture of $MnCl_2 \cdot 2thf$ and lithium bis(trimethylsilyl)amide in tetrahydrofuran. Unsymmetrically-substituted ("hemi") compounds of the type X–Mn–L (X = Cl, NO_3), which exhibit only moderate thermal stability, were also prepared.

Bis(methyldiphenylsilyl)amine forms a monomeric $Mn^{II}L_2$ complex, which does not give stable adducts due to the steric effects of the bulky phenyl group.

The $Mn^{III}L_3$ complex with HL = bis(trimethylsilyl)amine ligand is the first example of a well-characterized, three-coordinate manganese(III) complex.

Reference:

[1] Power, P. P. (Comments Inorg. Chem. **8** [1989] 177/202).

38.3.2 With Bis(trimethylsilyl)amine
$[(CH_3)_3Si]_2NH$ (= $C_6H_{19}NSi_2$ = hexamethyldisilazane = HL)

38.3.2.1 Manganese(II) Compounds

$\mathbf{[Mn(C_6H_{18}NSi_2)_2]_2}$ was first prepared by refluxing MnI_2 and sodium bis(trimethylsilyl)amide (1:2 mole ratio) in tetrahydrofuran for 2 h. Concentration and distillation of the raw product in vacuum (0.2 Torr) at 100°C yielded flesh-colored crystals, stable under nitrogen [1, 2]. A modification of this synthesis, using $MnCl_2$ and $Li(C_6H_{18}NSi_2)$ as starting materials, gave predominantly the $[Mn(C_6H_{18}NSi_2)_2(thf)]$ complex [3]. The removal of the tetrahydrofuran molecule can be achieved by heating the complex at 120°C in a stream of argon for about 1 h. Recrystallization from pentane yielded pink plate-like [4] or flesh-colored crystals [3]. By another modified preparative method, reported recently, $Li(C_6H_{18}NSi_2)_2 \cdot O(C_2H_5)_2$ and $MnCl_2$ (2:1 mole ratio) were stirred in THF at room temperature for 2 h. The solvent was removed in vacuum. The residue was heated at 120°C in a stream of argon for 1 h and then recrystallized from cold pentane (−30°C) to afford the crystalline compound. The complex melts at 55 to 60°C and can be distilled at 112 to 120°C (0.2 Torr) [4].

In view of the high volatility, the solid compound was originally assumed to be monomeric [1, 2]; however, X-ray structural data show the complex to be dimeric in the crystalline state [5, 6] and monomeric in the gas phase at 130 to 150°C and 1 Torr [4] or in organic solvents [2]. A dimeric structure in pentane at room temperature was found by [4]. Pink single crystals could be grown by cooling a concentrated solution in hexane to −30°C [6]. X-ray determination at room temperature showed that the complex crystallizes in the monoclinic system, space group $C2/c-C^6_{2h}$ (No. 15) [5, 6] with lattice constants a = 18.256(3), b = 14.982(1), c = 18.253(3) Å, β = 118.06(1)°; Z = 4; calculated density D = 1.13 g/cm³ [5]. At 140 K the parameters a = 17.997(2), b = 14.942(1), c = 18.636(2) Å, β = 121.24(1)°, and D_{calc} = 1.17 g/cm³ were found [6]. The structure was solved by the direct method from 3838 independent reflections to a final reliability of R = 0.0439 [5] or from 2672 reflections to R = 0.031 [6]. As shown in **Fig. 4**, the complex is a dimer with two bridging and two terminal bis(trimethylsilyl)amido groups, yielding an approximate D_2 molecular symmetry. Selected bond distances (in Å) in the coordination polyhedron at room temperature are: Mn(2)–N(3) 1.994, Mn(2)–N(1) 2.184, N(3)–Si(4,4′) 1.723, and N(1)–Si(1,2) 1.774. Selected bond angles are: N(1)–Mn(1)–N(1′) 98.9°, Mn(2)–N(1′)–Mn(2) 81.1°, Si(4)–N(3)–Si(4′) 120°, Si(1)– N(1)–Si(2) 112°. The Mn–Mn distance of 2.841(1) Å and Mn–N–Mn bridge angle of 81.1(1)° imply direct metal-to-metal interactions [5]. The structure at 140 K is similar to that found at room temperature, but shows a contraction along the C_2 axis connecting the atoms Mn(1), Mn(2), N(2), and N(3). The Mn(1)–Mn(2) distance is shortened from 2.841 to 2.811 Å. Bond distances and angles and positional and thermal parameters at 140 K are given in the paper [6]. Both metal atoms and the four nitrogen atoms are coplanar. The geometry at each of the three-coordinate metal centers is distorted trigonal-planar with each atom being bonded to one terminal and two bridging amido groups. The geometry at the terminal N(2) and N(3) atoms is nearly trigonal-planar with the Si(3)–N(2)–Si(3) angle being 120.9(2)° and Si(1)–N(1)–Si(2) being 112.2(2)°. The bond distances and angles suggest strong covalent bonding between the metal and the ligands [6].

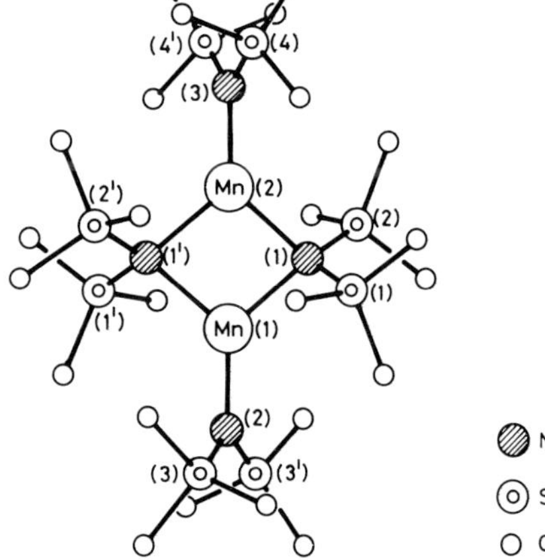

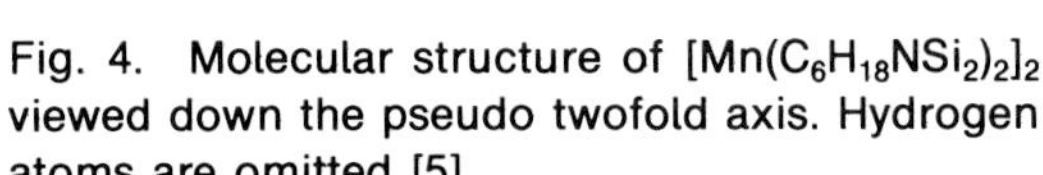
Fig. 4. Molecular structure of $[Mn(C_6H_{18}NSi_2)_2]_2$ viewed down the pseudo twofold axis. Hydrogen atoms are omitted [5].

The electron impact mass spectrum indicates that the complex is monomeric in the gas phase. The most intensive peak was that of the monomeric parent ion at m/e 375, with intense peaks at m/e 360 and 275. Gas electron diffraction data support the monomeric structure with the Mn–N bond distance of 1.95 Å and the N–Mn–N valence angle of 180°. Interligand strain is reduced by opening of the Si–N–Si angle to 132.1(2)° and rotation of the $Si(CH_3)_3$ group. Other interatomic distances, vibrational amplitudes, and valence angles are given in the paper. Comparison with the structure of the dimeric compound in the solid state shows that the dimer formation is accompanied by elongation of the terminal Mn–N bonds (0.05 Å) and compression of the Si–N–Si valence angle of the terminal ligands by 11° [4]. Both changes are interpreted as evidence for ligand-ligand repulsion and weak bridging in the dimer [4, 7]. SCF–MO calculations on high-spin $^6A_1$ $Mn(NH_2)_2$ yielded an average Mn–N bond distance of 1.93 Å, which is in close agreement with that of the monomeric $Mn(C_6H_{18}NSi_2)_2$. A high-spin configuration with very polar Mn–N bonding and of negligible Mn–N (d_π–p_π) bonding was suggested. Magnetic susceptibility measurements on the solid in the 5 to 290 K temperature range show that the compound does not follow the Curie-Weiss law and at higher temperatures the magnetism becomes temperature-independent [4]. The magnetic moments, $\mu_{eff} = 3.26$ and 3.34 μ_B, were calculated at 282 [4] and 296 K [5], respectively. Strong antiferromagnetic coupling, either directly between the two Mn^{2+} high-spin d^5 centers or by a super exchange, was supposed [4, 5]. The He(I) and He(II) photoelectron spectra of $[Mn(C_6H_{18}NSi_2)_2]_2$ recorded at 45°C showed bands with ionization energies at 7.9, 8.5, 9.1, and 11.4 eV, assigned to ionization from the 3 dσ orbitals, N lone pairs, Mn–N, and N–Si bond orbitals, respectively [4].

$[Mn(C_6H_{18}NSi_2)_2]_2$ is readily soluble in organic solvents [2, 3]. It is extremely oxygen- and moisture-sensitive and in contact with air turns black immediately. It is very reactive toward protic reagents and nucleophiles, supporting a weak dimer association [7]. Mixed ligand complexes $[Mn(C_6H_{18}NSi_2)_2 \cdot L'_2]$ were obtained with an excess of the ligand L' = thf, py, C_6H_6, or t-C_4H_9CN [3]. $[Mn(C_6H_{18}NSi_2)_2]_2$ reacts with *tert*-butyl alcohol at room temperature, but refluxing in benzene is desirable to obtain the pure alkoxide $Mn(OC_4H_9$-$t)_2$ as pale pink crystals [8]. An extremely oxygen- and water-sensitive surface compound between Mn^{II}/SiO_2 and $Mn(C_6H_{18}NSi_2)_2$, suitable for the purifications of gases, liquids, and polymerizable olefins, was reported in [9].

$[Mn(C_6H_{18}NSi_2)_2(thf)]$. The pink compound was observed first by [5]. It can be distilled from the reaction mixture of $MnCl_2 \cdot 2thf$ and lithium bis(trimethylsilyl)amide in tetrahydrofuran at 88°C and 10^{-3} Torr. According to [10] it was prepared by refluxing anhydrous $MnBr_2$ and the lithium salt of the ligand (1:2 mole ratio) in THF for 5 h. After removing the solvent in vacuum, the residue was extracted with hexane and LiBr was filtered off. The hexane was removed in vacuum and the residue distilled as a light rose-colored liquid at 90 to 100°C and 10^{-3} Torr, which crystallized as a pale flesh-colored solid upon standing at room temperature. The crystallographic analysis showed the structure to contain two independent molecules with very similar geometries. One molecule, a monomeric, three-coordinated species, is illustrated in **Fig. 5**. The main features of the molecular geometry are a planar MnN_2O unit with large N–Mn–N angles (145° and 150°) and small N–Mn–O angles (101° to 113°). The structure may arise from the attachment of the tetrahydrofuran ligand (followed by only a small rearrangement of the silylamide ligands) to $Mn(C_6H_{18}NSi_2)_2$, which has been shown to be a dimer with a linear D_{2d} symmetry. The Mn–O distance of 2.16(2) Å is in reasonable agreement with other Mn^{II}–O bond lengths, whereas the average Mn–N distance of 1.99(2) Å is extremely short. The average N–Si distance of 1.69(2) Å suggests only little or no Mn–N π interaction, but even increased N → Si π bonding. These distances are apparently due to the high thermal motion of the molecules in the crystals which melt at about 50°C [11], see also [12]. Susceptibility measurements yield a magnetic moment of $\mu_{eff} = 5.91\,\mu_B$, consistent with a high-spin three-coordinate $Mn^{II}(d^5)$ geometry [11].

The complex is extremely air-sensitive, but less than $[Mn(C_6H_{18}NSi_2)_2]_2$. It is thermally quite stable and may be distilled at 88°C and 10^{-3} Torr several times without losing the tetrahydrofuran molecule [3, 5]. The removal of the THF can be achieved by heating the adduct at 120°C in a stream of argon for ~1 h [5]. A series of novel air-sensitive alkoxides, $Mn(OR)_2$, including highly sterically-hindered derivatives, were prepared by alcoholysis of $[Mn(C_6H_{18}NSi_2)_2(thf)]$ in pentane [13]. Refluxing with $(t\text{-}C_4H_9)_2CHOH$ in a mixture of hexane and benzene for 2 h yields the trimer $[Mn_3\{OCH(C_4H_9\text{-}t)_2\}_6]$ by complete ligand exchange [14]. However, in the presence of lithium bis(trimethylsilyl)amide the ternary complex $[LiMn(C_6H_{18}NSi_2)\{OC(C_4H_9\text{-}t)_3\}_2]$ was obtained after 20 h of stirring [10]. The reaction with *tert*-butylcyanide or pyridine in pentane afforded complete replacement of tetrahydrofuran by two molecules of these reagents, whereas the reaction with benzene or toluene gave erratic results even after refluxing and long reaction times [3]. An extremely oxygen-sensitive $\geqslant Si\text{–}O\text{–}Mn(C_6H_{18}NSi_2)$ surface compound was obtained by reacting $[Mn(C_6H_{18}NSi_2)_2(thf)]$ with silcia gel in pentane [15].

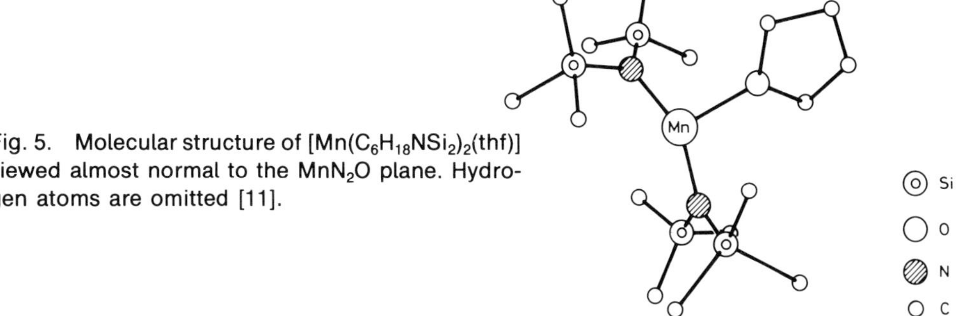

Fig. 5. Molecular structure of $[Mn(C_6H_{18}NSi_2)_2(thf)]$ viewed almost normal to the MnN_2O plane. Hydrogen atoms are omitted [11].

$[Mn(C_6H_{18}NSi_2)_2L'_2]$ (L′ = thf, py, $t\text{-}C_4H_9CN$, C_6H_6). The complex with tetrahydrofuran was obtained by refluxing $MnCl_2 \cdot 2thf$ and lithium bis(trimethylsilyl)amide (1:2 mole ratio) in tetrahydrofuran for 5 h, after which the solvent was removed at 50°C under vacuum. The residue was then extracted with cyclohexane, LiCl was filtered off, and the cyclohexane removed in vacuum at 50°C. The orange-brown residue, distilled at a pressure of 10^{-3} mbar gives at 75 to 95°C a yellow crystallizable forerun consisting of $[Mn(C_6H_{18}NSi_2)_2(thf)_2]$, followed by a rose-colored liquid main fraction of $[Mn(C_6H_{18}NSi_2)_2(thf)]$ at 90 to 120°C. By dissolving $[Mn(C_6H_{18}NSi_2)_2(thf)]$ in excess tetrahydrofuran at room temperature with stirring for several hours $[Mn(C_6H_{18}NSi_2)_2(thf)_2]$ is formed again. The solvent was removed in vacuum at 25°C and the residue was recrystallized twice from pentane to yield pale yellow crystals [3].

$[Mn(C_6H_{18}NSi_2)_2py_2]$ and $[Mn(C_6H_{18}NSi_2)_2(t\text{-}CH_9CN)_2]$ were prepared by adding pyridine or *tert*-butylcyanide dropwise to a pentane solution of $[Mn(C_6H_{18}NSi_2)_2]_2$. In the first case the solution turned yellow-green and upon addition of excess pyridine a yellow precipitate resulted, which was recrystallized from pentane and dried at 25°C at 10^{-3} mbar for 1 h. In the case of *tert*-butylcyanide the pale yellow solution deposited large pale rose crystals after several days of standing at −20°C; these were recrystallized from pentane and dried as above. The complex, $[Mn(C_6H_{18}NSi_2)_2(C_6H_6)_2]$, was prepared by refluxing $[MnCl(C_6H_{18}NSi_2)(thf)]$ (see below) in benzene for several hours. Evaporation of the benzene produced a residue from which yellow crystals could be isolated via pentane extraction. These were dried as above. It could be shown by X-ray structure analysis (data not published) that all the $[Mn(C_6H_{18}NSi_2)_2L'_2]$ complexes have tetrahedral configurations. The complexes are quite thermally stable and sensitive to air and moisture, but less so than $[Mn(C_6H_{18}NSi_2)_2]_2$ [3].

$[Mn(C_6H_{18}NSi_2)Cl]_n$ was obtained on heating $[Mn(C_6H_{18}NSi_2)Cl(thf)]$ (see p. 30) to 90°C at 10^{-3} mbar. After the $[Mn(C_6H_{18}NSi_2)_2(thf)_2]$, produced by disproportionation, was distilled off, the residue was refluxed with benzene for several hours and filtered while hot. Removal of benzene left a light yellow residue, which was extracted twice with pentane (50 mL) and dried at 50°C under high vacuum to give a beige-colored amorphous powder. The compound is most likely a coordination polymer, which upon heating to moderately higher temperatures rapidly disproportionates into $MnCl_2$ and $[Mn(C_6H_{18}NSi_2)_2]_2$. It is insoluble in pentane and reacts only slowly with oxygen [3].

$[Mn(C_6H_{18}NSi_2)(OC_4H_9\text{-}i)]$ was obtained by adding the lithium salt of *tert*-butyl alcohol in pentane to a solution of $[MnCl(C_6H_{18}NSi_2)(thf)]$ in tetrahydrofuran. The red-orange mixture was evaporated to dryness at room temperature under vacuum, the residue was dissolved in cyclohexane, and the LiCl was filtered off. The solvent was removed as before, giving a rust-red oily residue, which formed rose-colored crystals at −20°C after several days. The complex is more sensitive to oxygen than the pure Mn^{II} alkoxide and disproportionates into $Mn(OC_4H_9\text{-}i)_2$ and $Mn(C_6H_{18}NSi_2)_2$ upon storing at room temperature over longer periods of time [3].

$[Mn(C_6H_{18}NSi_2)X(thf)]$ ($X = Cl, NO_3$). The nitrato complex was prepared by stirring a solution containing equimolar amounts of $[Mn(C_6H_{18}NSi_2)_2(thf)_2]$ (see above) and $Mn(NO_3)_2 \cdot 3thf$ in tetrahydrofuran at room temperature for 3 d (protected from light). The solvent was then removed under vacuum at 20°C and the ocher-colored residue was dissolved in pentane. Evaporation of the solvent yielded a beige-colored product which was dried at 20°C and 10^{-3} mbar for 3 h and found to be very sensitive to light and oxygen. The chloro complex was obtained from $MnCl_2 \cdot 2thf$ and equimolar amounts of lithium bis(trimethylsilyl)amide in a tetrahydrofuran-hexane mixture. After 48 h the solvent was evaporated from the clear orange-red solution under vacuum at room temperature. The oily residue was extracted twice with ice-cold pentane, then dissolved in benzene by stirring at 25°C (15 h). After separation of the LiCl and removal of benzene, the yellow-brown oily residue was extracted with cold pentane and dried at 25°C and 10^{-3} mbar for 2 h. The chloro complex could also be obtained by reacting a solution of $[Mn(C_6H_{18}NSi_2)_2(thf)_2]$ in tetrahydrofuran with somewhat less than an equimolar amount of $MnCl_2$. The brilliant yellow product is sensitive to oxygen and light. It decomposes rapidly above 50°C with formation of $MnCl_2 \cdot 2thf$ and $[Mn(C_6H_{18}NSi_2)_2(thf)_2]$. The complex reacts with alcohols and phenols (ROH) to give $[MnCl(OR)(thf)_n]$ complexes and with alkyllithium (LiR) or lithium alkoxides (LiOR) to give compounds of the type $[Mn(C_6H_{18}NSi_2)(R)(thf)_n]$ or $[Mn(C_6H_{18}NSi_2)OR]_x$, respectively [3].

$[Li(thf)Mn(C_6H_{18}NSi_2)_3]$ was prepared as follows: Anhydrous $MnBr_2$ (3.9 g) and lithium bis(trimethylsilyl)amide (10 g) were added to 65 mL of chilled (−30°C) tetrahydrofuran. The mixture was refluxed for 5 h. After cooling to 20°C the solvent was removed in vacuum and the residue was distilled at 88 to 100°C and 4×10^{-3} Torr to give a pink oil which crystallized upon cooling to 20°C. The product was contaminated with a small amount of $[Mn(C_6H_{18}NSi_2)_2(thf)]$ but could be recrystallized from a concentrated hexane solution cooled to −30°C. The complex was also obtained by treating $[Mn(C_6H_{18}NSi_2)_2(thf)]$ with an equivalent amount of $Li(C_6H_{18}NSi_2)$ in tetrahydrofuran. Large well-shaped pale pink single crystals were grown by cooling a concentrated hexane solution to −30°C. After the crystals were removed from a Schlenk tube inside a dry-box, they were protected from air contamination by a layer of epoxy resin [6].

The complex melts at 125°C. X-ray diffraction studies at 140 K revealed a monoclinic lattice, space group $P2_1/n\text{-}C_{2h}^5$ (No. 14) with a = 11.678(2), b = 19.362(2), c = 17.020(3) Å, and β = 108.77(1)°; Z = 4. The structure was refined to R = 0.031 with anisotropic thermal parameters for the nonhydrogen atoms and isotropic for hydrogen. Atomic coordinates and temperature factors are given in the paper. The molecular structure is shown in **Fig. 6**.

The atoms N(1), N(2), N(3), Mn, Li, and O are essentially coplanar. The dihedral angle between the Mn, N(1), N(2), N(3) plane and the Mn, N(3), Si(5), Si(6) plane of 44.3° is believed to be a compromise of steric factors which favor an angle of 90°, and electronic (ligand-to-metal π-bonding) factors, which favor a zero dihedral angle. The Mn↔Li distance is similar to other structures described as strongly associated ion pairs, where the Li^+ ion acts as an electron-rich partner due to electron donation from a nearby donor molecule. Such excess charge permits electron transfer to the π-acceptor orbitals on the bridging ligands. $[Li(thf)Mn(C_6H_{18}NSi_2)_3]$ may also be considered an acid-base complex in which the lone electron pairs of the amido nitrogens can donate electrons either to the Mn or the Li atoms. The N–Si distances in the bridging amido groups (~1.73 Å) support a compromise between these two possibilities. The tetrahydrofuran molecule is not disordered. The density, D = 1.12 g/cm³, was calculated from the X-ray data.

Selected interatomic distances and angles are shown below [6]:

distances in Å		angles in °	
Mn–N(1)	2.143(3)	N(1)–Mn–N(2)	99.0(1)
Mn–N(2)	2.143(2)	N(1)–Mn–N(3)	133.2(1)
Mn–N(3)	2.023(3)	N(2)–Mn–N(3)	127.7(1)
Mn···Li	2.718(6)	N(1)–Mn–Li	49.6(1)
Li–N(1)	2.104(6)	N(2)–Mn–Li	49.5(1)
Li–N(2)	2.101(6)	N(3)–Mn–Li	173.6(1)

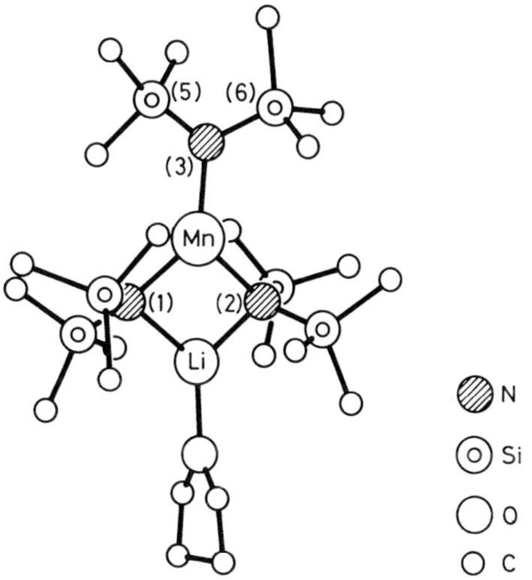

Fig. 6. Molecular structure of $[Li(thf)Mn(C_6H_{18}NSi_2)_3]$. Hydrogen atoms are omitted [6].

The complex is very soluble in hydrocarbons due to the ability of the $N[Si(CH_3)_3]_2^-$ groups in providing a hydrocarbon-like exterior. The solutions, however, are exceptionally air-sensitive and darken immediately to give violet or brown solutions, if small controlled amounts of O_2 are admitted. In the presence of excess O_2 the solutions turn brown to black rapidly and deposit solids which are dissolved only by mineral acids [6].

$[Li(C_{13}H_{27}O)_2Mn(C_6H_{18}NSi_2)]$, containing $(t\text{-}C_4H_9)_3CO^-$ groups ($=C_{13}H_{27}O^-$), was prepared by adding a solution of $(t\text{-}C_4H_9)_3COH$ (1.79 g) in tetrahydrofuran (10 mL) to a solution of $[Mn(C_6H_{18}NSi_2)_2(thf)]$ (2.00 g) and lithium bis(trimethylsilyl)amide (0.75 g) in hexane (30 mL). The mixture was stirred for 20 h and the volume was reduced to 15 mL. Slow cooling to −30°C yielded pale pink crystals melting at 158 to 160°C. Large, well-shaped, pale pink single crystals could be grown by cooling a concentrated hexane-THF solution to −30°C. X-ray diffractometer studies at 140 K show the complex to belong to the monoclinic system, space group Cc–C_s^4

(No. 9) with the lattice constants a = 21.243(4), b = 11.814(1), c = 20.334(3) Å, and β = 133.6(1)°; Z = 4. The structure was refined to R = 0.033. Atomic coordinates for nonhydrogen atoms are given in the paper. Selected bond distances and angles are given below [10]:

distances in Å		angles in °		angles in °	
Mn–O(1)	2.019(4)	O(1)–Mn–N	134.8(2)	N–Mn–Li	171.3(2)
Mn–O(2)	1.980(4)	O(2)–Mn–N	138.2(3)	Mn–O(1)–Li	87.1(4)
Mn–N	2.001(3)	O(1)–Mn–O(2)	86.4(2)	Mn–O(1)–C(1)	150.7(3)
Mn···Li	2.640(7)				

As observed for the complex [Li(thf)Mn($C_6H_{18}NSi_2$)$_3$], see above, the Mn atom is three-coordinated in a distorted trigonal planar geometry. The molecular structure is shown in **Fig. 7**.

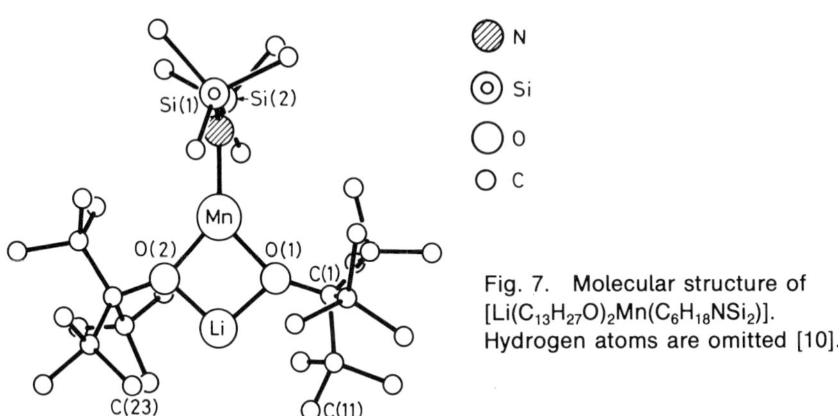

Fig. 7. Molecular structure of [Li($C_{13}H_{27}O$)$_2$Mn($C_6H_{18}NSi_2$)]. Hydrogen atoms are omitted [10].

The dihedral angle is 88.1°, indicating the dominant influence of packing effects in the crystal structure. The Mn–N distance of 2.001(3) Å is consistent with a normal terminal Mn–N[Si(CH_3)$_3$]$_2$ bond, but the Mn–OC(C_4H_9-*t*)$_3$ distances of 1.980 and 2.019 Å are considerably shorter than the Mn–OC_4H_8 distance of 2.16 Å in [Mn($C_6H_{18}NSi_2$)$_2$(thf)], see p. 28. The tertiary butyl groups in each alkoxo ligand are staggered with respect to each other, which disrupts the near C_2 axis through N, Mn, and Li. Such staggering allows weak interactions between Li and the methyl groups at C(11) and C(23) atoms. The steric bulk of the alkoxo groups prevents the addition of tetrahydrofuran to the formally two-coordinate lithium. Further evidence for steric congestion in [Li($C_{13}H_{27}O$)$_2$Mn($C_6H_{18}NSi_2$)] is that it does not form a trisalkoxo complex even on prolonged (20 h) treatment with excess (*t*-C_4H_9)$_3$COH in refluxing toluene.

The calculated density (D_{calc} = 1.12 g/cm^3) and the chemical behavior of the compound are the same as described for [Li(thf)Mn($C_6H_{18}NSi_2$)$_3$] above. [Li($C_{13}H_{27}O$)$_2$Mn($C_6H_{18}NSi_2$)] is also very soluble in hydrocarbons due to the (*t*-C_4H_9)$_3CO^-$ groups providing a hydrocarbon-like exterior [10].

References:

[1] Bürger, H.; Wannagat, U. (Monatsh. Chem. **95** [1964] 1099/102).
[2] Bürger, H. (Proc. 8th Intern. Conf. Coord. Chem., Vienna 1964, pp. 171/3).
[3] Horvath, B.; Möseler, R.; Horvath, E. G. (Z. Anorg. Allgem. Chem. **450** [1979] 165/77).

[4] Andersen, R. A.; Faegri, K.; Green, J. C.; Haaland, A.; Lappert, M. F.; Leung, W. P.; Rypdal, K. (Inorg. Chem. **27** [1988] 1782/6).
[5] Bradley, D. C.; Hursthouse, M. B.; Malik, K. M. A.; Möseler, R. (Transition Metal Chem. [Weinheim] **3** [1978] 253/4).
[6] Murray, B. D.; Power, P. P. (Inorg. Chem. **23** [1984] 4584/8).
[7] Power, P. P. (Comments Inorg. Chem. **8** [1989] 177/202).
[8] Bochmann, M.; Wilkinson, G.; Young, G. B.; Hursthouse, M. B.; Malik, K. M. A. (J. Chem. Soc. Dalton Trans. **1980** 1963/71).
[9] Möseler, R.; Horvath, B.; Lindenau, D.; Horvath, E. G.; Krauss, H. L. (Z. Naturforsch. **31b** [1976] 892/3).
[10] Murray, B. D.; Power, P. P. (J. Am. Chem. Soc. **106** [1984] 7011/5).

[11] Eller, P. G.; Bradley, D. C.; Hursthouse, M. B.; Meek, D. W. (Coord. Chem. Rev. **24** [1977] 1/95, 16/7).
[12] Bradley, D. C. (Chem. Brit. **11** [1975] 393/7).
[13] Horvath, B.; Möseler, R.; Horvath, E. G. (Z. Anorg. Allgem. Chem. **449** [1979] 41/51).
[14] Murray, B. D.; Hope, H.; Power, P. P. (J. Am. Chem. Soc. **107** [1985] 169/73).
[15] Horvath, B.; Strutz, J.; Möseler, R.; Horvath, E. G. (Z. Anorg. Allgem. Chem. **449** [1979] 5/24).

38.3.2.2 Manganese(III) Compounds

$Mn(C_6H_{18}NSi_2)_3$ was prepared by treating $[Mn^{II}(C_6H_{18}NSi_2)_2]_2$ (see p. 26) with an equimolar amount of $[(CH_3)_3Si]_2NBr$ (obtained from bis(trimethylsilyl)amine and N-bromosuccinimide in CCl_4 at 15°C) in toluene solution at ~0°C. The reaction mixture was reduced and the violet rod-like crystals formed after cooling at about −20°C [1]. Previous attempts to synthesize the complex by using $Li(C_6H_{18}NSi_2)$ and various Mn^{III} precursor salts, or through electrolytic oxidation of the divalent compound, were unsuccessful [2]. The complex melts at 108 to 110°C. The X-ray structure determination at 130 K reveals a trigonal lattice, space group $P\bar{3}1c-D_{3d}^2$ (No. 163) with a = 16.082(5), c = 8.436(2) Å; Z = 2. The structure was solved to R = 0.0695. As shown in **Fig. 8**, the complex is monomeric with a trigonal planar MnN_3 framework. The bond distances are Mn–N = 1.890(3) Å and Si–N = 1.755(2) Å; the Si–N–Si angle is 120.4(1)°. Consistent with a higher oxidation state of the metal, the Mn–N distance of 1.890(3) Å is significantly shorter

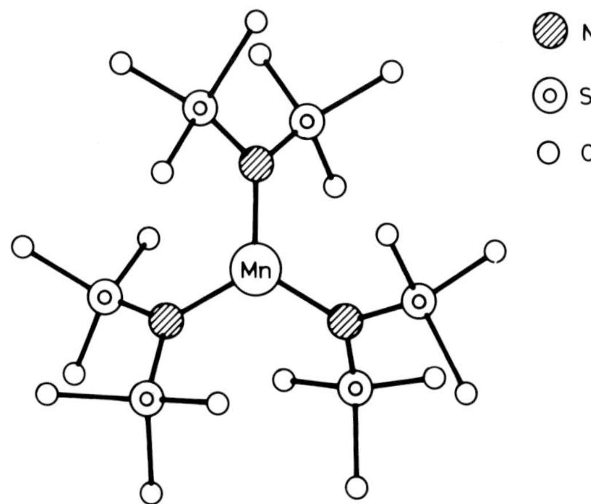

Fig. 8. Molecular structure of $Mn(C_6H_{18}NSi_2)_3$. Hydrogen atoms are omitted [1].

than the terminal Mn–N bond in $[Li(thf)Mn^{II}(C_6H_{18}NSi_2)_3]$ or in the dimer $[Mn(C_6H_{18}NSi_2)_2]_2$ complex, both of which involve three-coordination at the manganese(II) center. The NSi_2 plane forms a dihedral angle of 50° with the MnN_3 plane. The complex is isomorphous with analogous silylamide complexes of Al^{III}, Ga^{III}, Ti^{III}, V^{III}, Cr^{III}, and Fe^{III}.

The magnetic moment in C_6D_6/C_6H_6 solution is $\mu_{eff}=5.38$ μ_B at 297.3 K. The electronic spectrum in pentane solution shows band maxima (ε values in $L \cdot mol^{-1} \cdot cm^{-1}$ in parentheses) at 33800 (7120), 26740 (4220), 21280 (4160), and 17600 (3890) cm^{-1}. The lowest energy absorptions at 17600 and 21280 cm^{-1} were assigned to $^5E' \rightarrow {}^5E''$ and $^5E' \rightarrow {}^5A_1'$ transitions. An additional band at 33800 cm^{-1} was interpreted as evidence for luminescence. The ^1H NMR spectrum of the complex at 297.3 K displays a broad singlet at 27.3 ppm. The complex is freely soluble in relatively inert hydrocarbons such as pentane or toluene [1].

References:

[1] Ellison, J. J.; Power, P. P.; Shoner, S. C. (J. Am. Chem. Soc. **111** [1989] 8044/6).
[2] Alyea, E. C.; Bradley, D. C.; Copperthwaite, R. G. (J. Chem. Soc. Dalton Trans. **1972** 1580/4).

38.3.3 With Bis(methyldiphenylsilyl)amine $[CH_3(C_6H_5)_2Si]_2NH$ ($=C_{26}H_{27}NSi_2=HL$)

$Mn^{II}(C_{26}H_{26}NSi_2)_2$ is formed by the reaction of $Li(C_{26}H_{26}NSi_2)$ with manganese(II) iodide: An ice-cooled solution of the ligand (8 mmol in 5 mL THF) was treated dropwise with n-butyllithium (8 mmol in 5 mL hexane). After stirring for 1 h, MnI_2 (4 mmol) was added from a solid addition tube and the reaction mixture was stirred for 12 h at ambient temperature. The resulting pink solution was then evaporated to dryness under vacuum, the residue dissolved in toluene (25 mL), and filtered. The volume of the solution was halved under reduced pressure, and ca. 10 mL hexane was added. Filtration and a slight volume reduction resulted in the appearance of pink crystals, which were redissolved again. Cooling to −20°C over 24 h yielded pale pink crystals melting at 157 to 159°C (dec.).

The X-ray structure determination at 130 K reveals a monoclinic lattice, space group $P2_1/c-C_{2h}^5$ (No. 14), with a = 10.893(1), b = 15.399(6), c = 27.049(4) Å, β = 91.73(1)°; Z = 4. The structure was solved to R = 0.040. Selected bond distances and angles are given below [1]:

distances	in Å	angles	in °
Mn–N(1)	1.989(3)	N(1)–Mn–N(2)	170.7(1)
Mn–N(2)	1.988(3)	Si(1)–N(1)–Si(2)	127.7(2)
N(1)–Si(1)	1.719(3)		
N(1)–Si(2)	1.706(3)	Si(3)–N(2)–Si(4)	131.8(2)
N(2)–Si(3)	1.708(3)	Mn–N(1)–Si(1)	107.5(1)
N(2)–Si(4)	1.716(3)	Mn–N(2)–Si(3)	116.7(1)

The complex is monomeric, in the main, due to steric effects. Four of the eight phenyl rings shield the metal atom in an approximately octahedral arrangement [2]. A significant feature of the structure is the two-coordination of the manganese atom and a near linear geometry, see **Fig. 9**. As shown by the N(1)–Mn–N(2) angle (170.7°), the deviation from linearity is slight. The next closest atom to manganese is C(7) with an approach of 2.774(5) Å. The weak interactions are reflected in asymmetries of 13° and 14.8° in the Mn–N(1)–Si(1) and Mn–N(1)–Si(2) angles. There is only negligible asymmetry in the angles at the N(2) ligand center, which involves no metal-carbon interactions. It is probable that both the N(1)–Mn–N(2) and Si–N–Si angles are relatively easily distorted and that bending in the complex could be due to packing forces. The Mn–N distances are quite close to the terminal Mn–N distances observed in the $[Mn(C_6H_{18}NSi_2)_2]_2$ dimer (see p. 27).

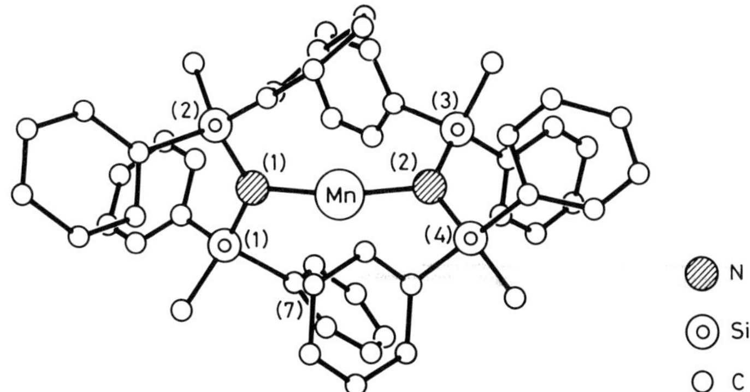

Fig. 9. Molecular structure of $Mn^{II}(C_{26}H_{26}NSi_2)_2$. Hydrogen atoms are omitted [1].

$Mn(C_{26}H_{26}NSi_2)_2$ is isomorphous with analogous complexes of Fe^{II} and Co^{II}; the M–N distances were found to decrease in the order Mn–N > Fe–N > Co–N. The magnetic moment is $\mu_{eff} = 5.72\ \mu_B$ at 300 K. The ESR spectrum recorded on a crystalline sample in the 7.6 to 9 K temperature range is characterized by a multiplicity of absorptions, suggesting a considerable number of low-lying energy states in the manganese system. Two major peaks, which are further split, were found near g values of 7.04 and 2.09. In the ^1H NMR spectrum in C_6D_6 no paramagnetically shifted peaks were observed, presumably due to the broadness of the resonances. The electronic spectrum in hexane is essentially featureless with an increase in intensity toward the short wavelengths. The complex does not form stable adducts with THF, due to the steric effects of the bulky phenyl rings [1].

References:

[1] Cheng, H.; Barlett, R. A.; Rasika Dias, H. V.; Olmstead, M. M.; Power, P. P. (J. Am. Chem. Soc. **111** [1989] 4338/45).

[2] Power, P. P. (Comments Inorg. Chem. **8** [1989] 177/202).

39 Complexes with Ligands Containing Phosphorus

Remark. Manganese phosphides, phosphates, thiophosphates, or selenophosphates have been described in "Manganese" C 9, 1983, pp. 1/232. In this chapter complexes with phosphanes are reported, followed by complexes with phosphane oxides, organyl phosphides, organyl phosphorus, oxo or thioxo acids and their derivatives, or complexes with heterocycles containing phosphorus.

39.1 Complexes with Monophosphanes

Survey. There are few data on manganese complexes with phosphane, PH_3, except for carbonyl phosphane complexes. These are not described here, but are reported along with organomanganese compounds. Reviews for carbonyl phosphane manganese complexes are given in [1 to 6].

This section treats predominantly manganese complexes with trialkyl-, triphenyl-, or mixed alkylphenyl phosphanes. Mainly phosphanes with CH_3 to C_4H_9 alkyl groups (short-chain phosphanes) are used as ligands, but complexes with phosphanes containing C_{12}, C_{14}, or C_{16} alkyl groups (long-chain phosphanes) have also been prepared. A large number of complexes, mostly of the $MnLX_2$ type, arises from the many possible combinations of the various phosphanes (L) and halides or pseudohalides (X). To aid the recognition of the R groups in the coordinated phosphanes, linearized formulas such as $Mn\{P(C_6H_5)(CH_3)_2\}X_2$ have been used rather than the empirical formulas.

$MnLX_2$ and MnL_2X_2 complexes are prepared by the reaction of the Mn^{II} halide with the phosphane in a noncoordinating solvent (e.g., toluene-dichloromethane) under absolutely anhydrous conditions. The successful preparation of these complexes was delayed for several years by the difficulty in procuring truly anhydrous MnX_2 and in identifying appropriate solvent systems.

The $MnLX_2$ complexes, which represent the most important group, are doubly halide-bridged oligomeric or polymeric compounds. The polymeric chains consist of alternating tetrahedral and octahedral subunits. The polymeric structure of the solid state is retained in noncoordinating solvents. It is broken by coordinating solvents, such as tetrahydrofuran, which add to the Mn to form monomeric octahedral complex molecules.

The $MnLX_2$ complexes and their tetrahydrofuran solvates are extraordinary in their ability to bind small molecules, especially gas molecules: O_2, CO, ethylene, and NO are bound reversibly, whereas SO_2, CS_2, and tetracyanoethylene are bound irreversibly. The ability to bind is attributed to the particular structure of the $MnLX_2$ complexes with vacant spaces above and below the tetrahedrally coordinated Mn atoms. The reactivity toward small molecules depends on the R groups at the phosphane ligands and on the halide. Equilibrium studies yielded reactivity sequences for the phosphanes and halides which are quite different for the individual gases. In the case of dioxygen the binding sequence for phosphanes is: $PR_3 > P(C_6H_5)R_2 > P(C_6H_5)_2R$. However, two extremes are observed: While $Mn\{P(C_6H_5)_3\}X_2$ complexes do not bind dioxygen at all, binding, which is strongest for trialkylphosphane complexes, is not reversible for $Mn\{P(CH_3)_3\}X_2$ complexes. These react irreversibly with dioxygen to give the Mn^{III} species, $Mn\{P(CH_3)_3\}_2X_3$. The binding ability of the $MnLX_2$ complexes is of possible importance for gas separations.

Of the adducts with the above small molecules, only those with O_2, SO_2, and NO are described in this section. Adducts containing Mn–C bonds will be treated along with organomanganese compounds. The vividly colored 1:1 adducts of $MnLX_2$ complexes with NO or O_2 are formed by exposure of the complexes to the gases in the solid state or in solution. While these adducts could not be isolated from the solution, yellow adducts containing 0.66 mol SO_2 per mol of $MnLX_2$ could be obtained from toluene. In addition to $Mn(NO)X_2$ complexes formed by the reaction of solid $MnLX_2$ complexes with NO gas, $Mn(NO)_2L_2X$ and $Mn(NO)_3L$ complexes are reported. Preparative procedures for the di- and trinitrosyl complexes are based in most cases on CO displacement from various manganese carbonyl complexes.

References:

[1] McAuliffe, C. A.; Levason, W. (Phosphine, Arsine, and Stibine Complexes of the Transition Elements; Studies in Inorganic Chemistry, Vol. I, Elsevier, Amsterdam – Oxford – New York 1979, pp. 90/1).
[2] Corbridge, D. E. C. (Phosphorus. An Outline of Its Chemistry, Biochemistry, and Technology; Studies in Inorganic Chemistry, Vol. 2, Elsevier, Amsterdam – Oxford – New York 1980, p. 444).
[3] Stelzer, O. (Top. Phosphorus Chem. **9** [1977] 1/229, 3/4).
[4] Fluck, E. (Fortschr. Chem. Forsch. **35** [1973] 1/64, 49/50).
[5] Robinson, S. D. (MTP [Med. Tech. Publ. Co.] Intern. Rev. Sci. Inorg. Chem. Ser. One **6** Pt. 2 [1972] 121/69, 123).
[6] Robinson, S. D. (Intern. Rev. Sci. Inorg. Chem. Ser. Two **6** [1975] 71/108, 73).

39.1.1 Complexes with PH_3

For the hypothetical dinuclear **$[Mn^{II}(PH_3)_2Cl_2]_2$** complex, which was assumed to have a triple Mn–Mn bond, the Mn–Mn bond energy, $\Delta E = -99$ kJ/mol, (and its various components due to steric, σ, π, and δ interactions) and the bond distance, $r_{Mn-Mn} = 1.92$ Å, were obtained by a Hartree-Fock-Slater calculation [1]. A previous simpler HFS calculation had given $\Delta E = -294.8$ kJ/mol and the same bond distance [2, 3].

Hypothetical **$Mn(PH_3)_4(N_2)Cl^{n+}$** complexes in various oxidation (Mn^0, Mn^I, and Mn^{II}) and spin states were studied by the CNDO method, assuming an equatorial arrangement of the PH_3 ligands and an "end-on" coordination of the N_2 ligand. The results show a significant delocalization of the Mn d orbitals and of the dinitrogen π^* orbitals; Mn to N_2 π back-bonding was found to be relatively low [4].

The formation of the **$[Mn^{VII}O_3(PH_3)(OH)X]^-$** species (X = Cl, Br, I) as intermediate complexes during the oxidation of PH_3 by $KMnO_4$ in the presence of X^- ions in aqueous solution (in a flowing gas reactor under isothermal conditions) is indicated by kinetic and potentiometric studies. The kinetic activity of the complexes increases in the order: Cl < Br < I. The corresponding complexes with AsH_3 (see p. 202) are more active than those with PH_3 [5, 6].

References:

[1] Ziegler, T.; Tschinke, V. (Polyhedron **6** [1987] 685/93).
[2] Ziegler, T. (J. Am. Chem. Soc. **106** [1984] 5901/8).
[3] Ziegler, T.; Tschinke, V.; Versluis, L. (NATO ASI Ser. C No. 176 [1986] 189/98).
[4] Pelikán, P.; Boča, R.; Magová, M. (J. Mol. Catal. **19** [1983] 243/55).
[5] Dorfman, Ya. A.; Polimbetova, G. S.; Mansurov, B. A.; Bikmukhametova, A. K.; Doroshkevich, D. M. (Koord. Khim. **14** [1988] 1219/23; C.A. **109** [1988] No. 238061).
[6] Bikmukhametova, A. K. (Avtoref. Diss. Kand. Nauk, Alma-Ata 1986 from Dorfman, Ya. A.; Polimbetova, G. S.; Mansurov, B. A.; Bikmukhametova, A. K.; Doroshkevich, D. M., Koord. Khim. **14** [1988] 1219/23, 1223).

39.1.2 Complexes with Triorganylphosphanes, PR_3 or PR_2R' (R and R' = Alkyl or Phenyl)

No.	ligand L	formula	No.	ligand L	formula
1	$P(CH_3)_3$	C_3H_9P	13	$P(C_6H_5)_3$	$C_{18}H_{15}P$
2	$P(C_2H_5)_3$	$C_6H_{15}P$	14	$P(C_6H_5)_2CH_3$	$C_{13}H_{13}P$
3	$P(C_3H_7)_3$	$C_9H_{21}P$	15	$P(C_6H_5)_2C_2H_5$	$C_{14}H_{15}P$
4	$P(C_4H_9)_3$	$C_{12}H_{27}P$	16	$P(C_6H_5)(CH_3)_2$	$C_8H_{11}P$
5	$P(C_4H_9\text{-}t)_3$	$C_{12}H_{27}P$	17	$P(C_6H_5)(C_2H_5)_2$	$C_{10}H_{15}P$
6	$P(C_5H_{11})_3$	$C_{15}H_{33}P$	18	$P(C_6H_5)(C_3H_7)_2$	$C_{12}H_{19}P$
7	$P(C_8H_{17})_3$	$C_{24}H_{51}P$	19	$P(C_6H_5)(C_4H_9)_2$	$C_{14}H_{23}P$
8	$P(C_{12}H_{25})_3$	$C_{36}H_{75}P$	20	$P(C_6H_5)(C_4H_9\text{-}i)_2$	$C_{14}H_{23}P$
9	$P(C_{14}H_{29})_3$	$C_{42}H_{87}P$	21	$P(C_6H_5)(C_{12}H_{25})_2$	$C_{30}H_{55}P$
10	$P(C_{16}H_{33})_3$	$C_{48}H_{99}P$	22	$P(C_6H_5)(C_{14}H_{29})_2$	$C_{34}H_{63}P$
11	$P(C_6H_{11}\text{-}c)_3$	$C_{18}H_{33}P$	23	$P(C_6H_5)(C_{16}H_{33})_2$	$C_{38}H_{71}P$
12	$P(CH_2\text{–}CH\text{=}CH_2)_3$	$C_9H_{15}P$	24	$P(C_6H_5)(CH_2CH_2C_6H_5)_2$	$C_{22}H_{23}P$

39.1.2.1 $Mn^{II}LX_2$ Compounds

39.1.2.1.1 Preparation

Absolutely crucial to the preparation of these complexes is the use of strictly anhydrous solvents and manganese(II) salts and the employment of dioxygen-free atmospheres [2].

Preparative procedures for $MnLX_2$ complexes with short-chain phosphane ligands, including the production of anhydrous MnX_2 samples, were developed and are reported in [1, 25]. For these complexes the MnX_2 compound and the phosphane have to be used in an exact 1:1 mole ratio [2]. As a typical example, the preparation of the "parent" compounds, $Mn\{P(C_6H_5)_3\}X_2$ with X = Cl, Br, I, NCS, will be described here: A solution of 4 mmol triphenylphosphane in 35 mL dichloromethane was added to 4 mmol of the MnX_2 salt in a 60:40 mixture of freshly-distilled toluene-dichloromethane (90 mL). The mixture was stirred under argon at room temperature for 5 days. Then the volume of the solution was reduced to ~25 mL in vacuum (heat was not applied). The resulting solid was separated by Schlenk techniques and dried under argon, then taken up in dichloromethane (50 mL). The solution was filtered and the volume again reduced (to ~10 mL). The resulting complexes were filtered and dried in vacuum. Yields are almost quantitative [1]. $Mn\{PR_3\}X_2$ and $Mn\{P(C_6H_5)R_2\}X_2$ with R = C_4H_9 for X = Cl, Br, I, and R = CH_3, C_2H_5, C_3H_7, C_4H_9 for X = NCS were prepared in the same way [2]. $Mn\{P(CH_2\text{–}CH\text{=}CH_2)_3\}Cl_2$ and $Mn\{P(C_6H_5)(CH_3)_2\}X_2$ with X = Cl, Br, I, NCS [5, 6], NCSe and $Mn\{P(C_6H_5)_3\}(NCSe)_2$ [4] were prepared in toluene [4 to 6]. For the preparation of the latter complex, the solvate $Mn(NCSe)_2 \cdot C_2H_5OH$ was used as the source of manganese [4]. During the attempts of other work groups to prepare $MnLX_2$ complexes with L = $P(C_3H_7)_3$ and $P(C_6H_5)(CH_3)_2$ for X = Cl, Br, I, or NCS [27, 35], satisfying results were obtained only for complexes with X = Br and I using diethyl ether as the solvent. Those with X = Cl and NCS were contaminated with unreacted halide. Use of toluene as the solvent led to greatly increased reaction times and lower purities. With toluene-dichloromethane the reaction time was comparable to that with ether, but inorganic halide was detected in the isolated solids [27]. In addition to toluene-dichloromethane and diethyl ether, pentane has also been used as the solvent system [26]. The choice of solvent is critical because solvents in which the $MnX_2\text{–}PR_3$ system is highly soluble enable the precipitation of the solvated manganese halide when the solvent volume is reduced [27]. $MnLBr_2$ complexes were prepared as films on KBr infrared windows by exposing $MnBr_2$ films to the appropriate phosphanes at their vapor pressure or at 400 Torr (in the case of $P(CH_3)_3$) for 30 to

120 min. The films formed were then subjected to evacuation at 10^{-6} Torr for 4 to 12 h at room temperature to remove all traces of excess phosphane [3, 8, 23]. Caution is urged in using the pyrophoric trimethylphosphane [29].

The compound $Mn\{P(C_6H_5)_3\}\{(CH_3)_3CCOO\}_2$, reported earlier in [37], was obviously prepared by the reaction of manganese(II) 2,2-dimethylpropanoate with triphenylphosphane in anhydrous benzene under Ar. The solution was concentrated at 50 to 60°C and 10 Torr until crystals appeared; these were washed with benzene-hexane and dried in vacuum (analogous to a method described for similar complexes in [38]).

In the preparation of $MnLX_2$ complexes with L = $P(C_{12}H_{25})_3$, $P(C_{14}H_{29})_3$, $P(C_{16}H_{33})_3$, $P(C_6H_5)(C_{12}H_{25})_2$, $P(C_6H_5)(C_{14}H_{29})_2$, or $P(C_6H_5)(C_{16}H_{33})_2$ for X = Cl, Br, or I, the mole ratio of MnX_2 compound and phosphane used need not be 1:1, in contrast to $MnLX_2$ complexes with short-chain phosphanes. The preparation of $Mn\{P(C_6H_5)(C_{16}H_{33})_2\}Br_2$ is described here as a typical example: A mixture of anhydrous $MnBr_2$ and $P(C_6H_5)(C_{16}H_{33})_2$ in pentane was refluxed at 40°C under Ar for 5 d. Then the mixture was cooled to room temperature. The solid complex was isolated by Schlenk techniques, washed with pentane, and dried in vacuum for 6 h (96% yield) [34].

The $MnLX_2$ complexes have different colors, depending on the halide or pseudohalide: white-pink for X = Cl or Br, pink-orange for X = I, yellow for X = NCS [1], and cream for X = NCSe [4].

39.1.2.1.2 Structure

$Mn\{P(C_6H_5)(CH_3)_2\}I_2$ crystallizes from ether in two forms: blue cubes and straw-colored needles, both with ~0.66 mol diethyl ether per Mn atom. The ether is lost from both crystal forms on removal from the mother liquor [27].

A single-crystal X-ray structure determination for this compound revealed a triclinic lattice, space group $P\bar{1}$–C_i^1 (No. 2) with the lattice parameters: a = 11.389(2), b = 15.227(3), c = 15.228(3) Å, α = 89.10(1)°, β = 97.51(1)°, γ = 89.53(1)°; Z = 2. The calculated density is 2.27 g/cm³. The structure was refined to a final reliability of R = 0.067. As shown in **Fig. 10**, eight $Mn\{P(C_6H_5)(CH_3)_2\}I_2$ units in the asymmetric unit cell form parts of infinite chains of Mn atoms

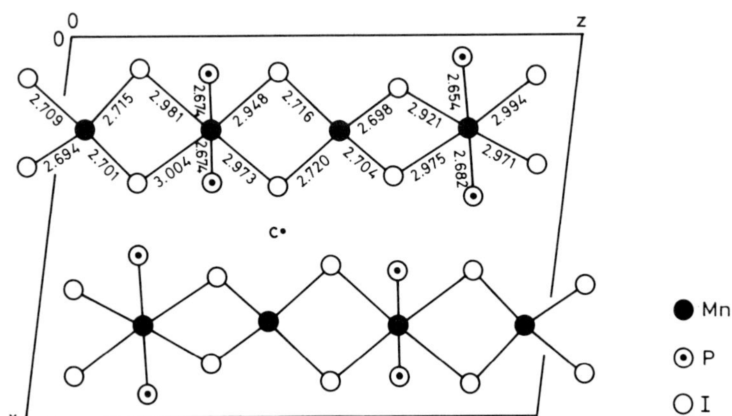

Fig. 10. Structure of $Mn\{P(C_6H_5)(CH_3)_2\}I_2$. View down y axis (methyl and phenyl groups omitted); bond lengths in Å [9].

References on pp. 47/9

linked by pairs of iodide bridges. The Mn atoms are alternately tetrahedrally coordinated by four I atoms and octahedrally coordinated by four equatorial I atoms and two axial P atoms (the P–Mn–P angle is almost exactly 180°). Alternate equatorial planes of I atoms are approximately perpendicular to each other. The Mn–P bond lengths (2.654 to 2.682 Å) are equivalent for each phosphane bound to an individual Mn atom. The Mn–I bond lengths in the octahedral units (2.921 to 3.004 Å) are much longer than those in the tetrahedral units (2.694 to 2.720 Å), causing increased electron density at the octahedral Mn centers. Adjacent chains are related by the center of symmetry at c. Bond lengths are shown in Fig. 10, p. 39; bond angles are not reported [9].

An infinite-chain polymeric structure can also be assumed for $Mn\{P(C_6H_5)_3\}(NCS)_2$ on the basis of IR studies (see p. 41), whereas these studies indicate only oligomeric structures (at least dimeric) for $Mn\{P(C_6H_5)_3\}Cl_2$ and $Mn\{P(C_6H_5)_3\}Br_2$ complexes [1]. Differences between the structures of $MnL(NCS)_2$ complexes with $L = P(C_3H_7)_3$ or $P(C_4H_9)_3$ (which are isomorphous) and those with $L = P(C_6H_5)R_2$, $P(C_6H_5)_2R$, or R_3PO are shown by the X-ray powder patterns for the two groups of compounds [2]. Thus, the $MnLX_2$ compounds with short-chain phosphane ligands are believed to be oligomeric or polymeric with the degree of polymerization dependent on the phosphane and X [1]. In contrast to complexes with short-chain phosphanes those with long-chain phosphanes have been assigned a dinuclear tetrahedral structure with two bridging X atoms, on the basis of ESR and IR studies [34]. A dinuclear cluster structure with four bridging carboxylate groups and axially arranged phosphane ligands is assumed for $Mn\{P(C_6H_5)_3\}\{(CH_3)_3CCOO\}_2$ on the basis of its magnetic properties [37].

39.1.2.1.3 Physical Properties

Melting points (in °C) and effective magnetic moments (in μ_B) of short-chain [1, 2, 29] and long-chain $MnLX_2$ complexes [34] are summarized below:

complex	m.p.	μ_{eff}	Ref.	complex	m.p.	μ_{eff}
$Mn\{P(CH_3)_3\}Cl_2$	—	5.8	[29]	$Mn\{P(C_{12}H_{25})_3\}Cl_2$	49	5.75
$Mn\{P(CH_3)_3\}Br_2$	—	5.6	[29]	$Mn\{P(C_{12}H_{25})_3\}Br_2$	38	5.70
$Mn\{P(CH_3)_3\}I_2$	—	5.4	[29]	$Mn\{P(C_{12}H_{25})_3\}I_2$	42	5.45
$Mn\{P(CH_3)_3\}(NCS)_2$	—	5.7	[2]	$Mn\{P(C_{14}H_{29})_3\}Cl_2$	82	5.80
$Mn\{P(C_2H_5)_3\}(NCS)_2$	—	5.9	[2]	$Mn\{P(C_{14}H_{29})_3\}Br_2$	88	5.70
$Mn\{P(C_3H_7)_3\}(NCS)_2$	—	6.0	[2]	$Mn\{P(C_{14}H_{29})_3\}I_2$	91	5.40
$Mn\{P(C_4H_9)_3\}(NCS)_2$	—	6.0	[2]	$Mn\{P(C_{16}H_{33})_3\}Cl_2$	92	5.80
$Mn\{P(C_6H_5)_3\}Cl_2$	96	5.8	[1]	$Mn\{P(C_{16}H_{33})_3\}Br_2$	58	5.65
$Mn\{P(C_6H_5)_3\}Br_2$	78	5.5	[1]	$Mn\{P(C_{16}H_{33})_3\}I_2$	52	5.45
$Mn\{P(C_6H_5)_3\}I_2$	194	5.4	[1]	$Mn\{P(C_6H_5)(C_{12}H_{25})_2\}Cl_2$	124	5.75
$Mn\{P(C_6H_5)_3\}(NCS)_2$	70	5.3	[2]	$Mn\{P(C_6H_5)(C_{12}H_{25})_2\}Br_2$	104	5.70
$Mn\{P(C_6H_5)(CH_3)_2\}(NCS)_2$	—	5.4	[2]	$Mn\{P(C_6H_5)(C_{12}H_{25})_2\}I_2$	64	5.40
$Mn\{P(C_6H_5)(C_2H_5)_2\}(NCS)_2$	—	5.4	[2]	$Mn\{P(C_6H_5)(C_{14}H_{29})_2\}Cl_2$	124	5.75
$Mn\{P(C_6H_5)(C_3H_7)_2\}(NCS)_2$	—	5.4	[2]	$Mn\{P(C_6H_5)(C_{14}H_{29})_2\}Br_2$	100	5.65
$Mn\{P(C_6H_5)(C_4H_9)_2\}(NCS)_2$	—	5.3	[2]	$Mn\{P(C_6H_5)(C_{14}H_{29})_2\}I_2$	86	5.40
$Mn\{P(C_6H_5)_2(CH_3)\}(NCS)_2$	—	5.3	[2]	$Mn\{P(C_6H_5)(C_{16}H_{33})_2\}Cl_2$	105	5.80
				$Mn\{P(C_6H_5)(C_{16}H_{33})_2\}Br_2$	92	5.65
				$Mn\{P(C_6H_5)(C_{16}H_{33})_2\}I_2$	58	5.40

The decrease of the magnetic moments for the $MnLX_2$ complexes in the order X = Cl > Br > I suggests that bridging Mn–X–Mn moieties lead to some significant spin-pairing in this order [1, 34]. $MnL(NCS)_2$ complexes fall into two classes: those containing phenyl-substituted phosphanes exhibit moments in the range of 5.2 to 5.4 μ_B, whereas the other complexes exhibit $\mu_{eff} = \sim 6.0$ μ_B. The low values of the first group indicate a considerable degree of spin-pairing in these complexes with only bridging NCS groups [2]. Varied temperature magnetic measurements (by the Faraday method) for $Mn\{P(C_6H_5)_3\}\{(CH_3)_3CCOO\}_2$ yielded the magnetic moment, $\mu_{eff} = 4.46$ μ_B, at 294 K which decreases to 2.19 μ_B at 78 K. The data are characteristic of antiferromagnetic dimeric compounds. The exchange constant, $J = -17$ cm^{-1}, and the g factor, 2.003, have been evaluated using the isotropic Hamiltonian for metal-metal interaction in dimeric molecules [37].

X-band ESR spectra of $Mn\{P(C_6H_5)_3\}X_2$ complexes with X = Cl, Br, I, NCS, or NCSe in the solid state or in dichloromethane [1, 4] and of $Mn\{P(C_6H_5)(CH_3)_2\}X_2$ complexes with X = Cl, Br, I, or NCS and $Mn\{P(CH_2{-}CH{=}CH_2)_3\}Cl_2$ in toluene [6] exhibit a broad single band at $g \approx 2$; no fine structure was observed [1, 4, 6]. $Mn\{P(C_6H_5)_3\}I_2$ in the solid state and in dichloromethane [1, 15], as well as all the long-chain phosphane complexes in the solid state and in toluene solution (frozen glasses at −160°C) [34], show a strong broad absorption at $g \approx 2$ and 4, together with additional lines down to $g \approx 6$. Comparison of the spectra with published data suggests an essentially tetrahedral environment of the Mn ions for all the complexes in the solid state and in dichloromethane or toluene [1, 15, 34]. Evaluation of the zero-field splitting parameters, D and E, for $Mn\{P(C_6H_5)_3\}I_2$ gave $D \approx 0.09$ cm^{-1} and λ (= D/E) ≈ 0.333, supporting the tetrahedral geometry for this complex in the solid state and in dichloromethane [1].

Far-IR data (in cm^{-1}) of short-chain phosphane complexes from mulls in Nujol or hexachlorobutadiene [1, 2] are given below:

complex	ν(Mn–P)	ν(Mn–X)	Ref.
$Mn\{P(CH_3)_3\}Cl_2$	420	232 [a]	[29]
$Mn\{P(CH_3)_3\}Br_2$	440	195 [a]	[29]
$Mn\{P(CH_3)_3\}I_2$	445	—	[29]
$Mn\{P(C_6H_5)_3\}Cl_2$	421	326 [b], 232 [a]	[1]
$Mn\{P(C_6H_5)_3\}Br_2$	420	229 [b], 192 [a]	[1]
$Mn\{P(C_6H_5)_3\}I_2$	417	—	[1]
$Mn\{P(C_6H_5)_3\}(NCS)_2$ [c]	418	256	[1]
$Mn\{P(C_6H_5)_3\}(NCSe)_2$	416	232	[4]
$Mn\{P(C_6H_5)(CH_3)_2\}(NCSe)_2$	417	240	[4]

[a] Bridging X atoms. – [b] Terminal X atoms. – [c] ν(Mn–S) at 304 cm^{-1}.

As shown in the table the far-IR spectra of $Mn\{P(C_6H_5)_3\}Cl_2$ and $Mn\{P(C_6H_5)_3\}Br_2$ show evidence of terminal and bridging Mn–X linkages [1]. This is also the case for all the long-chain phosphane complexes. Terminal and bridging ν(Mn–X) bands of these complexes are in the following ranges [34]:

X	Cl	Br	I
ν(Mn–X)	289 to 300	235 to 245	205 to 220
ν(Mn–X–Mn)	220 to 230	195 to 205	165 to 170

References on pp. 47/9

The ν(Mn–P) band of long-chain phosphane complexes occurs in the 380 to 405 cm^{-1} region increasing in energy in the order Cl<Br<I. This indicates that the Mn–P bond is strongest in the $MnLI_2$ complexes [34]. The IR spectrum of the $Mn\{P(C_6H_5)_3\}(NCS)_2$ complex shows a single ν(CN) band at 2140 cm^{-1} [1], whereas the other isothiocyanato or isoselenocyanato complexes studied exhibit two or three ν(CN) bands (2100 to 2140, 2080 to 2100, and 2040 to 2085 cm^{-1}). For these complexes an abrupt rise of 30 to 40 cm^{-1} in the high-energy ν(CN) band can be observed in going from those with PR_3 groups (R = alkyl) to those with $P(C_6H_5)_{3-x}R_x$ groups [2]. Graphs of the IR spectra are presented for $Mn\{P(CH_3)_3\}Br_2$ in the 3000 to 300 cm^{-1} range [8] and for $Mn\{P(C_2H_5)_3\}Br_2$ and $Mn\{P(C_6H_5)(CH_3)_2\}Br_2$ in the 1600 to 300 cm^{-1} range [3]; changes on formation of oxygenated species (see p. 53) are discussed, but no assignments are given for $MnLBr_2$ complexes [3, 8].

The electronic reflectance spectra recorded for $Mn\{P(C_6H_5)_3\}X_2$ with X = Cl, Br, I, and NCS show four bands in the 530 to 370 nm range, which were assigned to the $^4T_1(G)$; $^4T_2(G)$; $^4A_1$, $^4E(G)$; and $^4T_2(D)$ transitions from the $^6A_1$ ground state (assignments based essentially on T_d symmetry) [1].

39.1.2.1.4 Chemical Reactions

Once isolated, all the compounds, except for $Mn\{P(CH_3)_3\}X_2$, are quite stable and insensitive to all but very moist conditions. They can be stored indefinitely under dry inert conditions [1, 2]. Since the reaction of $Mn\{P(CH_3)_3\}X_2$ complexes with dioxygen is irreversible (see below) these compounds should be handled and stored under strictly inert conditions [29].

The complexes with short-chain alkylphosphanes, such as $P(C_4H_9)_3$ or $P(C_6H_5)(CH_3)_2$, are only very slightly soluble in noncoordinating solvents, such as toluene or dichloromethane [9, 34]. To improve their solubility, complexes with long-chain alkylphosphanes, such as $P(C_{12}H_{25})_3$, were prepared. These are moderately soluble in THF [36] and 1:1 toluene-dichloromethane (~10^{-3} mol/L), but only sparingly soluble in diethyl ether or pentane [34]. All the complexes retain the oligomeric or polymeric structures of the solid state (see p. 40) in solutions of noncoordinating solvents, as was shown by ESR and IR studies [1, 6, 34]. In contrast, breaking of the polymeric chains to give monomeric units can be assumed in coordinating solvents, such as tetrahydrofuran or diglyme (=1,1′-oxybis(2-methoxyethane) = $C_6H_{14}O_3$), based on ESR and molecular weight determinations [6, 9]. The $MnL(thf)_nX_2$ complexes are treated on p. 50. Poorly characterized $MnL(C_6H_{14}O_3)_nX_2$ complexes are reported in [6].

The complexes are distinguished by the peculiarity of adding small molecules, especially gas molecules. Of these, O_2, CO, ethylene, and NO are bound reversibly, whereas SO_2, CS_2, and tetracyanoethylene are bound irreversibly. The ability of individual $MnLX_2$ complexes to bind individual small molecules is related to the special structure of these complexes (see p. 39): there are vacant spaces above and below the T_d Mn atoms, which may be fine-tuned by changing the bridging halide or the steric requirements of the phosphane group [9, 30]. A survey on the coordination of small molecules by $MnLX_2$ complexes and their use in gas separations is given in [30]. Below, the reactions of $MnLX_2$ complexes with the various small molecules are treated in detail.

Reactions with Dioxygen. The nature of the coordinated phosphane is of prime importance in the reaction of $MnLX_2$ complexes with dioxygen. There are two extremes: The $Mn\{P(CH_3)_3\}X_2$ complexes with X = Cl, Br, or I react rapidly and irreversibly with O_2 to give oxidation products [3, 8, 18, 19, 29] (see p. 43), whereas $Mn\{P(C_6H_5)_3\}X_2$ complexes do not bind O_2, even at an O_2 pressure of 20 atm [19]. The $Mn(PR_3)X_2$ and $Mn\{P(C_6H_5)R_2\}X_2$ complexes with $R = C_2H_5$, C_3H_7, or C_4H_9 for X = Cl, Br, or I react (in the solid state and in solution) with O_2 under anhydrous conditions to give highly-colored O_2 adducts (see p. 53) which release the dioxygen under reduced pressure or at higher temperatures; i.e., these compounds react reversibly with oxygen [2, 14, 17]. Reversible reactions with O_2 were also observed for isothiocyanates, $Mn(PR_3)(NCS)_2$ with $R = CH_3$, C_2H_5, C_3H_7, or C_4H_9, whereas the $Mn\{P(C_6H_5)R_2\}(NCS)_2$ complex with R as above showed no ability to form O_2 adducts [2]. Studies with labeled dioxygen ($^{16}O_2$-$^{18}O_2$ mixtures) showed that the integrity of the dioxygen molecules was retained after desorption [14]. The compounds reacting reversibly with dioxygen have potential uses. For example, $Mn\{P(C_4H_9)_3\}(NCS)_2$ which binds O_2 at ambient partial pressure in air could possibly be used in a pressure-swing absorption process for separating O_2 from air. $Mn\{P(C_6H_5)(CH_3)_2\}Br_2$ binds O_2 at almost zero partial pressure and thus is suitable for scavenging traces of O_2 from inert gases [17].

For the solid $MnLX_2$ complexes reacting reversibly with dioxygen, the amount of initial O_2 uptake is dependent on the nature of the phosphane, ligand, halide or pseudohalide, O_2 partial pressure, temperature, light, and the surface area exposed [5, 7, 10, 14, 17]. Some dioxygen absorption isotherms are shown in **Fig. 11**. The hyperbolic curve for $Mn\{P(C_6H_5)(CH_3)_2\}Br_2$ (a) and the sigmoidal curves for $Mn\{P(C_6H_5)(C_2H_5)_2\}Br_2$ (b) and $Mn\{P(C_4H_9)_3\}(NCS)_2$ (c) indicate that the latter pair are analogs of haemoglobin, and the former compound is an analog of myoglobin [17]. From finely divided samples of the stated compounds, ~9% of the calculated molar equivalent of O_2 was rapidly absorbed on the surface. Further uptake of O_2 was diffusion-controlled and took several hours to reach the full stoichiometric amount [10, 17]. Sorption of dioxygen is accelerated by light. In many cases capacity and sensitivity of $MnLX_2$ complexes to dioxygen are also increased by irradiation or exposure to light [5, 7, 10, 11]. Other dioxygen absorption isotherm data (p_{O_2} and the corresponding weight increase) are presented for $MnLX_2$ complexes with $L = P(C_6H_5)(CH_3)_2$, $P(C_6H_5)(C_2H_5)_2$ for X = Br, with $L = P(C_3H_7)_3$, $P(C_6H_5)(CH_3)_2$ for X = I, and with $L = P(C_4H_9)_3$ for X = NCS in [5, 7]. Reversibility of the reaction $MnLX_2 + O_2 \rightleftharpoons MnL(O_2)X_2$ was studied by repeated oxygenation/evacuation cycles, using an appropriate O_2 partial pressure followed by a p_{O_2} drop. One cycle in the case of $Mn\{P(C_6H_5)(CH_3)_2\}I_2$ to 79 cycles for $Mn\{P(C_6H_5)(CH_3)_2\}Br_2$ were observed without deterioration or decomposition of the complexes. Absorption and desorption could be followed by means of the vivid colors of the $MnL(O_2)X_2$ complexes (see p. 54) [14, 17]. In practice the $MnLX_2$ compounds are distributed on the surface of a support material, for example, a mass of glass beads [5, 7]. In contrast to the results of [2, 14, 17, 19], other workers observed only a slow

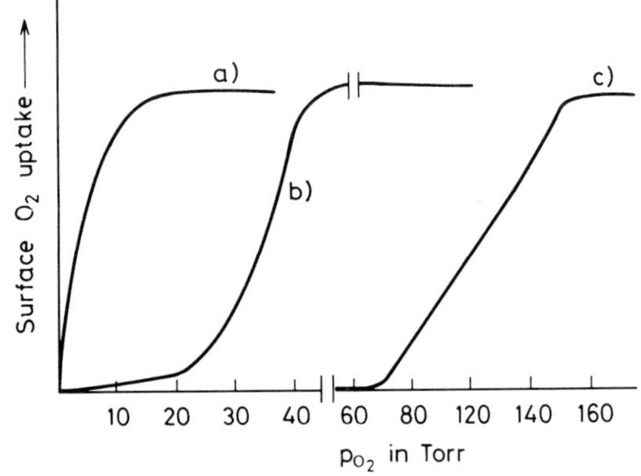

Fig. 11. Dioxygen absorption isotherms (a) for $Mn\{P(C_6H_5)(CH_3)_2\}Br_2$, (b) for $Mn\{P(C_6H_5)(C_2H_5)_2\}Br_2$, and (c) for $Mn\{P(C_4H_9)_3\}(NCS)_2$ [17].

weight change on exposure of solid $Mn\{P(C_6H_5)(CH_3)_2\}Br_2$ and $Mn\{P(C_6H_5)(CH_3)_2\}I_2$ to dioxygen. However, a slow loss of phosphane under vacuum made accurate work impossible. No reversibility could be detected under these conditions [27].

$Mn\{P(C_3H_7)_3\}Cl_2$ catalyzes the oxidation of acrolein by dioxygen with intermediate formation of the oxygenated Mn complex, as shown by a kinetic study of the reaction [31].

The irreversible reaction of solid $Mn\{P(CH_3)_3\}X_2$ compounds (X = Cl, Br, or I) with dry O_2 under strictly anhydrous conditions was studied by [3, 8, 29]. By magnetic, ESR, and IR spectral measurements it was shown that $Mn\{P(CH_3)_3\}X_2$ complexes are oxidized to highly-colored $Mn^{III}\{PCH_3)_3\}_2X_3$ complexes and an inorganic material having the composition $(MnI_2)_{0.2}(MnI_3)_{0.8} \cdot 10H_2O$ (in the case of X = I). This finding was supported by a single-crystal X-ray diffraction study of $Mn\{P(CH_3)_3\}_2I_3$ (see p. 59) [18, 29]. The results are in contrast to IR studies of the reaction between dioxygen and $Mn\{P(CH_3)_3\}Br_2$ prepared in situ as a thin film. In these studies the phosphane oxide complex, $Mn\{(CH_3)_3PO\}Br_2$, was the product and the dioxygen adduct, $Mn\{P(CH_3)_3\}Br_2 \cdot O_2$, an intermediate [3, 8].

In solution (toluene, dichloroethane, or THF) reversible reactions with O_2 were observed for a number of short-chain phosphane complexes (various studied in toluene are listed in Table 1, p. 45) [2, 10, 13] and for the long-chain phosphane $MnLX_2$ complexes with L = $P(C_{12}H_{25})_3$, $P(C_{14}H_{29})_3$, $P(C_{16}H_{31})_3$, $P(C_6H_5)(C_{12}H_{25})_2$, $P(C_6H_5)(C_{14}H_{29})_2$, and $P(C_6H_5)(C_{16}H_{33})_2$ for X = Cl, Br, or I [34]. Absorption and desorption reactions were followed by gas burette measurements [2, 10, 17], ESR [15], or electronic spectral measurements [2, 10, 12, 17, 34] using special vacuum devices [2]. The results indicate absorption of 1 mol O_2 per mol of complex, and the formation of 1:1 adducts, $Mn^{II}L(O_2)X_2$ was concluded [2, 15, 17, 34]. This is supported by molecular weight determinations of $Mn\{P(C_{12}H_{25})_3\}(O_2)Cl_2$ and $Mn\{P(C_6H_5)(C_{14}H_{29})_2\}(O_2)I_2$ [34]. Deviating results were obtained during volumetric O_2 absorption measurements on solutions of $Mn\{P(C_6H_5)(CH_3)_2\}I_2$ in THF [27]. The formation of additional species which are probably dimeric is discussed on the basis of ESR and spectral studies on $MnLX_2$ complexes with L = $P(C_3H_7)_3$, $P(C_4H_9)_3$, or $P(C_5H_{11})_3$ for X = Cl, Br, or I in toluene or THF [12, 15, 17]. By following the increase in intensity of the electronic spectra with increasing O_2 partial

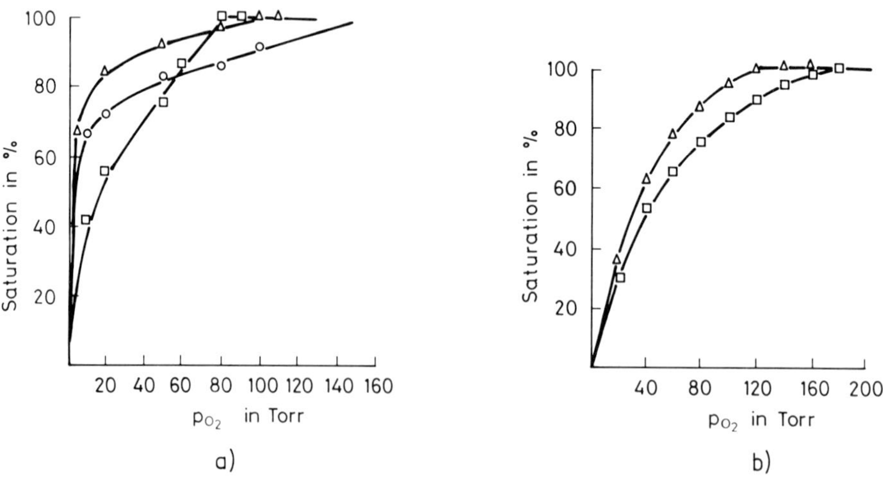

Fig. 12. Dioxygen absorption isotherms a) for $Mn\{P(C_4H_9)_3\}Br_2$ at 20°C, in $C_2H_4Cl_2$ (△), in toluene (○), and in THF (□); b) for $Mn\{P(C_6H_5)(C_4H_9)_2\}X_2$ in $C_2H_4Cl_2$, for X = Cl (△), for X = Br (□) [2].

pressure it has been possible to construct O_2-binding isotherms for the complexes in various solvents as shown for example in **Fig. 12**. Isotherms for other complexes are presented in [2, 10, 34]. From Fig. 12a) it can be seen that the shape of the curves in the noncoordinating solvents toluene and 1,2-dichloroethane are hyperbolic, while that for the coordinating solvent THF moves to the right. Its shape indicates a more complex interaction, presumably involving solvent participation [2, 6, 13]. Details of the dioxygen-binding reaction of (solvated) $MnLX_2$ complexes in THF are outlined on p. 51. For the reaction of dimeric long-chain phosphane $MnLX_2$ complexes with O_2 in toluene, rupture of the MnX_2Mn double bridge is assumed and addition of two O_2 molecules occurs to form two monomeric tetrahedral $MnL(O_2)X_2$ complexes [34]. From the O_2-binding isotherms equilibrium constants, K_{O_2} (in 10^{-3} $Torr^{-1}$) of the reaction $MnLX_2+O_2 \rightleftharpoons MnL(O_2)X_2$, were calculated using the Hill equation $y/1-y=[Mn(PR_3)(O_2)X_2]/[Mn(PR_3)X_2]=K_{O_2}\cdot(p_{O_2})^n=(p_{O_2})^n/p_{50}(O_2)$, where y is the fraction of the oxygenated sites [2, 10, 13, 34]. Values of K_{O_2} together with those of the O_2 partial pressure at 50% saturation (p_{50} in Torr) and of the Hill coefficient, n, are gathered in Table 1 for $MnLX_2$ complexes in toluene.

Table 1

Dioxygen-Binding Equilibrium Data for $MnLX_2$ Complexes with Short-Chain Phosphanes in Toluene at 20°C.

complex	p_{50}	K_{O_2}	n	Ref.
$Mn\{P(C_3H_7)_3\}Cl_2$	37	27.2	1.02	[10, 13]
$Mn\{P(C_3H_7)_3\}Br_2$	1.8	561	0.54	[10, 13]
$Mn\{P(C_4H_9)_3\}Br_2$	2.7	372	0.7	[2, 10, 13]
$Mn\{P(C_6H_5)(CH_3)_2\}Br_2$	7.13	140.3	0.92	[10]
	20.9	47.9	1.12	[13]
$Mn\{P(C_6H_5)(C_2H_5)_2\}Cl_2$	145.2	6.88	1.16	[10]
$Mn\{P(C_6H_5)(C_2H_5)_2\}Br_2$	125.6	7.96	1.22	[10]
$Mn\{P(C_6H_5)(C_4H_9)_2\}Cl_2$	165.6	6.04	1.2	[2, 10]
$Mn\{P(C_6H_5)(C_4H_9)_2\}Br_2$	199.5	5.01	1.3	[2, 10]

Equilibrium data of the complexes in dichloroethane are similar to those in toluene [2, 10, 13]. For long-chain phosphane $MnLX_2$ complexes with X = Cl, Br, or I, the K_{O_2} values vary between 5 and 50×10^{-3} $Torr^{-1}$ [34]. The data and also Fig. 12b) show that for the same solvent and the same coordinated phosphane, the affinity of $MnLX_2$ complexes for dioxygen varies with the halide, generally in the order Cl > Br > I. For a constant halide and solvent the ability to bind dioxygen is generally in the order $L=PR_3>P(C_6H_5)R_2>P(C_6H_5)_2R$ (R = alkyl) [13, 18, 34]. However, for long-chain phosphane complexes the affinity changes to I > Br at higher saturations (80 to 90%). For these complexes, Hill coefficients n > 1 (1.10 to 1.64) indicate cooperativity which may well be due to the dimeric structure proposed [34]. A diagram relating the dioxygen-binding constant, K_{O_2}, with the cone-angle and an electronic parameter of the phosphanes is presented in [19].

The reaction $MnLX_2+O_2 \rightleftharpoons MnL(O_2)X_2$ may be reversed by lowering the dioxygen partial pressure above the solution, by heating, or by bubbling an inert gas through the solution [2, 10]. The ability of the complexes repeatedly to absorb and desorb dioxygen (cycling), which is important in view of potential uses for gas separation purposes, is sensitive to the temperature and the dioxygen partial pressure above the solution [2]. Examples for the number of ab-

References on pp. 47/9

sorption/desorption cycles observed for $Mn\{P(C_4H_9)_3\}I_2$ in THF, for $Mn\{P(C_3H_7)_3\}Cl_2$ in THF, and for $Mn\{P(C_8H_{17})_3\}Br_2$ in polyethylene glycol diisopropyl ether are given in [2, 12], and [16], respectively.

The rate of oxygenation is dependent on the nature of the phosphane coordinated, on temperature (generally faster at −40°C than at 0°C), and on dioxygen partial pressure (faster for p_{O_2} = 760 Torr than for p_{O_2}=150 Torr). Usually, oxygenation is complete within 5 to 10 min at 0°C and 760 Torr [2, 10]. Dioxygen absorption isotherms (saturation vs. time) for $Mn\{P(C_2H_5)_3\}X_2$ (X = Cl, Br, I) and for $MnLBr_2$ (L = $P(C_2H_5)_3$, $P(C_6H_5)(CH_3)_2$, and $P(C_6H_5)(C_4H_9)_2$) all in THF at 0°C and 760 Torr are presented in [10].

Reactions with Nitric Oxide. $MnLX_2$ complexes with L = $P(C_3H_7)_3$, $P(C_4H_9)_3$, $P(C_6H_5)(CH_3)_2$, and $P(C_6H_5)(C_2H_5)_2$ for X = Cl, Br, I bind nitric oxide to form vividly colored $Mn(NO)LX_2$ complexes (see p. 61), both in the solid state and in THF. The reaction is reversible. From the variation of the electronic spectra with the partial pressure p_{NO} of complexes in THF, absorption isotherms were obtained (some example isotherms are presented in [21] and [32]). The isotherms have a hyperbolic shape similar to those on p. 44 for O_2 absorption isotherms. A saturation of 100% is reached at $p_{NO} \approx 150$ Torr for chloro complexes and at $p_{NO} \approx 200$ Torr for bromo complexes. NO binding constants at 25°C lie in the (14.5 to 21.3) $\times 10^{-3}$ $Torr^{-1}$ range for the chloro complexes and (4.0 to 6.5) $\times 10^{-3}$ $Torr^{-1}$ for the bromo complexes. The affinity of $MnLX_2$ complexes is in the order $P(C_3H_7)_3 > P(C_4H_9)_3 > P(C_6H_5)(CH_3)_2 \approx P(C_6H_5)(C_2H_5)_2$ for X = Cl and $P(C_6H_5)(C_2H_5)_2 > P(C_6H_5)(CH_3)_2 > P(C_3H_7)_3 > P(C_4H_9)_3$ for X = Br. The NO may be desorbed by purging the $Mn(NO)LX_2$ solutions with dry argon for several minutes or by applying a vacuum. At −78°C up to 15 complete 100% absorption/desorption cycles could be obtained for the chloro complexes, but usually only two for the iodo complexes. Graphs showing percentage NO uptake vs. cycle number for $Mn\{P(C_4H_9)_3\}X_2$ (X = Cl, Br, I) at −78, 0, and 25°C are presented in [32]. All systems are less stable at temperatures above 0°C [21, 32].

Reactions with Sulfur Dioxide. Solid samples of $Mn\{P(C_3H_7)_3\}Br_2$ and $Mn\{P(C_6H_5)(CH_3)_2\}$-Br_2, toluene slurries, and THF solutions of $MnLX_2$ complexes with L = $P(C_3H_7)_3$, $P(C_4H_9)_3$, $P(C_6H_5)_3$, $P(C_6H_5)(CH_3)_2$, $P(C_6H_5)(C_2H_5)_2$, $P(C_6H_5)(C_3H_7)_2$, $P(C_6H_5)(C_4H_9)_2$, and $P(C_6H_5)(C_4H_9\text{-}i)_2$ for X = Cl, Br, I were exposed to dry SO_2 in special vacuum devices at 0°C. While all of the iodo complexes (including that with L = $P(C_6H_5)_3$) formed SO_2 adducts, no SO_2 adduct of a chloro complex was obtained. For the bromo complexes only those with trialkylphosphanes or $P(C_6H_5)$-$(CH_3)_2$ were active. Reaction products had the composition $MnL(SO_2)_nX_2$ (see p. 57) with n = 0.66 for the solid complexes and toluene slurries and n = 0.66 to 0.76 for the THF solutions [24]. Only irreversible reactions were observed [19, 24]. Isotherm data for the absorption of SO_2 by solid $Mn\{P(C_3H_7)_3\}Br_2$ at 19°C are given in [5, 7]. In another study the complexes $MnLX_2$ with L = $P(CH_3)_3$, $P(C_2H_5)_3$, $P(C_6H_5)(CH_3)_2$ for X = Cl, Br, or I, were prepared as films on KBr windows and the interaction with SO_2 (p_{SO_2} = 15 to 120 Torr for 15 min to 72 h) followed by IR spectroscopy. At low SO_2 pressures and short exposure times reversible reactions were observed, while at high SO_2 pressures and long exposure times the complexes reacted irreversibly with SO_2. Products of the irreversible reactions had the composition $MnL(SO_2)_{0.5}X_2$. In the case of Mn-$\{P(C_6H_5)(CH_3)_2\}Cl_2$ only an irreversible reaction was observed [22, 23].

Reactions with Ethylene, Tetracyanoethylene, or *tert*-Butyl Isocyanide. $MnLX_2$ complexes with L = $P(C_3H_7)_3$, $P(C_4H_9)_3$, $P(C_6H_5)(CH_3)_2$, or $P(C_6H_5)(C_2H_5)_2$ for X = Cl and with L = $P(C_4H_9)_3$ for X = Br react with ethylene in THF at 0°C to give 1:1 adducts. The reaction appears to be reversible, desorption being effected by warming the solution to ~30°C under a static vacuum. The reversibility of binding was demonstrated by three absorption/desorption cycles for $Mn\{P(C_6H_5)(C_2H_5)_2\}Cl_2$ [28]. $MnLX_2$ complexes with L = $P(C_3H_7)_3$, $P(C_6H_5)(CH_3)_2$, or $P(C_6H_5)_3$ for X = Cl, Br, or I react with tetracyanoethylene in toluene slurries under Ar to give the complexes $Mn^{III}L(C_6N_4)X_2$ [33]. Treatment of $Mn\{P(C_6H_5)(CH_3)_2\}Br_2$ with *tert*-butyl isocyanide in toluene under strictly anhydrous conditions yields $Mn\{P(C_6H_5)(CH_3)_2\}(t\text{-}C_4H_9NC)Br_2$ [39]. A remark concerning the description of the mixed ligand complexes formed (with ethylene, tetracyanoethylene, or *tert*-butyl isocyanide) is given on p. 59.

Reactions with Carbon Monoxide or Carbon Disulfide. $MnLX_2$ complexes with $L = P(C_6H_5)(CH_3)_2$ for X = Cl, I and $L = P(C_3H_7)_3$, $P(C_6H_5)(C_2H_5)_2$ for X = Cl or Br react reversibly with carbon monoxide to give $Mn(CO)LX_2$ complexes. Absorption studies of solid complexes were made by exposing samples to one atmosphere of dry CO for one day or more. The shape of the absorption isotherms (**Fig. 13**) indicates that the CO uptake proceeds in three steps, which are explained by: 1) Formation of a monolayer of CO adducts on the surface of the complex, 2) molecular rearrangement and generation of a new set of coordination sites, and 3) occupation of the second series of coordination sites. Breaking of bridging Mn–X–Mn bonds is assumed to occur during the second step. The affinity for CO, following the sequence I > Br > Cl, may reflect the ease with which the Mn–X–Mn bonds are broken. In THF solution saturation vs. time curves (presented in the publication) show distinct plateaus, indicating the formation of intermediates. The curves show that the rate of absorption follows the order $P(C_6H_5)(CH_3)_2 > P(C_6H_5)(C_2H_5)_2 \approx P(C_3H_7)_3$ for X = Br and $P(C_3H_7)_3 > P(C_6H_5)(C_2H_5)_2 \approx P(C_6H_5)(CH_3)_2$ for X = Cl. The complexes were found to be capable of repeatedly absorbing and releasing CO without any sign of decomposition at 0°C for X = Cl, Br or at −40°C for X = I. The desorbed gas was shown to be CO by mass spectral methods [25]. $MnLX_2$ complexes with $L = P(C_3H_7)_3$, $P(C_4H_9)_3$, $P(C_6H_5)(CH_3)_2$, $P(C_6H_5)(C_2H_5)_2$, or $P(C_6H_5)(C_3H_7)_2$ for X = Cl, Br, or I react with excess CS_2 in diethyl ether to form insoluble $(MnLX_2)_2CS_2$ complexes. In THF, the solvated complexes, $\{MnLX_2(thf)_2\}_2CS_2$, are formed [26]. (A remark concerning carbon monoxide and carbon disulfide adducts of $MnLX_2$ complexes is given on p. 59.)

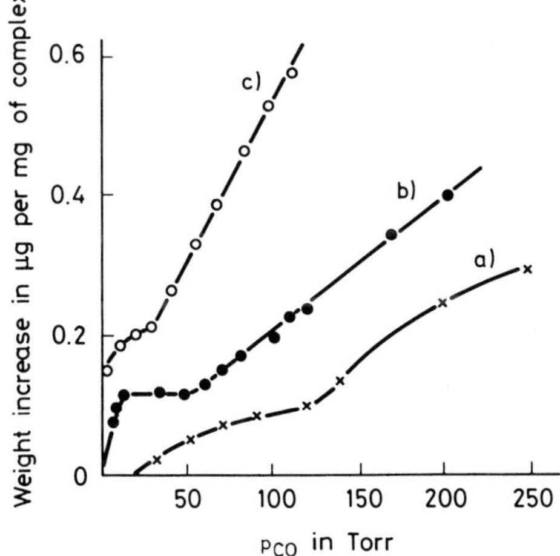

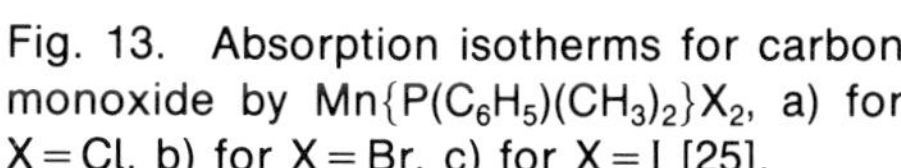
Fig. 13. Absorption isotherms for carbon monoxide by $Mn\{P(C_6H_5)(CH_3)_2\}X_2$, a) for X = Cl, b) for X = Br, c) for X = I [25].

References:

[1] Hosseiny, A.; Mackie, A. G.; McAuliffe, C. A.; Minten, K. (Inorg. Chim. Acta **49** [1981] 99/105).
[2] McAuliffe, C. A.; Al-Khateeb, H. F.; Barratt, D. S.; Briggs, J. C.; Challita, A.; Hosseiny, A.; Little, M. G.; Mackie, A. G.; Minten, K. (J. Chem. Soc. Dalton Trans. **1983** 2147/53).
[3] Newberry, V. F.; Burkett, H. D.; Worley, S. D.; Hill, W. E. (Inorg. Chem. **23** [1984] 3911/7).
[4] Benson, C. G.; Little, M. G.; McAuliffe, C. A. (Inorg. Chim. Acta **87** [1984] 169/75).
[5] McAuliffe, C. A.; McCullough, F. P.; Levason, W. (Belg. 874459 [1979] 1/19; C.A. **91** [1979] No. 195208).

[6] Hill, W. E.; McAuliffe, C. A. (U.S. 4442297 [1981/84]; C.A. **101** [1984] No. 40616).
[7] McAuliffe, C.A.; Levason, W.; McCullough, F. P. (U.S. 4251452 [1979/81]; C.A. **94** [1981] No. 157087).
[8] Burkett, H. D.; Newberry, V. F.; Hill, W. E.; Worley, S. D. (J. Am. Chem. Soc. **105** [1983] 4097/9).
[9] Beagley, B.; Briggs, J. C.; Hosseiny, A.; Hill, W. E.; King. T. J.; McAuliffe, C. A.; Minten, K. (J. Chem. Soc. Chem. Commun. **1984** 305/6).
[10] McAuliffe, C. A. (Spec. Publ. Roy. Soc. Chem. No. 39 [1981] 119/31; C.A. **95** [1981] No. 92442).

[11] McAuliffe, C. A.; Minten, K. (Symp. Papers Inst. Chem. Eng. North West. Branch **1982**; C.A. **96** [1982] No. 192249).
[12] Yatsimirskii, K. B.; Bratushko, Yu. I.; Ermokhina, N. I. (Teor. Eksperim. Khim. **18** [1982] 367/71; Theor. Exptl. Chem. [USSR] **18** [1982] 328/31).
[13] McAuliffe, C. A.; Al-Khateeb, H. (Inorg. Chim. Acta **45** [1980] L195/L196).
[14] Barber, M.; Bordoli, R. S.; Hosseiny, A.; Minten, K.; Perkin, C. R.; Sedgwick, R. D.; McAuliffe, C. A. (Inorg. Chim. Acta **45** [1980] L89/L90).
[15] McAuliffe, C. A.; Little, M. G.; Raynor, J. B. (J. Chem. Soc. Chem. Commun. **1982** 68/70).
[16] Richter, W.; Kummer, R.; Platz, R. (Ger. Offen. 2927734 [1979/81] 1/6; C.A. **95** [1981] No. 9411).
[17] McAuliffe, C. A.; Al-Khateb, H.; Jones, M. H.; Levason, W.; Minten, K.; McCullough, F. P. (J. Chem. Soc. Chem. Commun. **1979** 736/8).
[18] Beagley, B.; McAuliffe, C. A.; Minten, K.; Pritchard, R. G. (J. Chem. Soc. Chem. Commun. **1984** 658/9).
[19] McAuliffe, C. A. (Spec. Publ. Roy. Soc. Chem. No. 48 [1984] 263/74; C.A. **101** [1984] No. 194280).
[20] O'Sullivan, D. A. (Chem. Eng. News **56** No. 49 [1978] 24/6).

[21] Barratt, D. S.; McAuliffe, C. A. (J. Chem. Soc. Chem. Commun. **1984** 594/5).
[22] Newberry, V. F. (Diss. Abstr. Intern. B **45** [1985] 3814/5; C.A. **103** [1985] No. 63896).
[23] Hill, W. E.; Worley, S. D.; Paul, D. K.; Newberry, V. F. (Inorg. Chem. **24** [1985] 4429/30).
[24] Barratt, D. S.; Benson, C. G.; Gott, G. A.; McAuliffe, C. A.; Tanner, S. P. (J. Chem. Soc. Dalton Trans. **1985** 2661/4).
[25] McAuliffe, C. A.; Barratt, D. S.; Benson, C. G.; Hosseiny, A.; Little, M. G.; Minten, K. (J. Organometal. Chem. **258** [1983] 35/45).
[26] Barratt, D. S.; McAuliffe, C. A. (Inorg. Chim. Acta **97** [1985] 37/43).
[27] Wickens, D. A.; Abrams, G. (J. Chem. Soc. Dalton Trans. **1985** 2203/4).
[28] Gott, G. A.; McAuliffe, C. A. (J. Chem. Soc. Dalton Trans. **1987** 2241/3).
[29] Beagley, B.; McAuliffe, C. A.; Minten, K.; Pritchard, R. G. (J. Chem. Soc. Dalton Trans. **1987** 1999/2003).
[30] McAuliffe, C. A. (J. Mol. Catal. **44** [1988] 35/63, 41/50).

[31] Bratushko, Yu. I.; Ermokhina, N. I.; Yatsimirskii, K. B. (Zh. Neorgan. Khim. **33** [1988] 636/43; Russ. J. Inorg. Chem. **33** [1988] 357/61).
[32] Barratt, D. S.; McAuliffe, C. A. (J. Chem. Soc. Dalton Trans. **1987** 2497/501).
[33] Gott, G. A.; McAuliffe, C. A. (J. Chem. Soc. Chem. Commun. **1987** 1785/7).
[34] Barratt, D. S.; Gott, G. A.; McAuliffe, C. A. (J. Chem. Soc. Dalton Trans. **1988** 2065/70).
[35] Brown, R. M.; Bull, R. E.; Green, M. L. H.; Grebenik, P. D.; Martin-Polo, J. J.; Mingos, D. M. P. (J. Organometal. Chem. **201** [1980] 437/46).
[36] Barratt, D. S.; Gott, G. A.; McAuliffe, C. A. (Inorg. Chim. Acta **145** [1988] 289/98).

[37] Novotortsev, V. M.; Rakitin, Yu. V.; Pasynskii, A. A.; Kalinnikov, V. T. (Dokl. Akad. Nauk SSSR **240** [1978] 355/7; Dokl. Chem. Proc. Acad. Sci. USSR **238/243** [1978] 220/2).

[38] Pasynskii, A. A.; Idrisov, T. C.; Suvorova, K. M.; Kalinnikov, V. T. (Koord. Khim. **2** [1976] 1060/8; Soviet J. Coord. Chem. **2** [1976] 813/9).

[39] Beagley, B.; Benson, C. G.; Gott, G. A.; McAuliffe, C. A.; Pritchard, R. G.; Tanner, S. P. (J. Chem. Soc. Dalton Trans. **1988** 2261/6).

39.1.2.2 $Mn^{II}L_2X_2$ Compounds

$Mn\{P(C_2H_5)_3\}_2I_2$ was prepared by reaction of anhydrous MnI_2 in ether with a small excess of triethylphosphane. Recrystallization from ether gave pure pink needles in a 47% yield. An X-ray crystal structure determination shows an orthorhombic lattice, space group Pccn–D_{2h}^{10} (No. 56), with the lattice constants a=17.326(3), b=17.302(3), c=14.702(3) Å; Z=8. The experimental density is D_{exp}=1.643 g/cm³. Refinement of the structure yielded R=0.05. The molecular structure is shown in **Fig. 14**. The complex contains two crystallographically different molecules (1 and 2) with very similar structural parameters. Each Mn atom has four ligands. Two I and two P atoms are arranged in such a way that a crystallographic twofold axis passing through the metal bisects the angles I–Mn–I′ and P–Mn–P′. Thus, the structure is a distorted tetrahedron [1].

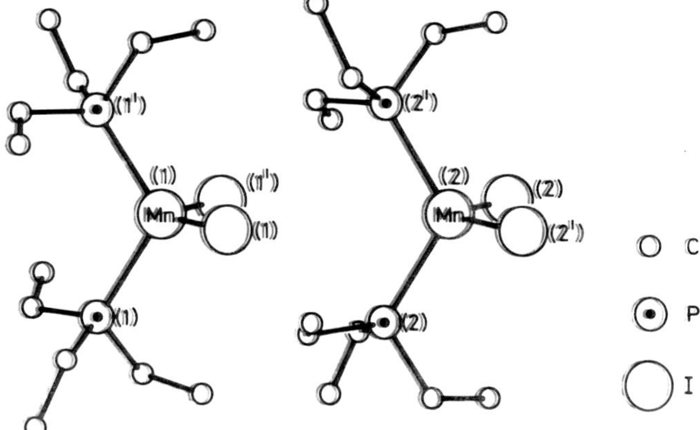

Fig. 14. Molecular structure of α-$Mn\{P(C_2H_5)_3\}_2I_2$ (hydrogen atoms are omitted) [1].

More recent investigations of MnL_2I_2 compounds showed that $Mn\{P(C_2H_5)_3\}_2I_2$ exists in two forms: α-$Mn\{P(C_2H_5)_3\}_2I_2$, which was described above, and β-$Mn\{P(C_2H_5)_3\}_2I_2$. Crystal determinations of β-$Mn\{P(C_2H_5)_3\}_2I_2$, $Mn\{P(C_6H_5)_3\}_2I_2$, and $Mn\{P(C_6H_5)(CH_2CH_2C_6H_5)_2\}_2I_2$ (no preparations given) reveal that in the three compounds the manganese atoms are tetrahedrally coordinated by two P and two I atoms, as is the case in α-$Mn\{P(C_2H_5)_3\}_2I_2$ [6]. Mn–I and Mn–P bond lengths (in Å) in the four MnL_2I_2 compounds are:

complex	Mn(1)–I(1)	Mn(2)–I(2)	Mn(1)–P(1)	Mn(2)–P(2)	Ref.
α-$Mn\{P(C_2H_5)_3\}_2I_2$	2.662(2)	2.670(2)	2.528(4)	2.539(4)	[1]
β-$Mn\{P(C_2H_5)_3\}_2I_2$	2.670(2)	2.675(2)	2.562(3)	2.568(4)	[7]
$Mn\{P(C_6H_5)_3\}_2I_2$	2.632(2)	2.642(2)	2.596(3)	2.617(3)	[7]
$Mn\{P(C_6H_5)(CH_2CH_2C_6H_5)_2\}_2I_2$	2.653(2)	2.653(2)	2.585(4)	2.585(4)	[7]

Bond angles in α-$Mn\{P(C_2H_5)_3\}_2I_2$ are: I(1)–Mn(1)–I(1′)=120.1(1); I(2)–Mn(2)–I(2′)=119.9(1); P(1)–Mn(1)–P(1′)=115.1(2); P(2)–Mn(2)–P(2′)=116.4(2); I(1)–Mn(1)–P(1′)=102.2(1); I(2)–Mn(2)–P(2)=105.5(1); I(1)–Mn(1)–P(1″)=105.8(1); I(2)–Mn(2)–P(2′)=105.1(1) [1].

The ^1H and ^{13}C NMR spectra (measured in C_6D_6) show the room temperature paramagnetic shifts: δ_{298}^{para} (in ppm) = −36.8 (CH_2), −654.2 (CH_2), −5.5 (CH_3), −158.8 (CH_3) (negative sign to high frequency). The shifts are very similar to those found for phosphane-stabilized manganese half-sandwich compounds that contain 5 unpaired electrons. The electron-spin delocalization most probably occurs through σ bonds with a hyperconjugative contribution to the large α-carbon shift [1].

The compound is soluble in benzene, ether, and THF. With small amounts of oxygen the benzene solution turns dark green, while in THF a blue-violet color is observed. The originally pink solutions can be regenerated by bubbling argon through them. This is indicative of a reversible oxygen binding [1]. $Mn\{P(C_2H_5)_3\}_2I_2$ reacts with bis(η-methylcyclopentadienyl)manganese, $(C_6H_7)_2Mn$, or $(C_6H_7)Na$ under inert conditions to give dinuclear $[(C_6H_7)Mn\{P(C_2H_5)_3\}I]_2$ [2].

Proposed $Mn\{P(C_6H_5)_3\}_2X_2$ compounds with X = Cl, Br, I, which had been prepared by action of $P(C_6H_5)_3$ on the manganese(II) halides in THF under anhydrous conditions [3], proved to be the corresponding complexes with phosphane oxide, $[Mn(C_{18}H_{15}OP)_2X_2]$ (see p. 90) [4, 5].

References:

[1] Hebendanz, N.; Koehler, F. H.; Müller, G. (Inorg. Chem. **23** [1984] 3043/4).
[2] Koehler, F. H.; Hebendanz, N.; Thewaldt, U.; Kanellakopulos, B.; Klenze, R. (Angew. Chem. **96** [1984] 697/9).
[3] Naldini, L. (Gazz. Chim. Ital. **90** [1960] 1337/42).
[4] Negoiu, D. (Analele Univ. Bucuresti Ser. Stiint. Nat. Chim. **14** [1965] 145/50; C.A. **67** [1965] 60439).
[5] Casey, S.; Levason, W.; McAuliffe, C.A. (J. Chem. Soc. Dalton Trans. **1974** 886).
[6] Beagley, B.; McAuliffe, C.A.; Pritchard, R. G.; White, E. W. (Acta Chem. Scand. A **42** [1988] 544/53).
[7] Beagley, B.; Briggs, J. C.; Chalitta, A.; Gott, G. A.; McAuliffe, C.A.; Pritchard, R. G.; Russell, D. R. (unpublished work from [6]).

39.1.2.3 $MnL(thf)_nX_2$ Compounds

Complexes in Solution. Solutions of complexes with X = Cl or Br for L = $P(C_3H_7)_3$, $P(C_4H_9)_3$, $P(C_6H_5)_3$, $P(C_6H_5)(CH_3)_2$, or $P(C_6H_5)(C_2H_5)_2$ in tetrahydrofuran (THF) were prepared under rigorously anhydrous conditions by adding a limited amount of dry THF and an equimolar amount of the required phosphane to anhydrous MnX_2 and stirring the resulting slurry under Ar for about a week. After this time dry THF was added to give a solution of the required concentration. The procedure for the preparation of complexes with X = I was essentially the same as that for X = Cl or Br, except that sufficient THF was added to give a solution of the required concentration before the addition of the phosphane [1]. The preparation of THF solutions of complexes with X = Cl for L = $P(CH_2{-}CH{=}CH_2)_3$, with X = Cl or Br for L = $P(C_4H_9)_3$, and with X = Br or I for L = $P(C_3H_7)_3$ or $P(C_6H_5)(CH_3)_2$ by dissolving the corresponding $MnLX_2$ complexes in THF is reported in [2].

X-band ESR spectra of complexes with X = Br or I for L = $P(C_4H_9)_3$ [3] and with X = NCS [3, 4] or NCSe [5] for L = $P(C_6H_5)_3$ [3 to 5] in THF at room temperature [4] or in frozen THF at 77 K [3], at 93 K [5], or at −60°C [3] show an absorption at g_{eff} = 2 and additional lines down to g_{eff} = 6

[3 to 5]. Evaluation of the spectra yielded the distortion parameters $D \sim 0.1$ and $\lambda < 0.1$ and hyperfine coupling constants between 75 and 95 G [2 to 5]. The ESR data reported in [3 to 5] and molecular weight measurements strongly suggest monomeric complexes with an essentially octahedral structure [6]. The findings suggest that the species present are $MnL(thf)_3X_2$ complexes with the X ions in *trans* positions. It follows that the polymeric structure of $MnLX_2$ complexes is broken on dissolution in THF and that there must be a phosphane transfer from one of the octahedral sites to a tetrahedral site [6]. X-band ESR spectra of complexes with long-chain phosphanes in THF are very similar to those with short-chain phosphanes and thus may be assigned a similar structure (octahedral with 3 THF molecules coordinated, X in *trans* positions) [7, 8]. In contrast, ESR spectra of $[MnL(thf)X_2]_2$ complexes in toluene at 77 K are almost identical to those in the solid state (see below), suggesting retention of the dimeric structure of the solid compounds (see p. 52) [8].

$MnL(thf)_3X_2$ complexes in tetrahydrofuran and $[MnL(thf)X_2]_2$ complexes in toluene react with dioxygen in a 1:1 mole ratio to give the dioxygen adducts. This is supported by molecular weight determinations of the adducts $Mn\{P(C_{14}H_{29})_3\}(thf)(O_2)Br_2$ and $Mn\{P(C_{16}H_{33})_3\}(thf)(O_2)$-$Br_2$ in toluene. Thus, in toluene, on reaction with dioxygen the Mn–thf bond of the dimer is not broken but the MnX_2Mn double bridge is, and two O_2 molecules are added to give the monomeric penta-coordinate adducts. On the other hand, in THF one of the coordinated THF molecules, presumably that in *trans* position to the phosphane [9], is substituted by O_2 to give the monomeric octahedral adduct [8]. From absorption isotherms constructed as described on p. 44, dioxygen-binding equilibrium data were calculated using the Hill equation (see p. 45). In the table below dioxygen-binding equilibrium data for complexes with short-chain phosphanes in THF are gathered (symbols are explained on p. 45).

complex	t in °C	p_{50}	K_{O_2}	n	Ref.
$Mn\{P(C_2H_5)_3\}(thf)_3Cl_2$	20	11.83	84.5	0.86	[10]
$Mn\{P(C_2H_5)_3\}(thf)_3Br_2$	20	11.3	88.3	0.88	[10]
$Mn\{P(C_2H_5)_3\}(thf)_3I_2$	−40	12.9	84.5	1.62	[10]
$Mn\{P(C_3H_7)_3\}(thf)_3Cl_2$	20	509	19.6	1.35	[10, 11]
$Mn\{P(C_3H_7)_3\}(thf)_3Br_2$	20	11.8	84.5	0.86	[10, 11]
$Mn\{P(C_3H_7)_3\}(thf)_3I_2$	−40	22.9	43.5	1.77	[10]
$Mn\{P(C_4H_9)_3\}(thf)_3Br_2$	20	20.9	47.9	1.12	[10 to 12]
$Mn\{P(C_4H_9)_3\}(thf)_3I_2$	−40	29.7	33.7	1.8	[10, 12]
$Mn\{P(C_6H_5)(CH_3)_2\}(thf)_3I_2$	−40	27.0	37.1	1.43	[10]

Equilibrium data for long-chain phosphane $[MnL(thf)X_2]_2$ complexes with X = Cl, Br, or I in toluene at 20°C [8] and of $MnL(thf)_3X_2$ complexes with X = Cl or Br at 20°C and with X = I at −40°C in THF [7, 8] are presented in the papers [7, 8]. (It was found necessary to perform measurements on iodo complexes in THF at −40°C, because at higher temperatures the dioxygen adducts decomposed before equilibrium was reached [8].) In both solvents the dioxygen affinity is in the order: X = Cl > Br > I. However, the relative affinity referred to the phosphane is different for the halides in the two solvents. Sequences of the phosphanes for the three halides in toluene and in THF are presented in [8]. Hill coefficients, n > 1 (1.10 to 1.64), for the complexes in toluene suggest some cooperativity in this solvent. The chloro and bromo complexes in THF have n ~ 1 (0.75 to 1.30) while the iodo complexes in THF have n > 1 (1.30 to 1.55), indicating some cooperativity as well, although the reason for this is unclear [7, 8].

The reaction of (solvated) $MnLX_2$ complexes in THF with nitric oxide, sulfur dioxide, ethylene, carbon monoxide, and carbon disulfide is described on pp. 46/7.

$[MnL(thf)X_2]_2$ complexes with long-chain phosphanes were prepared by the reaction of anhydrous $MnCl_2$ with an equimolar amount of the phosphane in dry THF under rigorously inert conditions. The solution was stirred under Ar for 10 d. After this time the THF was removed by application of a vacuum. Dry pentane was added and the mixture stirred for 1 h. The resulting slurry was filtered in a Schlenk apparatus and the solid complex dried in vacuum for 6 h. The chloro and bromo complexes are pale pink, the iodo complexes are pale orange. Below are summarized the physical data of $[MnL(thf)X_2]_2$ complexes: melting points (m.p. in °C), magnetic moments (μ_{eff} in μ_B), and IR data (from mulls, in cm^{-1}) [8]:

complex	m.p.	μ_{eff}	$\nu(COC)_{thf}^{a)}$	ν(Mn–O)	ν(Mn–P)	ν(Mn–X)
$[Mn\{P(C_{12}H_{25})_3\}(thf)Cl_2]_2$	32	5.70	1040, 880	425	380	295
$[Mn\{P(C_{14}H_{29})_3\}(thf)Cl_2]_2$	34	5.68	1040, 880	420	385	300
$[Mn\{P(C_{16}H_{33})_3\}(thf)Cl_2]_2$	38	5.71	1040, 880	425	380	300
$[Mn\{P(C_6H_5)(C_{12}H_{25})_2\}(thf)Cl_2]_2$	48	5.65	1040, 880	420	385	295
$[Mn\{P(C_6H_5)(C_{14}H_{29})_2\}(thf)Cl_2]_2$	141	5.70	1040, 880	425	390	300
$[Mn\{P(C_6H_5)(C_{16}H_{33})_2\}(thf)Cl_2]_2$	125	5.70	1040, 880	425	390	295
$[Mn\{P(C_{12}H_{25})_3\}(thf)Br_2]_2$	47	5.59	1042, 878	430	390	240
$[Mn\{P(C_{14}H_{29})_3\}(thf)Br_2]_2$	33	5.60	1042, 878	430	390	245
$[Mn\{P(C_{16}H_{33})_3\}(thf)Br_2]_2$	44	5.65	1042, 880	435	395	240
$[Mn\{P(C_6H_5)(C_{12}H_{15})_2\}(thf)Br_2]_2$	92	5.60	1040, 878	435	400	235
$[Mn\{P(C_6H_5)(C_{14}H_{29})_2\}(thf)Br_2]_2$	116	5.65	1040, 878	435	400	235
$[Mn\{P(C_6H_5)(C_{16}H_{33})_2\}(thf)Br_2]_2$	102	5.60	1040, 878	435	400	235
$[Mn\{P(C_{12}H_{25})_3\}(thf)I_2]_2$	54	5.45	1040, 880	445	400	220
$[Mn\{P(C_{14}H_{29})_3\}(thf)I_2]_2$	45	5.40	1040, 880	445	400	210
$[Mn\{P(C_{16}H_{33})_3\}(thf)I_2]_2$	28	5.45	1040, 880	440	400	210
$[Mn\{P(C_6H_5)(C_{12}H_{33})_2\}(thf)I_2]_2$	44	5.40	1040, 880	445	405	210
$[Mn\{P(C_6H_5)(C_{14}H_{29})_2\}(thf)I_2]_2$	32	5.40	1040, 880	445	405	210
$[Mn\{P(C_6H_5)(C_{16}H_{33})_2\}(thf)I_2]_2$	58	5.40	1040, 880	450	410	205

a) Noncoordinated THF absorbs at 1060 and 900 cm^{-1}.

The ESR spectra of the compounds show a broad strong absorption at $g_{eff}=2$ (~3300 G) and other much weaker bands at ~1000, 1500, and 5400 G. The magnetic moments (5.40 to 5.71 μ_B) give evidence for a halide-bridged structure. The IR data show the presence of coordinated THF and terminal-bonded halide (however, the far-IR spectra in the bridging ν(Mn–X) region were not recorded). On the basis of the physical data the complexes were assigned a dinuclear trigonal-bipyramidal structure with two bridging X atoms and one terminal X atom in equatorial sites and the THF and PR_3 ligands in the axial sites [8].

The complexes are quite soluble in toluene and THF, but they are only sparingly soluble in pentane or diethyl ether [8].

$MnL(thf)_nX_2$. Poorly defined and characterized compounds with n=2 [13] or varying from 0 to 3 [2] are reported to have been prepared by the reaction of anhydrous MnX_2 with a short-chain phosphane in dry THF for 10 d [2, 13] or by dissolving an $MnLX_2$ compound in THF [2] under rigorously inert conditions [13].

References:

[1] Gott, G. A.; McAuliffe, C. A. (J. Chem. Soc. Dalton Trans. **1987** 2241/3).
[2] Hill, W. E.; McAuliffe, C. A. (U.S. 4442297 [1981/84]; C.A. **101** [1984] No. 40616).
[3] McAuliffe, C. A.; Little, M. G.; Raynor, J. B. (J. Chem. Soc. Chem. Commun. **1982** 68/70).
[4] Hosseiny, A.; Mackie, A. G.; McAuliffe, C. A.; Minten, K. (Inorg. Chim. Acta **49** [1981] 99/105).
[5] Benson, C. G.; Little, M. G.; McAuliffe, C. A. (Inorg. Chim. Acta **87** [1984] 169/75).
[6] Beagley, B.; Briggs, J. C.; Hosseiny, A.; Hill, W. E.; King, T. J.; McAuliffe, C. A.; Minten, K. (J. Chem. Soc. Chem. Commun. **1984** 305/6).
[7] Barratt, D. S.; Gott, G. A.; McAuliffe, C. A. (J. Chem. Soc. Dalton Trans. **1988** 2065/70).
[8] Barratt, D. S.; Gott, G. A.; McAuliffe, C. A. (Inorg. Chim. Acta **145** [1988] 289/98).
[9] Barratt, D. S.; McAuliffe, C. A. (J. Chem. Soc. Dalton Trans. **1987** 2497/501).
[10] McAuliffe, C. A. (Spec. Publ. Roy. Soc. Chem. No. 39 [1981] 119/31; C.A. **95** [1981] No. 92442).

[11] McAuliffe, C. A.; Al-Khateeb, H. F. (Inorg. Chim. Acta **45** [1980] L195/L196).
[12] McAuliffe, C. A.; Al-Khateeb, H. F.; Barratt, D. S.; Briggs, J. C.; Challita, A.; Hosseiny, A.; Little, M. G.; Mackie, A. G.; Minten, K. (J. Chem. Soc. Dalton Trans. **1983** 2147/53).
[13] Barratt, D. S.; McAuliffe, C. A. (Inorg. Chim. Acta **97** [1985] 37/43).

39.1.2.4 Dioxygen Adducts, $Mn^{II}L(O_2)X_2$ and $Mn^{II}L(thf)_n(O_2)X_2$

Complexes in Solution. Molecular weight determinations and ESR and electronic spectral measurements indicate the existence of $MnL(O_2)X_2$ and $MnL(thf)(O_2)X_2$ species in noncoordinating solvents such as toluene and of $MnL(thf)_2(O_2)X_2$ species in tetrahydrofuran [1, 3, 7, 11 to 14]. The oxygen adducts are formed by the exposure of solutions of the nonoxygenated complexes to extremely dry oxygen [1 to 3, 7, 11, 13, 14]. Thus, 3.2 mmol $Mn\{P(C_4H_9)_3\}I_2$ in 100 mL tetrahydrofuran was exposed to oxygen (p_{O_2} > 30 Torr) at −20°C with stirring of the solution in a special vacuum device. During O_2 uptake the nearly colorless solution changed to deep green-purple. After 6 min the complex had absorbed 70.8 mL O_2 (corrected for blank THF absorption) corresponding to 98.7% of the theoretical 1:1 absorption. Details of the apparatus and procedure used are described in [1]. Under the conditions employed absorption was complete after 5 to 10 min, depending on stirrer speed and the nature of the complex [1]. A number of short-chain phosphane $MnLX_2$ and $MnL(thf)_3X_2$ complexes capable of forming oxygen adducts in toluene or THF is listed in Table 1, p. 45, and in the table on p. 51, respectively. In addition, oxygen adducts of long-chain phosphane $MnLX_2$ and $[MnL(thf)X_2]_2$ complexes with $L=P(C_{12}H_{25})_3$, $P(C_{14}H_{29})_3$, $P(C_{16}H_{33})_3$, $P(C_6H_5)(C_{12}H_{25})_2$, $P(C_6H_5)(C_{14}H_{29})_2$, and $P(C_6H_5)(C_{16}H_{33})_2$, all for X = Cl, Br, or I, in toluene or THF were prepared [13, 14]. A controversy about the existence of $MnL(O_2)X_2$ complexes is discussed in [6, 7]. Molecular weight determinations of several long-chain phosphane $MnL(O_2)X_2$ and $MnL(thf)(O_2)X_2$ complexes in toluene show that the complexes are monomeric [13, 14].

X-band ESR spectra were recorded at 77 K for the oxygen adducts of short-chain phosphane complexes listed on p. 54. The compounds were prepared at the temperatures given in the table. The spectra consist of several groups of six lines, together with broad features in which hyperfine coupling was not resolved. The line positions given are the magnetic field strengths at the centers of the sextuplets or broad lines [11]:

complex	solvent	color	t in °C	ESR line positions in G
$Mn\{P(C_4H_9)_3\}(thf)_2(O_2)Cl_2$	THF	pink	−60	numerous lines at 0 to 3000 [a)], 3300 s, 5500 m
	THF	purple	0	numerous lines at 0 to 3000 [a)], 3300 vs, 5500 m
$Mn\{P(C_4H_9)_3\}(thf)_2(O_2)Br_2$	THF	deep pink	−60	1125 s [a)], 1600 m [a)], 2200 m [a)], 3300 m, 5700 m, 8500 w
	THF	blue	0	same as at −60°C
$Mn\{P(C_4H_9)_3\}(thf)_2(O_2)I_2$	THF	purple-brown	−60	950 s [a)], 1400 m [a)], 3300 m, 5350 m
$Mn\{P(C_5H_{11})_3\}(O_2)Cl_2$	toluene	purple	0	1650 w, 3300 s
$Mn\{P(C_5H_{11})_3\}(O_2)Br_2$	toluene	blue	0	1500 s, 3300 s [a)]
$Mn\{P(C_5H_{11})_3\}(O_2)I_2$	toluene	green	0	1600 s, 3310 s

[a)] Split into six lines by hyperfine coupling to the nuclear spin (I = 5/2) of ^{55}Mn.

The complexes with X = Br or I in THF exhibit a strong sextuplet at g ≈ 6. The shape and position of this line is characteristic of $g_\perp$ signals observed in six-coordinate high-spin (S = 5/2) complexes with axial symmetry. This is consistent with the well-resolved hyperfine couplings in this region (80 to 90 G). Thus, the most likely structure for oxygen adducts of the short-chain phosphane complexes in tetrahydrofuran is one which involves two coordinated THF molecules and the halogens both in *trans* positions [11].

A similar structure is assumed for oxygen adducts of long-chain phosphane complexes in tetrahydrofuran [13, 14]. In toluene, a pseudotetrahedral structure is suggested for $MnL(O_2)X_2$ complexes (both short-chain and long-chain phosphane) [11, 13] and a monomeric five-coordinate structure for long-chain phosphane $MnL(thf)(O_2)X_2$ complexes [14]. The O_2 molecule is presumably bonded in a "bent end-on" configuration (see below) so as to remove the degeneracy of the d_{xz} and d_{yz} orbitals which would not be possible with an "end-on" bonded O_2 molecule [11]. The ESR spectra of $Mn\{P(C_3H_7)_3\}(O_2)Cl_2$ at −182 and 20°C are presented in [10].

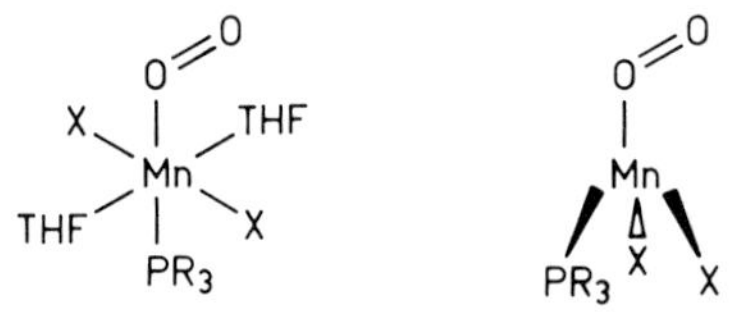

The electronic spectra of the oxygen adducts of short-chain phosphane complexes in toluene or THF which were measured at 20°C in a specially-designed apparatus [1, 2] show maxima at the following wavelengths (λ in nm); ε in $L \cdot mol^{-1} \cdot cm^{-1}$:

complex	solvent	λ_1	ε	λ_2	Ref.
$Mn\{P(C_2H_5)_3\}(O_2)Cl_2$	toluene	535	2600	401	[12]
$Mn\{P(C_2H_5)_3\}(O_2)Br_2$	toluene	574.5	25600	419.5	[12]
$Mn\{P(C_2H_5)_3\}(O_2)I_2$	toluene	625 [a)]	25750	412	[12]

complex	solvent	λ_1	ε	λ_2	Ref.
$Mn\{P(C_3H_7)_3\}(O_2)Cl_2$	THF	530	—	395	[1]
		534	>200	398 [b)]	[10]
$Mn\{P(C_3H_7)_3\}(O_2)Br_2$	THF	570	—	414	[1]
$Mn\{P(C_3H_7)_3\}(O_2)I_2$	THF	620	—	455	[1]
$Mn\{P(C_4H_9)_3\}(O_2)Br_2$	THF	569 [c)]	—	414 [c)]	[3]
$Mn\{P(C_6H_5)(CH_3)_2\}(O_2)Br_2$	THF	550	—	410	[9]

a) A third band was observed at 458 nm. – b) $\varepsilon > 90\ L \cdot mol^{-1} \cdot cm^{-1}$. – c) At 0°C.

Absorption maxima shift to lower energy on going from Cl to I, with simultaneous change of color [1, 10]. The colors of the oxygen adducts appear to depend more upon the nature of the halide than that of the phosphane, and also to some extent depend on the solvent. The oxygen adducts are purple (X = Cl), blue (X = Br), green (X = I, toluene), or red-brown (X = I, THF). The positions of the band maxima for each halide series show only small variations with change in phosphane ligand, usually by 5 to 10 nm. Average positions of the band maxima (λ in nm) and their molar absorption coefficients (ε in $L \cdot mol^{-1} \cdot cm^{-1}$) for long-chain phosphane $MnL(O_2)X_2$ complexes are listed below [13]:

X	solvent	λ_1	ε	λ_2	ε
Cl	toluene	520 ± 10	3800 ± 500	380 ± 10	220 ± 400
	THF	540 ± 10	1600 ± 300	400 ± 5	1200 ± 100
Br	toluene	545 ± 10	5700 ± 500	410 ± 5	3200 ± 400
	THF	560 ± 10	1800 ± 300	415 ± 5	1600 ± 100
I	toluene	545 ± 20	4100 ± 400	390 ± 10	5700 ± 500
				370 ± 5	440 ± 400
	THF	620 ± 10	2800 ± 300	520 ± 20	2700 ± 200

The absorption coefficients vary with the solvents, being largest in noncoordinating solvents [1].

Because of the linear relationship between the magnitude of the absorption coefficient and the degree of oxygenation, which was found for all the complexes studied, measurements of electronic spectra could be used to follow the O_2 absorption of $MnLX_2$ complexes [1, 2, 8] (see "Chemical Reactions of $MnLX_2$ Complexes" p. 44). Strong evidence for two different $MnL(O_2)X_2$ species (see magnetic properties, p. 56) was obtained from following the electronic spectrum during oxygenation of $Mn\{P(C_4H_9)_3\}Br_2$ in THF at 0°C: The two absorption maxima which develop do so at vastly different rates [3].

The stability of the complexes in solution (THF, toluene, 1,2-dichloroethane) is dependent on the O_2 partial pressure above the solution and on the temperature. For example, it was possible to store $Mn\{P(C_4H_9)_3\}(O_2)I_2$ in THF for 24 h under $p_{O_2} = 100$ Torr at −20°C with no detectable deterioration, whereas under the same p_{O_2}, but at 0°C, the complex began to deteriorate perceptibly after only 6 h. As the major decomposition product of $Mn\{P(C_4H_9)_3\}(O_2)I_2$, the white-yellow phosphane oxide complex, $[Mn^{II}(C_{12}H_{27}OP)I_2]$ (see p. 87), was isolated [1]. Decomposition at temperatures >0°C was also observed for long-chain phosphane $MnL(O_2)I_2$ complexes in THF, and phosphane oxides were isolated from solution [13, 14]. The chloro and bromo complexes are more stable than the iodo complexes [1, 13, 14]. The stability of an

$MnL(O_2)X_2$ complex and its dependence on temperature and O_2 partial pressure is also reflected in the number of oxygen absorption/desorption cycles (cycling) which are possible for the individual complexes. Thus, for $Mn\{P(C_4H_9)_3\}I_2$ in THF 2, 5, and >400 cycles were possible at 0°C (760 Torr), −20°C (760 Torr), and −20°C (150 Torr), respectively [1]. Details of cycling are given for $Mn\{P(C_3H_7)_3\}Cl_2$ in THF [10] and for $Mn\{P(C_8H_{17})_3\}Br_2$ in polyethylene glycol diisopropyl ether [15]. The oxygen adducts are nonconducting in THF and toluene [1].

Solid Compounds. $MnL(O_2)X_2$ complexes in the solid state were obtained by the exposure of the corresponding solid $MnLX_2$ compounds to extremely dry oxygen for several hours or days [1, 3, 7]. For example, when 4.48 g of $Mn\{P(C_6H_5)(CH_3)_2\}I_2$ (0.01 mol) is allowed to stand in an atmosphere of dry oxygen for 2 d, the weight of the green $Mn\{P(C_6H_5)(CH_3)_2\}(O_2)I_2$ complex obtained is 4.80 g (0.01 mol). No further weight gain is observed over long periods [7]. The $MnL(O_2)Br_2$ complexes with $L=P(CH_3)_3$ [4, 5], $P(C_2H_5)_3$, and $P(C_6H_5)(CH_3)_2$ [5] were prepared as thin films on KBr windows in a specially designed infrared cell [4, 5]. Colors of the solid complexes are the same as those of the complexes in toluene (see p. 54) [2]. Magnetic moments, μ_{eff} (in μ_B), and ν(CN) bands (in cm^{-1}) of some $MnL(O_2)(NCS)_2$ complexes are gathered below [1]:

complex	μ_{eff}	ν(CN)
$Mn\{P(CH_3)_3\}(O_2)(NCS)_2$	6.8	2100, 2080, 2040
$Mn\{P(C_2H_5)_3\}(O_2)(NCS)_2$	6.8	2115, 2100, 2080
$Mn\{P(C_3H_7)_3\}(O_2)(NCS)_2$	6.8	2108, 2085, —
$Mn\{P(C_4H_9)_3\}(O_2)(NCS)_2$	6.9	2100, 2085, —

Magnetic moments of ~6.8 μ_B for the $MnL(O_2)X_2$ complexes indicate systems with six unpaired electrons [1, 3]. The most plausible explanation for this is that there are two types of $MnL(O_2)X_2$ complexes coexisting in a 50:50 mixture: 50% exists as $Mn^{II}L(^3O_2)X_2$ ($^3O_2$ = triplet oxygen) having a total of 7 unpaired electrons and 50% exists as $Mn^{II}L(^1O_2)X_2$ ($^1O_2$ = singlet oxygen) having a total of 5 unpaired electrons, thus giving an average magnetic moment equivalent to 6 unpaired electrons. The ESR and IR spectra of solid $MnL(O_2)X_2$ complexes are consistent with this theory. In surface oxygenation of solid $MnLX_2$ complexes the ESR spectra remain quenched until 50% of the oxygen is added, and after this the ESR signal grows as the remaining 50% is taken up. Also the ESR signal is typical of Mn^{II}, and there is no evidence of a superoxide signal [3].

Each IR spectrum of an $MnL(O_2)X_2$ complex shows the following band (in cm^{-1}), which disappears on deoxygenation and which is assignable to the O–O stretching mode of the coordinated neutral O_2 molecule [3]:

complex	ν(O–O)	complex	ν(O–O)
$Mn\{P(C_3H_7)_3\}(O_2)Cl_2$	1430	$Mn\{P(C_4H_9)_3\}(O_2)I_2$	1403
$Mn\{P(C_3H_7)_3\}(O_2)Br_2$	1409	$Mn\{P(C_6H_5)(CH_3)_2\}(O_2)Cl_2$	1416
$Mn\{P(C_3H_7)_3\}(O_2)I_2$	1402	$Mn\{P(C_6H_5)(CH_3)_2\}(O_2)Br_2$	1409

Apart from ν(O–O) bands, IR spectra of $MnL(O_2)(NCS)_2$ complexes were found to be almost identical to those of $MnL(NCS)_2$ complexes. No bands assignable to $Mn-O_2$ vibrations were observed [1]. On the basis of the physical properties the $MnL(O_2)X_2$ complexes were assigned a monomeric tetrahedral structure [3]. In contrast, thin films of $MnL(O_2)Br_2$ with $L=P(CH_3)_3$ or $P(C_2H_5)_3$ showed a band at ~1130 cm^{-1} that shifted to 1095 cm^{-1} for the $^{18}O_2$ adduct. This band was assigned to the O–O stretching mode in a hyperoxide (O_2^-) species of $Mn^{III}L(O_2)Br_2$ with the

Mn^{III} providing the chromophore [5]. The solid complexes may be stored in a dry atmosphere for a long time [3]. Thin films of $Mn\{P(C_2H_5)_3\}(O_2)Cl_2$ and $Mn\{P(C_2H_5)_3\}(O_2)Br_2$ are reported to react with SO_2 (8 Torr) to give the phosphane oxide complexes and $Mn\{P(C_2H_5)_3\}(SO_2)_{0.5}X_2$ (p. 58), as derived from IR studies [16].

References:

[1] McAuliffe, C. A.; Al-Khateeb, H. F.; Barrat, D. S.; Briggs, J. C.; Challita, A.; Hosseiny, A.; Little, M. G.; Mackie, A. G.; Minten, K. (J. Chem. Soc. Dalton Trans. **1983** 2147/53).

[2] McAuliffe, C. A. (Spec. Publ. Roy. Soc. Chem. No. 39 [1981] 119/31; C.A. **95** [1981] No. 92442).

[3] McAuliffe, C. A.; Al-Khateeb, H. F.; Jones, H. H.; Levason, W.; Minten, K.; McCullough, F. P. (J. Chem. Soc. Chem. Commun. **1979** 736/8).

[4] Burkett, H. D.; Newberry, V. F.; Hill, W. E.; Worley, S. D. (J. Am. Chem. Soc. **105** [1983] 4097/9).

[5] Newberry, V. F.; Burkett, H. D.; Worley, S. D.; Hill, W. E. (Inorg. Chem. **23** [1984] 3911/7).

[6] Brown, R. M.; Bull, R. E.; Green, M. L. H.; Grebenik, P. D.; Martin-Polo, J. J.; Mingos, D. M. P. (J. Organometal. Chem. **201** [1980] 437/46).

[7] McAuliffe, C. A. (J. Organometal. Chem. **228** [1982] 255/64).

[8] McAuliffe, C. A.; Minten, K. (Symp. Papers Inst. Chem. Eng. North West. Branch **1981/82**; C.A. **96** [1982] No. 192249).

[9] Hosseiny, A.; McAuliffe, C. A.; Minten, K.; Parrott, M. J.; Pritchard, R.; Tames, J. (Inorg. Chim. Acta **39** [1980] 227/31).

[10] Yatsimirskii, K. B.; Bratushko, Yu. I.; Ermokhina, N. I. (Teor. Eksperim. Khim. **18** [1982] 367/71; Theor. Exptl. Chem. [USSR] **18** [1982] 328/31).

[11] McAuliffe, C. A.; Little, M. G.; Raynor, J. B. (J. Chem. Soc. Chem. Commun. **1982** 68/70).

[12] Beagley, B.; McAuliffe, C. A.; Minten, K.; Pritchard, R. G. (J. Chem. Soc. Dalton Trans. **1987** 1999/2003).

[13] Barratt, D. S.; Gott, G. A.; McAuliffe, C. A. (J. Chem. Soc. Dalton Trans. **1988** 2065/70).

[14] Barratt, D. S.; Gott, G. A.; McAuliffe, C. A. (Inorg. Chim. Acta **145** [1988] 289/98).

[15] Richter, W.; Kummer, R.; Platz, R. (Ger. 2927734 [1979/81]).

[16] Newberry, V. F.; Hill, W. E.; Worley, S. D. (Phosphorus Sulfur **27** [1986] 253/60).

39.1.2.5 Sulfur Dioxide Adducts, $Mn^{II}L(SO_2)_nX_2$

Complexes with n = 0.66 and $L = P(C_3H_7)_3$, $P(C_4H_9)_3$, or $P(C_6H_5)(CH_3)_2$ for X = Br or I, and $L = P(C_6H_5)_3$, $P(C_6H_5)(C_2H_5)_2$, $P(C_6H_5)(C_4H_9)_2$, or $P(C_6H_5)(i\text{-}C_4H_9)_2$ for X = I were prepared by exposing toluene slurries of $MnLX_2$ complexes to an atmosphere of dry sulfur dioxide (760 Torr) for 48 h with stirring. $Mn\{P(C_6H_5)_3\}(SO_2)_{0.66}I_2$ and $Mn\{P(C_6H_5)(CH_3)_2\}(SO_2)_{0.66}I_2$ were readily obtained as yellow solids, whereas the remaining complexes were isolated initially as yellow oils that subsequently solidified on standing. $Mn\{P(C_3H_7)_3\}(SO_2)_{0.66}Br_2$ and $Mn\{P(C_6H_5)(CH_3)_2\}(SO_2)_{0.66}Br_2$ were also obtained by exposing solid samples of the corresponding $MnLX_2$ complexes to one atmosphere of sulfur dioxide in a special tube connected to a high-vacuum line. Complexes of the $MnL(SO_2)_nX_2$ type with n = 0.65 to 0.76 (L and X as stated above) were obtained by exposing THF solutions of the corresponding $MnLX_2$ complexes to SO_2–Ar mixtures with increasing SO_2 partial pressures (initial $p_{SO_2} = 80$ Torr) at 0°C in the special high-vacuum apparatus [3] described in [4]. X-band ESR spectra were taken from $MnL(SO_2)_nX_2$ (n = 0.65 to 0.76, L and X as stated above) in frozen THF at 93 K (spectra presented for

$Mn\{P(C_3H_7)_3\}Br_2$ and $Mn\{P(C_4H_9)_3\}I_2$ in the paper). The spectra show an increase in the g=6 peak and changes in the fine structure of the g=2 peak compared to the $MnLX_2$ complexes, suggesting a structure with axial symmetry for the SO_2 adducts. The $MnL(SO_2)_{0.66}X_2$ complexes show an IR band in the 1140 to 1110 cm^{-1} range which was assigned to the ν(SO) vibration. No reversible coordination of SO_2 could be found for these complexes. The sulfur dioxide could be removed neither by heating solid samples for 3 h at 50°C nor by exposing solutions for 3 h to a static vacuum. Solutions in THF exhibited nonelectrolytic behavior. The complexes were recovered unchanged from THF [3].

Thin films of $Mn\{P(C_6H_5)(CH_3)_2\}Br_2$ were prepared on KBr windows, as described on p. 38, and exposed to SO_2 (50 Torr) for 24 h at 298 K, followed by brief evacuation of the specially designed vacuum IR cell. The spectrum showed the following bands: 1075, 1002, 995 cm^{-1} assigned to ν(SO); 525 cm^{-1} to ν(Mn–O); and 445, 412 cm^{-1} to ν(Mn–S) (assignment was made by isotopic shifts using $S^{18}O_2$). The bands declined markedly upon evacuation at 10^{-6} Torr. After 12 h evacuation, the sulfur dioxide was essentially removed. Cycling of the intensities of these bands by SO_2 absorption/desorption was observed for up to three cycles. However, on exposure to SO_2 at 120 Torr for 72 h or prolonged evacuation a changed spectrum with new bands at 982 and 625 cm^{-1} was obtained; this could not be reversed by lengthy evacuation. Similar results were obtained for $Mn\{P(C_2H_5)_3\}Br_2$ films. The conditions of SO_2 exposure in this case were: 15 Torr for 15 min to give the adduct with reversibly-coordinated SO_2 and 50 Torr for 24 h to give the adduct with irreversibly-bonded SO_2 [2, 5]. Studies of the interactions of $MnLX_2$ films (L=$P(CH_3)_3$ for X=Cl, Br, or I and L=$P(C_2H_5)_3$ and $P(C_6H_5)(CH_3)_2$ for X=Cl or I) with SO_2 showed results analogous to those obtained for $Mn\{P(C_6H_5)(CH_3)_2\}Br_2$ and $Mn\{P(C_2H_5)_3\}Br_2$. Initially, adducts were formed which could be cycled by exposure to and evacuation of SO_2. Subsequently, each film decomposed to an adduct which was not affected by evacuation. An exception to this generalization was $Mn\{P(C_6H_5)(CH_3)_2\}Cl_2$, which converted directly to the adduct of the second kind by exposure to 50 Torr of SO_2 for 24 h. Bulk analyses (data given for complexes with X=Br or I for L=$P(CH_3)_3$ and X=Br for L=$P(C_2H_5)_3$ or $P(C_6H_5)(CH_3)_2$) indicate the composition $MnL(SO_2)_{0.5}X_2$ for the irreversibly-formed adducts [5].

The IR spectra of all the adducts show a band in the 1120 to 1115 cm^{-1} range which was assigned to ν(P–C_{aryl}). The complexes with reversibly-bonded SO_2 show two bands in the 1080 to 1050 and 1010 to 975 cm^{-1} ranges, which were assigned to ν(SO) vibrations, while the bands for the $MnL(SO_2)_{0.5}X_2$ complexes appear in the 980 to 900 and 900 to 800 cm^{-1} ranges (in the paper specified for the individual complexes) [5]. The IR spectra together with the complex stoichiometries indicate that the two forms of each complex contain SO_2 in a bridging mode with S and O atoms involved in the bridge [2, 5].

The reaction with dioxygen of the reversibly-formed SO_2 adduct of $Mn\{P(C_6H_5)(CH_3)_2\}Br_2$ was studied. After three exposures to O_2, a band was observed at 1140 cm^{-1}, attributable to the ν(P–O) vibration in the $Mn\{(C_6H_5)(CH_3)_2PO\}Br_2$ complex. This indicates substitution of the SO_2 by O_2 and formation of the phosphane oxide complex [5].

References:

[1] Newberry, V. F. (Diss. Abstr. Intern. B **45** [1985] 3814/5; C.A. **103** [1985] No. 63896).
[2] Hill, W. E.; Worley, S. D.; Paul, D. K.; Newberry, V. F. (Inorg. Chem. **24** [1985] 4429/30).
[3] Barratt, D. S.; Benson, C. G.; Gott, G. A.; McAuliffe, C.A.; Tanner, S. P. (J. Chem. Soc. Dalton Trans. **1985** 2661/4).
[4] Newberry, V. F.; Burkett, H. D.; Worley, S. D.; Hill, W. E. (Inorg. Chem. **23** [1984] 3911/7, 3912).
[5] Newberry, V. F.; Hill, W. E.; Worley, S. D. (Phosphorus Sulfur **27** [1986] 253/60).

39.1.2.6 Other Adducts of $Mn^{II}LX_2$ Complexes

Nitric oxide adducts are treated along with the "Nitrosyl Compounds" on p. 61. The formation of ethylene, tetracyanoethylene, *tert*-butyl isocyanide, carbon monoxide, and carbon disulfide adducts of $Mn^{II}LX_2$ complexes is briefly described together with the chemical reactions of $Mn^{II}LX_2$ complexes on pp. 46/7. These adducts containing Mn–C bonds are not included in the coordination compounds of manganese but are treated later, along with the organomanganese compounds.

39.1.2.7 $[L_2Mn^{II}(NCS)_2Hg(OCN)_2]_n$

The bimetallic complex was synthesized by the reaction of $[Mn(NCS)_2Hg(OCN)_2]_n$ with $L=P(C_6H_5)_3$ (1:5 mole ratio) in acetone, followed by the usual workup. The IR spectrum of the complex (in polyethylene disks) shows a number of bands (cm^{-1}): 2203, $\nu(CN)_{OCN}$; 2133, $\nu(CN)_{NCS}$; 452, 438, δ(NCS); 261, ν(Mn–N); 201, ν(HgS); 181, ν(Mn–L); 163, δ(N–Mn–N). The IR data indicate the presence of bridged thiocyanato groups, but of terminal cyanato groups. These data, together with considerations of charge densities at the metal ions and at the ends of the coordinating groups, suggest a polymeric structure with NCS bridges and tetrahedral geometries for the two complex units. Presumably, the Mn ion is coordinated by two N and two P atoms and the Hg ion by two S and two O atoms.

Reference:

Ojha, T. N.; Sharma, S. B. (Chem. Era **19** [1983] 34/7; C.A. **101** [1984] No. 16211).

39.1.2.8 Manganese(III) Compounds

$Mn^{III}\{P(CH_3)_3\}_2X_3$ (X = Cl, Br, I). As an example, the preparation of the iodo compound is described: A solution of $Mn^{II}\{P(CH_3)_3\}I_2$ in diethyl ether was stirred under an atmosphere of dry oxygen for ~1 h. The diethyl ether was then evaporated and the green powder formed was purified by sublimation at 120°C and 0.1 Torr to give a dark green sublimate, $Mn\{P(CH_3)_3\}_2I_3$, and a flesh-colored residue. The yield was ~40%. The compound could also be obtained by the reaction of solid $Mn^{II}\{P(CH_3)_3\}I_2$ with dry oxygen for 5 to 7 d. The green powder which formed was sublimed as above. Yields were 45 to 52%. In both preparative routes rigorously anhydrous conditions must be applied. Crystals of $Mn\{P(CH_3)_3\}_2I_3$ suitable for an X-ray crystallographic study were grown from anhydrous diethyl ether. They are orthorhombic, space group Pnma–D_{2h}^{16} (No. 62), with the lattice parameters a = 10.509(1), b = 11.167(8), c = 14.398(2) Å; Z = 4. The calculated density is D_{calc} = 2.31 g/cm³. The structure was solved to R = 0.081. In the trigonal-bipyramidal $Mn\{P(CH_3)_3\}_2I_3$ molecule, the phosphane ligands are axially and the iodine atoms are equatorially arranged around the Mn atom (see **Fig. 15**, p. 60). The Mn–I and the P–C bonds are staggered. The packing of the $Mn\{P(CH_3)_3\}_2I_3$ molecules in the unit cell shows the I atoms well-removed from the Mn atoms of neighboring molecules. Bond distances (in Å) and bond angles (in °) are:

Mn–I(1)	2.605(6)	I(1)–Mn–I(2)	121.3(1)	I(1)–Mn–P(2)	89.4(3)
Mn–I(2)	2.635(3)	I(2)–Mn–I(2′)	117.5(1)	I(2)–Mn–P(1)	90.3(2)
Mn–P(1)	2.43(1)	I(1)–Mn–P(1)	89.1(3)	I(2)–Mn–P(2)	90.5(2)
Mn–P(2)	2.44(1)				

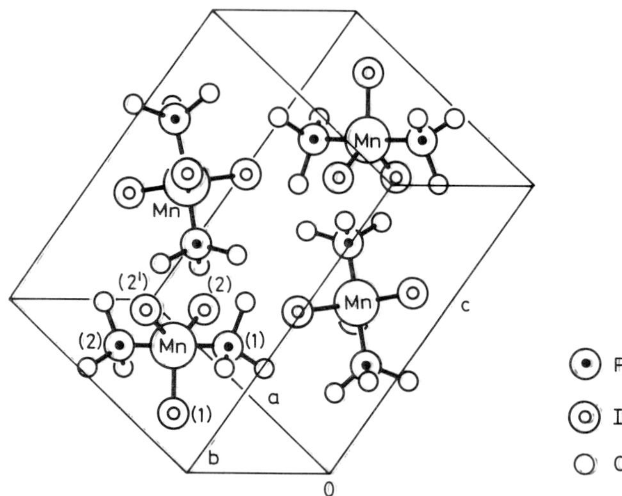

Fig. 15. Packing diagram of $Mn^{III}\{P(CH_3)_3\}_2I_3$ in the unit cell (hydrogen atoms are omitted) [1].

Magnetic moments (μ_{eff} in μ_B), bands in the far-IR spectrum of the solid complexes in Nujol (ν in cm^{-1}), and bands in the electronic spectrum of the complexes in toluene (λ in nm, ε in $L \cdot mol^{-1} \cdot cm^{-1}$) are shown below:

complex	color	μ_{eff}	ν(Mn–P)	ν(Mn–X)	λ_1 (log ε)	λ_2
$Mn\{P(CH_3)_3\}_2Cl_3$	crimson red	4.8	425	324	485.5 (5.26)	388
$Mn\{P(CH_3)_3\}_2Br_3$	purple	4.7	445	229	514 (5.17)	406.5
$Mn\{P(CH_3)_3\}_2I_3$	dark green	4.8	445	192	687, 585 (4.90)	452, 360.5

The electronic reflectance spectra of solid $Mn\{P(CH_3)_3\}_2X_3$ complexes show an additional 1 to 3 bands. $Mn\{P(CH_3)_3\}_2I_3$ is soluble in tetrahydrofuran, toluene, and diethyl ether. Neither these solutions nor the pure solids show an ESR signal. This together with the magnetic moment of 4.8 μ_B strongly point to the presence of a $Mn^{III}(d^4)$ species [1]; preliminary publication [2].

$[Mn^{III}\{P(C_6H_5)_3\}(T4\text{-}MPP)]PF_6$ (5, 10, 15, 20-tetrakis(4-methylphenyl)porphyrin = H_2T4-MPP). The black-green complex was prepared by refluxing a solution of Mn(T4-MPP)Cl and $P(C_6H_5)_3$ (1 : 4 mole ratio) in toluene for 2 h. The toluene was removed in vacuum and the free phosphane in high vacuum at 80 to 100°C (~5 h). The remaining crystals were washed with hexane, then recrystallized from CH_2Cl_2. To the recrystallization solution containing the complex, NH_4PF_6 (1 : 1.5 mole ratio) in acetone was added. In the IR spectrum (KBr) bands at 1440, 995, and 525 cm^{-1} were assigned to the coordinated phosphane, and bands at 840 and 565 cm^{-1} to the PF_6^- anion. The electronic spectrum of the complex in CH_2Cl_2 shows seven bands (λ in nm, log ε in parentheses): 610 (3.99), 575 (3.98), 520 (3.78), 479 (4.76), 434 sh (4.49), 407 (4.72), 388 (4.80) [3].

References:

[1] Beagley, B.; McAuliffe, C. A.; Minten, K.; Pritchard, R. G. (J. Chem. Soc. Dalton Trans. **1987** 1999/2003).

[2] Beagley, B.; McAuliffe, C. A.; Minten, K.; Pritchard, R. G. (J. Chem. Soc. Chem. Commun. **1984** 658/9).

[3] Buchler, J. W.; Dreher, C.; Lay, K. L. (Chem. Ber. **117** [1984] 2261/74).

39.1.2.9 Nitrosyl Compounds

Remark: Other nitrosyl compounds are reported on pp. 68, 71, 73, and 79.

$Mn(NO)LX_2$. Vividly-colored complexes of this type with X = Cl (deep red-purple), X = Br (deep blue), and X = I (brown) for L = $P(C_3H_7)_3$, $P(C_4H_9)_3$, $P(C_6H_5)(CH_3)_2$, and $P(C_6H_5)(C_2H_5)_2$ were obtained by exposing solid samples of the appropriate $MnLX_2$ complexes to NO gas at temperatures below 0°C. Room temperature magnetic moments of the $Mn(NO)LX_2$ complexes range from 4.85 to 5.25 μ_B, indicative of four unpaired electrons. No ESR spectra could be observed [1]. The IR spectra show a band around 1600 cm^{-1} that was assigned to ν(NO) vibrations. The ν(NO) frequencies (cm^{-1}) of some solid $Mn(NO)LX_2$ complexes are tabulated below [18]:

complex	ν(NO)
$Mn(NO)\{P(C_3H_7)_3\}Cl_2$	1610
$Mn(NO)\{P(C_3H_7)_3\}Br_2$	1600
$Mn(NO)\{P(C_6H_5)(CH_3)_2\}Cl_2$	1605
$Mn(NO)\{P(C_6H_5)(CH_3)_2\}Br_2$	1595
$Mn(NO)\{P(C_6H_5)(C_2H_5)_2\}Cl_2$	1607
$Mn(NO)\{P(C_6H_5)(C_2H_5)_2\}Br_2$	1598

The table shows that the bromo complexes absorb at lower frequencies than the chloro complexes and that the dialkylphenylphosphane complexes absorb at slightly lower frequencies than the trialkylphosphane complexes. Thus, there appears to be more Mn → NO π back bonding in the bromo and dialkylphenylphosphane complexes than in the chloro and trialkylphosphane complexes. The large shift of the ν(NO) band from that of free nitric oxide (ν_{NO} at 1840 cm^{-1}) proves that the NO group is chemically bonded. The complexes are indefinitely stable below 0°C. At room temperature they undergo a slow decomposition, the major decomposition products being the Mn^{II} phospane oxide complexes, as shown by the IR and ESR spectra of the products both in the solid state and in THF solution. Upon exposure to a vacuum the chloro and bromo complexes lost their color, but the iodo complexes remained colored, even when left under vacuum for 24 h [18].

THF solutions of $Mn(NO)LX_2$ complexes were obtained by treating $MnLX_2$ complexes in THF with NO at −78°C. THF solutions of $Mn(NO)LBr_2$ complexes did not show ESR or ^{31}P NMR spectra. The visible electronic spectra of $Mn(NO)LX_2$ complexes in THF exhibit intense bands at 535, 410, 385 nm (for X = Cl) and 550, 395, 370 nm (for X = Br). The iodo complexes were too unstable to be studied at room temperature. The variation of electronic spectra with p_{NO} was used to construct NO absorption isotherms of $MnLX_2$ complexes [1, 18]. These together with the equilibrium constants and reversibility of the $Mn(NO)LX_2$ formation are treated on p. 46.

$Mn(NO)_2L_2X$. In most cases preparative procedures are based on CO displacement from various manganese(II) phosphane carbonyl complexes. The ligand introduced is the NO molecule or the halide atom. All operations were performed under an inert atmosphere. Preparation methods for the individual complexes are reported following the table below. The table compiles the complexes prepared, their melting points (m.p. in °C), and their ν(NO) frequencies in the IR spectra (races A_1 and E, respectively, in cm^{-1}) measured in various media:

No.	complex	color	m.p. in °C	ν(NO) in cm^{-1}	medium	Ref.
1	$Mn(NO)_2$-$\{P(C_2H_5)_3\}_2I$	brown [2] (oil)	—	1705, 1662[a)] [3]	C_6H_{12}	[2, 3]
2	$Mn(NO)_2$-$\{P(C_2H_5)_3\}_2NO_2$	—	—	—	—	[4]

No.	complex	color	m.p. in °C	ν(NO) in cm^{-1}	medium	Ref.
3	$Mn(NO)_2$-$\{P(C_6H_5)(CH_3)_2\}_2Cl$	brick red (prisms)	120 to 122	1706, 1660	CH_2Cl_2	[5]
4	$Mn(NO)_2$-$\{P(C_6H_5)_3\}_2H$	yellow	153 to 154	1684, 1640	b)	[2]
5	$Mn(NO)_2$-$\{P(C_6H_5)_3\}_2Cl$	copper (plates)	178 to 181	1712, 1664	CH_2Cl_2	[5]
		orange	—	1703, 1659	Fluorolube c)	[6]
6	$Mn(NO)_2$-$\{P(C_6H_5)_3\}_2Br$	ochre [2]	—	1715, 1670 [7]	CH_2Cl_2	[2, 7]
		orange	—	1702, 1658	Fluorolube c)	[6]
		—	—	1708, 1660	b)	[3]
7	$Mn(NO)_2$-$\{P(C_6H_5)_3\}_2I$	red-brown	167 to 168	1717, 1673	CH_2Cl_2	[7]
		—	—	1714, 1670 d)	CCl_4	[3]
8	$Mn(NO)_2$-$\{P(C_6H_5)_3\}_2CN$	brown [2]	—	1720, 1676 [7]	CH_2Cl_2	[2, 7]
		—	—	1727, 1681	b)	[3]
9	$Mn(NO)_2$-$\{P(C_6H_5)_3\}_2NCO$	orange	—	1698, 1659	Fluorolube c)	[6]

a) Valence force constant, 12.47 mdyn/Å. – b) Not given. – c) Polydichlorotrifluoroethylene. – d) Valence force constant, 12.60 mdyn/Å.

Complexes 1, 3, and 5 to 7 were prepared by passing dry NO gas through refluxing solutions of the corresponding $Mn(CO)_3L_2X$ complexes in benzene (3 h for complexes 1 and 7, 4 h for complex 3, and 6.5 h for complex 6). After hot filtration, most of the solvent was removed [2, 5]. Complexes 6 and 7 precipitated on addition of ethanol or pentane. They were recrystallized from benzene-alcohol or benzene-petroleum ether and dried in high vacuum. Complex 1 (no analysis given) was obtained only as a brown oil and was identified by IR spectroscopy [2]. The oily residue of complex 3 was recrystallized from ether-hexane to give the compound in 52% yield. Complex 5 was recrystallized from dichloromethane-methanol and was obtained in 75% yield [5].

Complex 2 (no analysis given) was obtained as a by-product of $Mn(NO)_3P(C_2H_5)_3$ (see p. 64) in the reaction of $Mn(NO)_3 \cdot THF$ with excess $P(C_2H_5)_3$ in THF for 3 h at room temperature. Evaporation of the solvent under vacuum, dissolution of the residue in dry degassed pentane, and cooling to −10°C yielded the crystalline complex [4].

Complex 4 was prepared by the reaction of complex 6 in THF with $NaBH_4$. The solution was refluxed and portions of $NaBH_4$ repeatedly added. The solvent was removed from the yellow solution and the residue was recrystallized from benzene-ethanol or benzene-pentane [2].

Complexes 6 to 8 (no analysis given for complex 8) were prepared by treating benzene solutions of $Mn(CO)_3(NO)P(C_6H_5)_3$ with neat Br_2 or I_2, or a solution of BrCN in benzene in a 1:3 (for complexes 6 and 7) or 1:4 mole ratio (for complex 8). After 1 h the solutions were concentrated. Complexes 6 and 7 precipitated on addition of ethanol. Only a few crystals of complex 8 precipitated on addition of pentane. It was identified by its IR spectrum [2]. In another procedure, $[C_6H_7Mn(CO)(NO)P(C_6H_5)_3]BF_4$ ($C_6H_7 = \eta^5$-indenyl) was treated with: a 1.5-fold excess of KBr in THF for 5.5 h at 20°C, a slight excess of KI in THF for 84 h at 20°C, or a ca. threefold excess of KCN in alcohol for 15 min at 50 to 60°C. Evaporation of the solvent, extraction of the residue with toluene, evaporation of the extract, and repeated washing of this residue with pentane gave the bromo complex in a 24% yield. The reaction mixture containing the iodo complex was kept standing for 10 days. Then the solvent was removed in vacuum and the residue purified by thin-layer chromatography on Al_2O_3 with benzene as the eluent. The yield was 43%. Complex 8 was obtained in a 9% yield by chromatography with chloroform as the eluent [7].

Complexes 5, 6, and 9 (no analyses given) were obtained by the reaction of [Mn(CO)(NO)-$P(C_6H_5)_3]PF_4$ with $[\{(C_6H_5)_3P\}N\{P(C_6H_5)_3\}]X$ (X = Cl, Br, or NCO) in acetone at 25°C [6].

A trigonal-bipyramidal complex geometry was shown by an X-ray structure determination for complex 2 [4] and various spectroscopic studies for complexes 3 [5], 6, 7, and 8 [7]. Crystals of complex 2 are monoclinic, space group $P2_1$–D_2^2 (No. 4) with the lattice constants a = 8.745(9), b = 15.157(16), c = 7.584(13) Å, β = 92.69(6)°; Z = 2. The structure was solved to an R factor of 9.8%. The experimental density is 1.30 g/cm³, the calculated density is 1.32 g/cm³. The two phosphane ligands are in axial positions. The nitrito ligand is unidentate and oxygen-bonded (see **Fig. 16**). Average bond distances (in Å) and angles (in °) are: Mn–P = 2.33, $Mn–N_{NO}$ = 1.67, $Mn–O_{NO_2}$ = 2.08, $N–O_{NO}$ = 1.17, $N–O_{NO_2}$ = 1.19; $Mn–N–O_{NO}$ = 165, $Mn–O–N_{NO_2}$ = 123, $O–N–O_{NO_2}$ = 118, P–Mn–P = 167 [4].

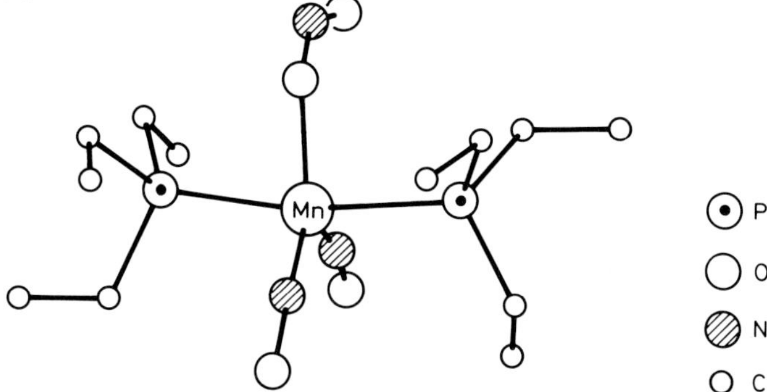

Fig. 16. Molecular structure of $Mn(NO)_2\{P(C_2H_5)_3\}_2NO_2$ (hydrogen atoms are omitted) [4].

While from X-ray photoelectron and ^1H NMR spectroscopic studies a structure with the phosphane ligands in axial positions was also suggested for complex 3 [5], the location of one phosphane ligand in an equatorial position and of the other phosphane ligand in an axial position was inferred from ^{13}C and ^{31}P NMR measurements for complexes 6, 7, and 8 [7]. Differences in frequency and intensity of the strong ν(NO) bands in the IR spectra of complexes 1, 6, and 7 (see table on pp. 61/2) are typical of NO groups in *cis* positions [3]. It is assumed that complexes 5 to 7 are isostructural [5]. The positions of the ν(NO) bands indicate considerable Mn $\rightarrow$ NO π back bonding (i.e., occupation of antibonding NO orbitals by Mn d electrons) [3].

Measurements of the magnetic susceptibility of complex 4 in the range between 77 and 292 K show that the complex is diamagnetic. This is consistent with the finding that no ESR signal is observed for the complex in the solid state or in solution [2]. The ^1H NMR spectrum of complex 3 in perdeuterated acetone solution shows a distorted triplet with a chemical shift of τ = 1.86 ppm and a separation of J* = 10.2 Hz for the outer peaks [5]. The ^{13}C NMR spectrum of complex 7 in dichloromethane solution exhibits two multiplets of the phenyl groups at the P atom with the following chemical shifts, δ (in ppm; the coupling constant, $J_{^{13}C-^{31}P}$ (in Hz) is given in parentheses): 134.62 (4.4), 134.42 (4.4), 132.55 (25), 131.57 (25), 130.89, 128.64 (4.4), 128.45 (4.4). The ^{31}P NMR spectrum displays two singlets with δ = 51.74 and 43.16 ppm. The spectrum did not exhibit any $^{31}P_1$–$^{31}P_2$ spin-spin splitting, even at −70°C [7]. From X-ray photoelectron spectroscopic measurement, the N_{1s} and P_{2p} binding energies were obtained: 400.0 and 130.6 eV for complex 3 and 400.1 and 131.4 eV for complex 5, calculated with respect to 284.0 eV for the C_{1s} electrons [5].

Complexes 3 and 5 are reasonably air-stable compounds that required an inert atmosphere for extended storage [5]. On heating complex 4 at 250°C for 50 h, the complex decomposed, with N_2 and H_2 as the gaseous decomposition products. Complex 4 is soluble in benzene, carbon disulfide, and dimethyl sulfoxide, but insoluble in pentane and ethanol. The electric dipole moment in benzene is $\mu = 4.07 \pm 0.11$ D. The properties indicate the existence of the compound as the hydride with the H atom occupying the coordination position of the Br atom in its precursor compound, $Mn(NO)_2\{P(C_6H_5)_3\}_2Br$ [2]. Conductivity measurements for complexes 3, 5 [5], and 1, 4, 6, 7 [2] in acetone show that they are nonelectrolytes in this solvent.

$[Mn(NO)_2\{P(C_6H_5)(CH_3)_2\}_3]BF_4$. The compound was prepared by first treating $Mn(NO)_2\{P(C_6H_5)(CH_3)_2\}_2Cl$ (complex 3 in the above section) with $AgBF_4$ (1:1 mole ratio) in degassed acetone, followed by removal of the AgCl precipitate, and treating the filtrate with $P(C_6H_5)(CH_3)_2$ in acetone (1:1 mole ratio). The solvent was removed and the resulting red oil left to stand under methanol at 0°C. The solid which slowly separated was recrystallized from dichloromethane-methanol. It formed red prisms of the desired product in 57% yield. The complex with the melting point 146 to 149°C is less stable than its neutral precursor, both in solid state and in solution, requiring storage under N_2 atmosphere. The ^1H NMR spectrum shows a distorted triplet with a chemical shift of $\tau = 1.84$ ppm and $J^* = 9.4$ Hz (J^* = separation of outer peaks) and a doublet with $\tau = 1.22$ ppm and a coupling constant of $J_{P-H} = 8.0$ Hz, indicating a trigonal-bipyramidal structure. The triplet is due to the methyl resonances of the two strongly-coupled *trans* axial ligands, whereas the doublet is ascribed to the single equatorial phosphane ligand. The IR spectrum of the complex in dichloromethane shows the ν(NO) bands at 1731 and 1687 cm^{-1}. The shift to higher wavenumbers (with respect to $Mn(NO)_2\{P(C_6H_5)(CH_3)_2\}_2Cl$) is accounted for by a decrease of the electron density on the Mn atom of the cationic complex. From X-ray photoelectron spectroscopic measurements the binding energy of the N_{1s} electron, 400.0 eV, and of the P_{2p} electrons, 130.6 eV, were obtained, calculated with respect to 284.0 eV for the C_{1s} electrons. The molar conductivity of the complex in dichloromethane, $\Lambda = 147\ cm^2 \cdot \Omega^{-1} \cdot mol^{-1}$, confirms its cationic nature [5].

$Mn(NO)_3L$. The dark green complexes were prepared by one of the following methods:

A) Reaction of $Mn(NO)_3 \cdot thf$ with excess ligand (threefold [8]) in pentane or tetrahydrofuran for several hours [4, 8, 9]. (The $Mn(NO)_3 \cdot thf$ had been prepared by photo-nitrosylation of $Mn_2(CO)_{10}$ in tetrahydrofuran [4, 8, 9], and this solution was used directly [4, 9] or after isolation of the $Mn(NO)_3 \cdot thf$ [8].)

B) Reaction of $Mn(NO)_3CO$ with equimolar amounts of the ligand in hexane, benzene, or p-xylene [10, 11]. (The $Mn(NO)_3CO$ solution had been obtained by photo-nitrosylation of $Mn_2(CO)_{10}$ in hexane [10] or by reaction of $Mn(CO)_5I$ or $[Mn(CO)_4I]_2$ with NO gas [11].)

C) Passing NO gas through a hot or boiling solution of $Mn(CO)_4L$ or $Mn(CO)_4LI$ in benzene or cyclohexane for several hours [2, 11 to 13].

D) Reaction of $Mn_2(CO)_8L$ (only for $L = P(C_6H_5)_2C_2H_5$) with $Co(NO)_2Br$ in tetrahydrofuran under UV irradiation [14].

The solvent was removed and the residue purified by recrystallization [2, 9 to 11, 13] or chromatography [9]. Some of the complexes were not isolated from solution and some were only obtained as an oil. On p. 65 are summarized the complexes, methods, and conditions of preparation, the solvents used for recrystallization, and the states of the products obtained:

ligand L	method	reaction medium[a)]	t in °C	time	solvent used for recrystallization	state of product	Ref.
$P(CH_3)_3$	A	pentane	r.t.	—	—	solution	[8]
$P(C_2H_5)_3$	A	THF	r.t.	3 h	—	—	[4]
	C	benzene	80	3 h	—	oil	[2]
$P(C_4H_9)_3$	B	p-xylene	r.t.	—	—	solution	[15]
$P(C_4H_9\text{-}t)_3$	B	hexane	r.t.	1 h	hexane	crystals	[10]
$P(C_6H_{11})_3$	C	C_6H_{12}	70	1.5 h	C_6H_{12}-petroleum ether	crystals	[2]
$P(C_6H_5)_2(C_2H_5)$	D	THF	r.t.	50 min	pentane-toluene[b)]	oil	[14]
$P(C_6H_5)_3$	A	THF	r.t.	3 h	pentane	crystals	[4]
	A	THF	r.t.	>8 h	—	solution	[9]
	B	benzene	r.t.	24 h	methanol	crystals	[11]
	C	p-xylene	100	2 h	methanol	crystals	[11]
	C	C_6H_{12}	80	2 to 5 h	—	crystals	[12]
	C	benzene	80	6 h	ethanol	crystals	[13]

[a)] C_6H_{12} = cyclohexane. – [b)] Solvent used for chromatography on Florisil.

All operations were carried out under an inert gas (N_2 or Ar) with careful exclusion of oxygen, using anhydrous degassed solvents [9, 10, 14, 15]. A kinetic study of the reaction $Mn(NO)_3CO + L \rightarrow Mn(NO)_3L + CO$ for $L = P(C_4H_9)_3$ and $P(C_6H_5)_3$ is reported in [15].

The melting point of $Mn(NO)_3\{P(C_6H_5)_3\}$ is 119°C (dec.) [11] or 128 to 130°C (dec). [13]. The compound is monomeric in benzene [11, 13].

A single-crystal X-ray diffraction study was carried out for $Mn(NO)_3\{P(C_6H_5)_3\}$. The complex crystallizes in the triclinic system, space group $P1\text{–}C_i^1$ (No. 2), with the lattice constants a = 10.411(5), b = 11.070(7), c = 10.189(6) Å, $\alpha = 77.88(3)°$, $\beta = 115.52(3)°$, $\gamma = 118.67(2)°$; Z = 2. The structure was solved to R = 7.3%. The calculated density is 1.456 g/cm³, the measured density is 1.46 g/cm³. As shown in **Fig. 17**, the molecule possesses C_3 geometry with essentially linear

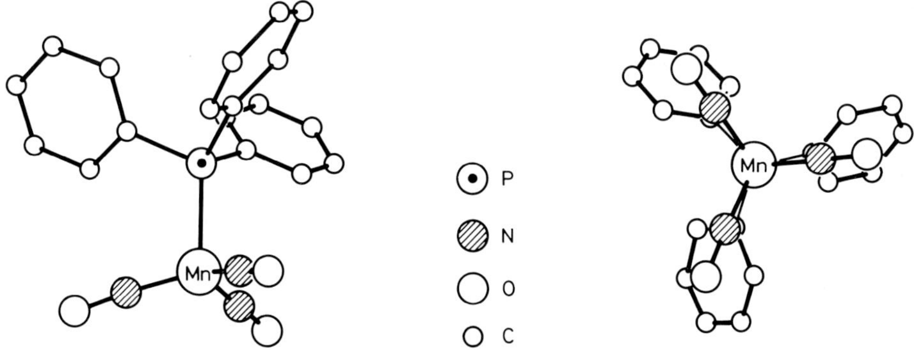

Fig. 17. Structure of the $Mn(NO)_3P(C_6H_5)_3$ molecule (hydrogen atoms omitted). a) View perpendicular to the Mn–P axis, b) view along the Mn–P axis [4].

nitrosyl groups and a nearly eclipsed configuration of the phenyl and NO groups (the average N–Mn–P–C dihedral angle is 5.3°). The nitrosyl groups are bent away from the idealized tetrahedral positions around the Mn ion toward the bulky phenyl groups. The phenyl rings are arranged in a helical way and are roughly perpendicular to each other. Average bond distances (in Å) and angles (in °) are:

Mn–P	2.315(2)	N–O	1.165(10)	Mn–N–O	177.2(7)	N–Mn–N	114.7(4)
Mn–N	1.686(7)	P–C	1.815(4)	P–Mn–N	103.6(2		

The opening up of the N–Mn–N angles is consistent with a pair of electrons in the d_{z^2} orbital which would have a density along the C_3 axis between the nitrosyl ligands. The eclipsed configuration is also consistent with MO theory: The relatively high-lying HOMO 1e′ orbital (xy, x^2-y^2, π^*(NO)) localizes electron density in nonbonding orbitals of trigonal symmetry in the xy plane. The eclipsed geometry should minimize the repulsions between the electrons in 1e′ and those in the P–C and Mn–N bonds [4].

Measurements of the magnetic susceptibility show that $Mn(NO)_3P(C_6H_5)_3$ is diamagnetic [11]. ^1H, ^{31}P, and ^{55}Mn NMR measurements on solutions of some $Mn(NO)_3L$ complexes at room temperature yielded the following chemical shifts (δ in ppm, $\delta<0$ means upfield shift) and coupling constant ($J={}^3J(HCC^{31}P)$ in Hz):

complex	solvent	$\delta(^{55}Mn)$ a)	$\delta(^1H)$ b)	$\delta(^{31}P)$ c)	J	ΔMn d)	Ref.
$Mn(NO)_3\{P(CH_3)_3\}$	pentane	−1240	—	—	—	—	[8]
$Mn(NO)_3\{P(C_4H_9\text{-}t)_3\}$	benzene	−1180	1.08	103.6 e)	12.8	41.1	[10]
$Mn(NO)_3\{P(C_6H_5)_2(C_2H_5)\}$	THF	−1220	—	—	—	—	[14]

a) Relative to concentrated aqueous $KMnO_4$ as external standard. – b) Relative to TMS. – c) Relative to 85% H_3PO_4 as external standard. – d) $\Delta Mn=\delta(^{31}P)_{complex}-\delta(^{31}P)_{ligand}$ with $\delta(^{31}P)_{ligand}=$ 62.5 ppm. – e) In deuterobenzene.

From the chemical shifts conclusions may be drawn about the ligand strengths of the phosphanes, tabulated on p. 38 [10].

In the IR spectra, the $Mn(NO)_3L$ complexes show two ν(NO) bands indicating C_{3v} symmetry of the complex molecule [3]. The ν(Mn–N) band was located at 665, the δ(Mn–NO) bands at 625 and 540 cm^{-1} [10]. As expected, the ratio of intensities of the higher-energy (race A_1) to the lower-energy NO band (race E), $\nu(A_1)/\nu(E)$, is ~2:10. The difference between the wavenumbers of the two bands is ~100 cm^{-1}. The position of the bands (between 1790 and 1670 cm^{-1}) indicates that coordination involves the σ electrons of the nitrogen and the unpaired electron from the antibonding NO_π orbital. It also indicates considerable metal-ligand π back bonding (i.e., occupation of antibonding NO orbitals by metal d electrons) [3]. Below are summarized frequencies of the ν(NO) bands and the valence force constants, f_{N-O}, of the N–O bonds of three $Mn(NO)_3L$ complexes in cyclohexane, together with the basicities (pK_a) of the L ligands in order of increasing f_{N-O} (in mdyn/Å):

complex	$\nu_{NO}(A_1)$	$\nu_{NO}(E)$	f_{N-O}	pK_a	Ref.
$Mn(NO)_3\{P(C_6H_{11})_3\}$	1773.8	1679.6	12.89	9.70	[3]
$Mn(NO)_3\{P(C_2H_5)_3\}$	1778.5	1685	12.97	8.69	[3]
$Mn(NO)_3\{P(C_6H_5)_3\}$	1781.5	1691	13.04	2.73	[3]
	1781	1688	—	—	[11]
	1780	1697	—	—	[12]

The results show increasing N–O valence force constants with increasing π acceptor ability and decreasing σ-donor ability (as given by the basicity) of the L ligands [3]. The table below lists ν(NO) bands of $Mn(NO)_3L$ complexes in various media:

complex	medium	$\nu_{NO}(A_1)$	$\nu_{NO}(E)$	f_{N-O}	Ref.
$Mn(NO)_3\{P(CH_3)_3\}$	pentane	1748	1688	—	[8]
$Mn(NO)_3\{P(C_2H_5)\}$	CCl_4	1782	1688.5	—	[3]
$Mn(NO)_3\{P(C_4H_9)\}$	p-xylene	1774	1680	—	[15]
$Mn(NO)_3\{P(C_4H_9\text{-}t)_3\}$	Nujol	1773	1679	—	[10]
$Mn(NO)_3\{P(C_6H_{11})_3\}$	CCl_4	1771.6	1677.4	—	[16]
	$CHCl_3$	1769.7	1674.8	—	[16]
$Mn(NO)_3\{P(C_6H_5)_2(C_2H_5)\}$	THF	1772	1682(sh), 1675	—	[14]
$Mn(NO)_3\{P(C_6H_5)_3\}$	KBr	1790	1680	—	[13]
	hexane	1784	1693	—	[17]
	CCl_4	1782	1688.5	13.02	[3]
	p-xylene	1780	1690	—	[15]
	THF	1776	1686	—	[9]

Investigations of solvent effects on ν(NO) vibrations show decreasing frequencies and increasing half-widths of the ν(NO) bands on going from nonpolar to polar solvents; e.g., for $Mn(NO)_3\{P(C_6H_{11})_3\}$ the sequence is: $C_6H_{12}>CCl_4>CHCl_3$ (see the tables above). The solvent effects increase in the direction of decreasing π acceptor ability of the L ligands. For the low π acceptors a higher occupation of the antibonding NO π orbitals by d_π metal electrons raises the polar character of the N–O bond and consequently the dipole-dipole interaction with the solvent [16]. From the ratio of the intensities, $I_E/I_{A_1}=9.51$, the bond angle between the equivalent NO groups was calculated as $\beta=111°$ for $Mn(NO)_3\{P(C_6H_5)_3\}$ [17].

The UV-visible absorption spectrum was measured for $Mn(NO)_3\{P(C_4H_9\text{-}t)_3\}$ in benzene. Absorptions were observed at 280, 310, 405, 480, and 651 nm [10].

The compounds with $L=P(C_2H_5)_3$, $P(C_6H_{11})_3$, and $P(C_6H_5)_3$ are thermally more stable and more stable to air oxygen than $Mn(NO)_3CO$ [2, 11, 12]. $Mn(NO)_3\{P(C_4H_9\text{-}t)_3\}$ could be stored at −10°C under Ar for a long time [10]. All the compounds are soluble in the usual solvents, such as pentane, cyclohexane, benzene, petroleum ether, and dichloromethane [2, 8, 10, 12, 13], and are nonelectrolytes [2, 11, 12]. The green solution of $Mn(NO)_3\{P(CH_3)_3\}$ in pentane was found to be stable at room temperature [8].

References:

[1] Barratt, D. S.; McAuliffe, C. A. (J. Chem. Soc. Chem. Commun. **1984** 594/5).
[2] Hieber, W.; Tengler, H. (Z. Anorg. Allgem. Chem. **318** [1962] 136/54, 137/8, 149/50).
[3] Beck, W.; Lottes, K. (Chem. Ber. **98** [1965] 2657/73, 2659, 2662, 2664/71).
[4] Wilson, R. D.; Bau, R. (J. Organometal. Chem. **191** [1980] 123/32, 125).
[5] Copperthwaite, R. G.; Reimann, R. H.; Singleton, E. (Inorg. Chim. Acta **28** [1978] 107/11).

[6] Haymore, B. L. (J. Organometal. Chem. **137** [1977] C11/C15).
[7] Kolobova, N. E.; Lobanova, I. A.; Zdanovich, V. I.; Petrovskii, P. V. (Izv. Akad. Nauk SSSR Ser. Khim. **1981** 935/8; Bull. Acad. Sci. USSR Div. Chem. Sci. **30** [1981] 707/10).
[8] Rehder, D.; Ihmels, K.; Wenke, D.; Oltmanns, P. (Inorg. Chim. Acta **100** [1985] L11/L12).
[9] Herberhold, M.; Razavi, A. (J. Organometal. Chem. **67** [1974] 81/6).
[10] Schumann, H.; Meissner, M. (Z. Naturforsch. **35b** [1980] 863/8).

[11] Barraclough, C. G.; Lewis, J. (J. Chem. Soc. **1960** 4842/6).
[12] Hieber, W.; Beck, W.; Tengler, H. (Z. Naturforsch. **15b** [1960] 411).
[13] Lambert, R. F.; Johnston, J. D. (Chem. Ind. [London] **1960** 1267/8).
[14] Oltmanns, P.; Rehder, D. (J. Organometal. Chem. **281** [1985] 263/72).
[15] Wawersik, H.; Basolo, F. (J. Am. Chem. Soc. **89** [1967] 4626/30).
[16] Beck, W.; Lottes, K. (Z. Naturforsch. **19b** [1964] 987/94).
[17] Beck, W.; Melnikoff, A.; Stahl, R. (Chem. Ber. **99** [1966] 3721/7).
[18] Barratt, D. S.; McAuliffe, C. A. (J. Chem. Soc. Dalton Trans. **1987** 2497/501).

39.1.2.10 Benzenediazo Complexes

$Mn(N_2C_6H_5)_2\{P(C_6H_5)_3\}_2X$ and **$Mn(NO)(N_2C_6H_5)\{P(C_6H_5)_3\}_2X$** (X = Cl, Br, or NCO). The burgundy red or red-brown compounds were prepared by the reaction of $[Mn(CO)(N_2C_6H_5)_2\{P(C_6H_5)_3\}_2]PF_6$ or $[Mn(NO)(CO)(N_2C_6H_5)\{P(C_6H_5)_3\}_2]PF_6$, respectively, with $[\{(C_6H_5)_3P\}N\{P(C_6H_5)_3\}]X$ complexes in acetone at 25°C under N_2. Resonance interactions of the ν(NN) vibrations with one or two weak phenyl vibrational modes were observed in the IR spectra. The decoupled values were determined with the aid of ^{15}N isotopic substitution. Characteristic bands of the complexes in Fluorolube mulls (ν in cm^{-1}) were assigned as follows:

complex[a)]	ν(NN)	complex[a)]	ν(NO)	ν(NN)
$Mn(N_2C_6H_5)L_3Cl$	1630, 1611, 1569, 1542, 1471	$Mn(NO)(N_2C_6H_5)L_3Cl$	1676	1630, 1564, 1476
$Mn(N_2C_6H_5)L_3Br$	1632, 1607, 1570, 1543, 1471	$Mn(NO)(N_2C_6H_5)L_3Br$	1677	1631, 1565, 1476
$Mn(N_2C_6H_5)L_3NCO$	1629, 1612, 1570, 1540, 1471	$Mn(NO)(N_2C_6H_5)L_3NCO$	1680	1627, 1560, 1476

a) $L = P(C_6H_5)_3$

The ν(NCO) bands for the isocyanato complexes were located at 2225 and ~1435 cm^{-1}. In the mixed nitrosyl-benzenediazo complexes, the ν(NO) and ν(NN) vibrations seem to be weakly coupled, thus increasing the observed ν(NO) and decreasing the observed ν(NN) frequencies by 2 to 5 cm^{-1}. The spectroscopic data are all consistent with a trigonal-bipyramidal geometry about the metal with axial phosphane ligands and singly-bent benzenediazo ligands. Considerable π back bonding of the benzenediazo ligands is assumed.

Reference:

Haymore, B. L. (J. Organometal. Chem. **137** [1977] C11/C16).

39.1.3 Complexes with Cyanoethyl- or Cyanoethyl(phenyl)phosphanes

$Mn^{II}\{P(C_6H_5)_{3-n}(CH_2CH_2CN)_n\}X_2$ (n=1 to 3; X=Cl, Br, I, or NCS). All the complexes were synthesized by the same general method. All manipulations were performed under an atmosphere of dry argon [1]. Solvents and Mn^{II} salts were carefully dried by previously published methods for $Mn^{II}\{P(C_6H_5)_3\}X_2$ complexes [2]. The synthesis of $Mn\{P(C_6H_5)(CH_2CH_2CN)_2\}I_2$ is described here as a typical example: A mixture of dry manganese(II) iodide (5 mmol) and solid $P(C_6H_5)(CH_2CH_2CN)_2$ (5 mmol) in diethyl ether (75 cm^3) was stirred at room temperature for 5 d. The volume of solvent was then reduced to ~25 cm^3 by application of a vacuum and the resulting slurry filtered in a Schlenk apparatus, washed with dry diethyl ether, and dried in vacuum. The colors of the complexes are similar to those of $Mn^{II}LX_2$ complexes with L=PR_3 (R=alkyl or phenyl): off-white for X=Cl, pink for X=Br, pink-orange for X=I, and yellow for X=NCS. Molecular weight determinations for $Mn\{P(CH_2CH_2CN)_3\}I_2$ (985; 1004 calculated for the dimer) and for $Mn\{P(C_6H_5)(CH_2CH_2CN)_2\}I_2$ (1020; 1050 calculated for the dimer) in dichloromethane show that these compounds are dimeric in the solid state and in dichloromethane. Melting points (m.p. in °C), room temperature magnetic moments (μ_{eff} in μ_B), and bands (in cm^{-1}) observed in the Nujol mull IR spectra of the complexes are as follows [1]:

complex	m.p.	μ_{eff}	ν(CN) [a]	ν(Mn–X)	ν(Mn–X–Mn)	ν(Mn–P)	ν(Mn–N)
$Mn\{P(CH_2CH_2CN)_3\}Cl_2$	191	5.8	2270, 2240	290	220	390	250
$Mn\{P(CH_2CH_2CN)_3\}Br_2$	124	5.75	2275, 2240	230	190	392	250
$Mn\{P(CH_2CH_2CN)_3\}I_2$	125	5.7	2280, 2240	218	160	394	245
$Mn\{P(CH_2CH_2CN)_3\}(NCS)_2$	77	5.45	2270, 2240	—	—	388	280, 250
$Mn\{P(C_6H_5)(CH_2CH_2CN)_2\}Cl_2$	185	5.8	2285	300	225	385	255
$Mn\{P(C_6H_5)(CH_2CH_2CN)_2\}Br_2$	143	5.7	2290	240	185	395	260
$Mn\{P(C_6H_5)(CH_2CH_2CN)_2\}I_2$	135	5.6	2300	205	170	400	255
$Mn\{P(C_6H_5)(CH_2CH_2CN)_2\}$-$(NCS)_2$	80	5.4	2280	—	—	390	285, 250
$Mn\{P(C_6H_5)_2(CH_2CH_2CN)\}Cl_2$	162	5.75	2230	300	230	400	—
$Mn\{P(C_6H_5)_2(CH_2CH_2CN)\}Br_2$	102	5.7	2230	240	185	400	—
$Mn\{P(C_6H_5)_2(CH_2CH_2CN)\}I_2$	92	5.6	2230	205	155	410	—
$Mn\{P(C_6H_5)_2(CH_2CH_2CN)\}$-$(NCS)_2$	62	5.35	2230	—	—	390	285

[a] ν(CN) for free ligands: $P(CH_2CH_2CN)_3$, 2240; $P(C_6H_5)(CH_2CH_2CN)_2$, 2270; $P(C_6H_5)_2(CH_2CH_2CN)$, 2230 cm^{-1}.

The ν(CN) bands of the terminal and bridging NCS groups are 2080 and 2210 cm^{-1}, respectively, for the tris(2-cyanoethyl)phosphane complex and 2060 and 2120 cm^{-1}, respectively, for the other complexes [1].

The magnetic moments of all the complexes are lower than the spin-only value (5.92 μ_B) expected for high-spin Mn^{II}. The values decrease in the order Cl>Br>I>NCS, indicating significant spin pairing in this order in the complexes. The ESR spectra of solid tris- and bis(2-cyanoethyl)phosphane complexes exhibit broad bands at g_{eff}=6 and 2, suggesting a pseudooctahedral environment of the Mn^{II} centers. A broad band at g_{eff}=6 points to a similar structure of these complexes in dichloromethane, whereas the spectra of the (2-cyanoethyl)diphenylphosphane complexes consist of two major bands in the $g_{eff} = 4$ and 2 regions, implying tetrahedral geometry about the Mn^{II} centers [1].

The IR spectra of the tris(2-cyanoethyl)phosphane complexes in the solid state and in dichloromethane show two ν(CN) bands, one at the position of the free ligand band and one at a higher frequency, indicating coordination of some of the CN groups. The spectra indicate only coordinated CN groups for bis(2-cyanoethyl)phenylphosphane complexes and only noncoordinated CN groups for (2-cyanoethyl)diphenylphosphane complexes. On the basis of the physical data the tris(2-cyanoethyl)phosphane complexes were assigned a dimeric halide- or pseudohalide-bridged structure in the solid state and in dichloromethane (as shown below).

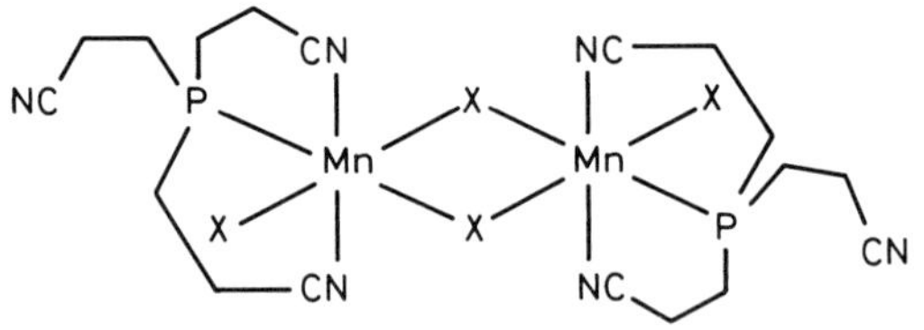

The bis(2-cyanoethyl)phenylphosphane complexes were assigned a quite similar structure. However, from IR evidence it is clear that the (2-cyanoethyl)diphenylphosphane complexes are at least dimeric, possibly polymeric in the solid state and in dichloromethane. The IR spectra of all the complexes in tetrahydrofuran show only one ν(CN) band in the same region as that of the free ligand. The ESR spectra are similar to that of $Mn\{P(C_4H_9)_3\}X_2$, indicating monomeric pseudooctahedral complexes with three THF molecules coordinated and the X atoms in axial positions [1].

THF solutions of the 2-cyanoethylphosphane complexes did not show any reaction with dioxygen. In a ligand exchange experiment, solutions of $Mn\{P(CH_2CH_2CN)_3\}I_2$ and $Mn\{P(C_4H_9)_3\}I_2$ in THF were prepared. Equimolar quantities of the complementary PR_3 ligand were then added and the solutions stirred for several days. In both cases $Mn\{P(CH_2CH_2CN)_3\}I_2$ was isolated. This shows that for manganese(II) iodide, $P(CH_2CH_2CN)_3$ is bonded in preference to $P(C_4H_9)_3$, possibly because of the better π-acceptor properties of the former. The manganese(II) complexes with the former ligand would therefore be expected to be more electron-deficient at the metal center. This deficiency would, in turn, mean that the Mn atom would bond more strongly to other σ donors, such as THF molecules, and that they cannot be so easily substituted by dioxygen molecules as can complexes with the sterically similar $P(C_4H_9)_3$ ligand [1].

References:

[1] Barratt, D. S.; Hosseiny, A.; McAuliffe, C. A.; Stacey, C. (J. Chem. Soc. Dalton Trans. **1985** 135/9).

[2] Hosseiny, A.; Mackie, A. G.; McAuliffe, C. A.; Minten, K. (Inorg. Chim. Acta **49** [1981] 99/105).

39.1.4 Complexes with Amino- or Pyridinylphosphanes

No.	ligand L	formula
1	$P\{N(CH_3)_2\}_3$	$C_6H_{18}N_3P$
2	$P(2\text{-}C_5H_4N)_3$	$C_{15}H_{12}N_3P$
3	$(C_6H_5)_2P{-}NH{-}(2\text{-}C_5H_4N)$	$C_{17}H_{15}N_2P$
4	$P(CH_2{-}2\text{-}C_5H_4N)_3$	$C_{18}H_{18}N_3P$

$Mn(NO)_3(C_6H_{18}N_3P)_3$ was prepared by treating $[Mn(CO)_4(C_6H_{18}N_3P)_3]_2$ in tetrahydrofuran for 4 h with nitric oxide (purified by passing through a trap containing molecular sieves at −78°C). Solvent was then removed from the resulting deep green solution at ~25°C/35 Torr and the resulting residue extracted with pentane. The pure crystalline compound was isolated by concentrating the pentane solution and cooling it in a −78°C bath. The green complex decomposed appreciably after a few hours at room temperature (color change to brown). It was therefore characterized by its mass spectrum (m/e values, relative intensities, and ions assigned in the paper). The IR spectrum of the complex (in cyclohexane) shows strong bands at 1785 and 1690 cm^{-1}, which were assigned to $\nu_{NO}(A_1)$ and $\nu_{NO}(E)$ vibrations, respectively [1].

$[Mn^{II}(C_{15}H_{12}N_3P)_2](ClO_4)_2$. Manganese(II) perchlorate hexahydrate was refluxed for 1 to 12 h in 95% ethanol–2,2-dimethoxypropane (1:1 v/v) under nitrogen before adding tris(2-pyridinyl)-phosphane (1:2 mole ratio) dissolved in 95% ethanol. Precipitation occurred within a few minutes. The reaction mixture was refluxed under nitrogen for 1 to 2 h and then cooled. The precipitate was collected, washed with ethanol, and then refluxed again in 95% ethanol–2,2-dimethoxypropane (1:1 v/v) under nitrogen. It was collected again and dried at 90 to 100°C in vacuum. The yield was >80%. The IR spectrum of the complex (in mineral oil or hexachlorobutadiene mulls) shows bands at 1575, 1550, 1425, and 1008 cm^{-1}, which were assigned to the pyridyl group. Other bands, at 1110 to 1080 and at 617 cm^{-1}, were ascribed to the ClO_4^- anion. The shift of the pyridine bands with respect to the free ligand bands reveals coordination of the ligand by the pyridine N atoms. The molar conductivity of a 10^{-3} M solution in acetonitrile ($\Lambda = 311\ cm^2 \cdot \Omega^{-1} \cdot mol^{-1}$) indicates a 1:2 electrolyte, i.e., the complex has the same composition in solution as in the solid state [2].

$[Mn^{II}(C_{17}H_{15}N_2P)_3]Br_2$. The complex with diphenyl(2-pyridinylamino)phosphane was prepared by the reaction of $MnBr_2$ with the ligand in methanol. Stoichiometry and properties of the compound (no details given) indicate the formation of five-membered chelate rings, i.e., the ligand is bidentate with the pyridyl nitrogen and the phosphorus as the donor atoms [3].

$[Mn^{II}(C_{18}H_{18}N_3P)Cl]Cl$ was prepared by the reaction of $MnCl_2$ with ligand 4 (no solvent given). The golden yellow compound is assumed to have a trigonal-bipyramidal structure with the three pyridine N atoms arranged in equatorial positions and the P and one Cl atom in axial positions. The assumption is consistent with results of conductivity measurements ($\Lambda_{eq} \approx 80\ cm^2 \cdot \Omega^{-1} \cdot mol^{-1}$ of a 10^{-3} M complex solution in nitromethane) [4].

References:

[1] King, R. B.; Korenowski, T. F. (J. Organometal. Chem. **17** [1969] 95/105, 98/9, 101/3).
[2] Boggess, R. K.; Zatko, D. A. (J. Coord. Chem. **4** [1975] 217/24).
[3] Seidel, W.; Schoeler, H. (Z. Chem. [Leipzig] **7** [1967] 431).
[4] Chiswell, B. (Australian J. Chem. **20** [1967] 2533/4).

39.1.5 Complex with Tris(diethoxyphosphoryl)phosphane $P\{P(O)(OC_2H_5)_2\}_3 = C_{12}H_{30}O_9P_4$

$[Mn^{II}(C_{12}H_{30}O_9P_4)_2](AsF_6)_2$. In a carefully dried round-bottomed flask, $Mn(SO_2)_2(AsF_6)_2$ (1.7 mmol) and $P\{P(O)(OC_2H_5)_2\}_3$ (3.4 mmol) were treated with a stream of N_2 and cooled to −190°C, then the flask was evacuated and 15 mL SO_2 condensed onto the mixture. The flask was warmed to room temperature, the yellow solution stirred for 18 h, and then the SO_2 was removed. Recrystallization from SO_2-petroleum ether (40 to 80°C) 1:1 (v/v) yielded the complex as yellow crystals. The yield was 94%.

The compound crystallizes in the trigonal system, space group $P\bar{3}1c$–D_{3d}^2 (No. 163). The lattice constants are: a = 12.047 and c = 21.865 Å at 20°C. The diffractometric study shows that the Mn complex and the corresponding Co complex are isomorphous with the Zn complex, for which a complete X-ray diffractometric study had been performed. As shown in **Fig. 18**, the structure consists of two molecules of tris(diethoxyphosphoryl)phosphane arranged sandwich-like around the metal atom, which is coordinated by the six phosphoryl oxygen atoms.

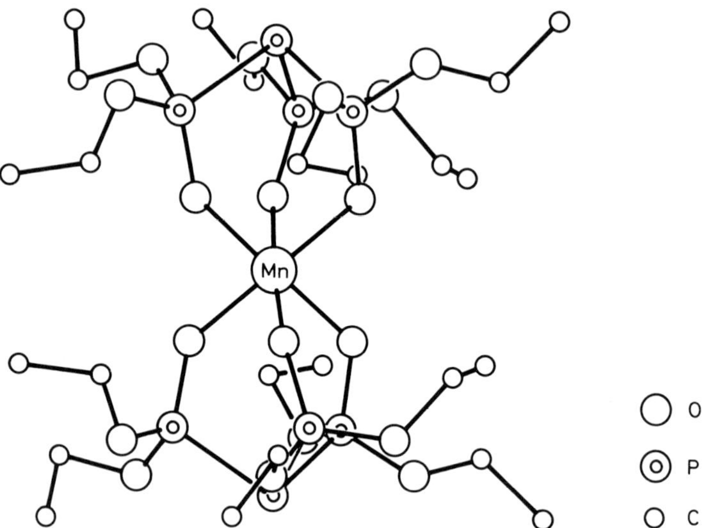

Fig. 18. Structure of the $[Mn(C_{12}H_{30}O_9P_4)_2]^{2+}$ ion (hydrogen atoms are omitted).

Magnetic measurements by the Faraday method between 3.6 and 293 K reveal Curie behavior with $\mu_{eff} = 6.0(1)$ μ_B. The ^{31}P NMR spectrum of the complex (in CH_2Cl_2) shows no signal.

Characteristic bands (cm^{-1}) in the IR spectrum were assigned as follows (free ligand bands in parentheses): 1185 (1260) to ν(P=O); 1165 (1172) to ν_{as}(POC); 1106 (1103) to ν_s(POC); 1020 (1030) to $\nu_{as}(PO_2)$; 950 (978) to ν(C–C); 699 to $\nu(AsF_6)$; and 399 to $\delta(AsF_6)$. The large shift of the ν(P=O) band (by -75 cm^{-1}) indicates the considerably lowered fraction of π bonding in the ν(P=O) bond caused by the coordination of the phosphoryl oxygen atoms. In the mass spectrum the peak of the $(Mn–AsF_6)^+$ cation was observed with 100% intensity. This indicates a great stability of the cation.

The complex is stable in air. The Mn–O bonds are resistant to hydrolysis, and the P–P bonds are not attacked by a metal-catalyzed SO_2 oxidation (as observed for free diphosphanes). The high stability of the complex is attributed to the chelate effect observed on coordination of the tripod ligand.

Reference:

Roesky, H. W.; Gries, T.; Dhathathreyan, K. S.; Lueken, H. (Z. Anorg. Allgem. Chem. **547** [1987] 199/204).

39.1.6 Complexes with Trimethylsilyl-, -germanyl- or -stannanylphosphanes

$\{(CH_3)_3M\}_nP(C_4H_9\text{-}t)_{3-n}$ with n = 1, 2, or 3 and M = Si, Ge, or Sn.

No.	ligand L	formula
1	$(CH_3)_3SiP(C_4H_9\text{-}t)_2$	$C_{11}H_{27}PSi$
2	$\{(CH_3)_3Si\}_2P(C_4H_9\text{-}t)$	$C_{10}H_{27}PSi_2$
3	$\{(CH_3)_3Si\}_3P$	$C_9H_{27}PSi_3$
4	$(CH_3)_3GeP(C_4H_9\text{-}t)_2$	$C_{11}H_{27}GeP$
5	$\{(CH_3)_3Ge\}_2P(C_4H_9\text{-}t)$	$C_{10}H_{27}Ge_2P$
6	$\{(CH_3)_3Ge\}_3P$	$C_9H_{27}Ge_3P$
7	$(CH_3)_3SnP(C_4H_9\text{-}t)_2$	$C_{11}H_{27}PSn$
8	$\{(CH_3)_3Sn\}_2P(C_4H_9\text{-}t)$	$C_{10}H_{27}PSn_2$
9	$\{(CH_3)_3Sn\}_3P$	$C_9H_{27}PSn_3$

$Mn(NO)_3L$. The dark green complexes were prepared under strictly O_2-free and anhydrous conditions by the addition of the appropriate ligand (1.6 mmol) to a hexane solution of $Mn(CO)(NO_3)_3$ (1 mmol) (obtained by photo-induced nitrosylation of Mn_2CO_{10}) and stirring the mixture for 1 h. Then the solvent was removed and the product purified by recrystallization from hexane at −10°C. Yields, which depend on the solubility of the respective compound, were 50 to 90% based on the phosphane [1]. Results of NMR measurements and wavelengths, λ, of the lowest-energy band in the UV-visible spectra of $Mn(NO)_3L$ complexes (in benzene, if not stated otherwise) are tabulated below (chemical shifts, δ, in ppm; λ in nm; δ<0 means upfield shift) [1, 2]:

L	1	2	3	4	5
$\delta(^{55}Mn)$ a)	−1191	−1169	−1145	−1191	−1164
$\delta(^{55}Mn)$ b)	−20	2	26	−20	7
$\delta(^{31}P)$ c)	36.7	−71.5	−218.4	48.2	−45.9
$\Delta(^{31}P)$ d)	39.9	36.9	32.8	33.9	36.4
λ	648	652	668	648	658

L	6	7	8	9
$\delta(^{55}Mn)$ a)	−1153	−1180	−1154	−1086 e)
$\delta(^{55}Mn)$ b)	18	−9	17	85
$\delta(^{31}P)$ c)	−191.2	44.2	−84.7	−293.3 f)
$\Delta(^{31}P)$ d)	37.3	23.5	26.4	35.3
λ	669	650	663	684

a) Relative to concentrated aqueous $KMnO_4$ as external standard [1]. – b) Relative to $Mn(CO)(NO)_3$ [2]. – c) In C_6D_6; δ relative to 85% H_3PO_4 as external standard. – d) $\Delta(^{31}P) = \delta(^{31}P)_{complex} - \delta(^{31}P)_{ligand}$. – e) In CH_2Cl_2. – f) In CH_2Cl_2-C_6D_6.

Chemical shifts of the ^{55}Mn signals and the lowest-energy electron transition, ΔE, can be considered as a measure of the σ donor and π acceptor ability of the L ligands, i.e., the ligand strength is highest for $(CH_3)_3SiP(C_4H_9\text{-}t)_2$ (ligand 1) and lowest for $P\{Sn(CH_3)_3\}$ (ligand 9). The chemical shift, $\delta(^{55}Mn)$, and the reciprocal value of the lowest-energy electron transition

($1/\Delta E \sim \lambda$) show the expected linear relation (plot of $\delta(^{55}Mn)$ against λ presented) [1]. ^{119}Sn NMR measurements of organylstannanylphosphane complexes in C_6D_6 yielded the following results [3]:

complex	$\delta(^{119}Sn)$ a)	$^1J(P-Sn)$	$^2J(H-Sn)$
$Mn(NO)_3(CH_3)_3SnP(C_4H_9\text{-}t)_2$	1.2	137	53.0
$Mn(NO)_3\{(CH_3)_3Sn\}_2P(C_4H_9\text{-}t)$	29.3	286	53.8
$Mn(NO)_3P\{(CH_3)_3Sn\}_3$	62.0 b)	399 b)	54.5

a) Relative to $Sn(CH_3)_4$ as external standard. – b) In C_6D_6-$CHCl_3$ (2:3 v/v).

The data are discussed with respect to those of the free organylstannanylphosphanes and compared with those of the related complexes of the pseudocarbonylnickel series (Fe(CO)-$(NO)_2$L, $Co(CO)_2(NO)$L, and $Ni(CO)_3$L) [3].

The IR spectra of the complexes with ligands 1 to 8 (in Nujol) show bands at ~1770 cm^{-1} assigned to $\nu_{NO}(A_1)$ and around 1678 cm^{-1} assigned to $\nu_{NO}(E)$. Lower frequencies for the complex with ligand 9 (in Nujol-$CHCl_3$) (at 1766 and 1667 cm^{-1}) indicate a lower π-acceptor strength of this ligand. Three bands in the range between 665 and 535 cm^{-1} were assigned to ν(Mn–N) and δ(Mn–NO) vibrations. The UV-visible spectra of the complexes in benzene show two distinct bands, at ~480 nm and in the 650 to 690 nm range. Another three higher-energy bands appear as shoulders on the strong charge-transfer bands with maxima between 270 and 280 nm [1].

All the complexes can be stored under Ar at −10°C for a longer period of time. They decompose at temperatures between 150 and 200°C. Except for $Mn(NO)_3P\{Sn(CH_3)_3\}_3$, which is practically insoluble in hexane, the compounds readily dissolve in hexane, benzene, and dichloromethane, and the solubility increases in this order. The solubility increases in the order $n = 0 < 1 < 2 < 3$ and $M = Sn < Ge < Si$ [1].

References:

[1] Schumann, H.; Meißner, M. (Z. Naturforsch. **35b** [1980] 863/8).
[2] Schumann, H.; Meißner, M.; Kroth, H.-J. (Z. Naturforsch. **35b** [1980] 639/41).
[3] Meißner, M.; Kroth, H.-J.; Köhricht, K.-H.; Schumann, H. (Z. Naturforsch. **36b** [1981] 904/6).

39.1.7 Complexes with Other Triorganylphosphanes

The complex $\mathbf{Mn\{(C_6H_5)_2PCH_2CH_2Si(OCH_3)_3\}I_2}$ was prepared by the reaction of powdered anhydrous MnI_2 with the ligand in pentane under strictly anhydrous conditions [1] (as were described in [2]). The reaction mixture was stirred for 7 d, filtered using Schlenk techniques, washed with dry pentane, and dried in vacuum. The yield was quantitative [1].

A silica-supported complex, $\mathbf{Mn\{(C_6H_5)_2PCH_2CH_2 \cdot sil\}I_2}$, was prepared by the reaction of $Mn\{(C_6H_5)_2PCH_2CH_2Si(OCH_3)_3\}I_2$ with dehydrated silica or by treating dehydrated silica with $(C_6H_5)_2PCH_2CH_2Si(OCH_3)_3$ and then with MnI_2. On exposure to sulfur dioxide the off-white color of $Mn\{(C_6H_5)_2PCH_2CH_2 \cdot sil\}I_2$ rapidly turned to deep yellow. After 20 min exposure elemental analysis gave 2.6% S, i.e., there was an ~1:1 Mn:SO_2 interaction. Upon application of a vacuum the yellow color was rapidly lost and the off-white color returned. The procedure appeared to be repeatable almost indefinitely. However, not all the SO_2 was desorbed upon application of a vacuum. After ~15 cycles a constant S content of ~1% was found. $Mn\{(C_6H_5)_2PCH_2CH_2Si(OCH_3)_3\}I_2$ does not reversibly bind SO_2 [1].

Polymer-supported complexes $\mathbf{MnL_nX_2}$, where L is a fragment of a polystyrene-divinylbenzene copolymer with one diphenylphosphanyl group bound to a phenyl group of the polymer chain and n=1 for X=Cl or Br and n=1.5 for X=I, were prepared by the reaction of rigorously anhydrous Mn^{II} salt with the required molar quantity of the polymer-supported ligand in THF-pentane under strictly anhydrous and oxygen-free conditions. (The polymer-supported ligand had been obtained by brominating beads of a 1% styrene-divinylbenzene copolymer in nitromethane and subsequent treatment with $LiP(C_6H_5)_2$ in THF to give samples containing 9.4% P.) The complexes are gray for X=Cl, buff for X=Br, and light pink for X=I. All the compounds irreversibly absorb SO_2, either in the solid state or suspended in toluene-dichloromethane (60:40 v/v), the colors changing to bright yellow for X=Cl and Br, and to deep red for X=I. For the SO_2 adducts obtained by treatment of the suspended complexes with SO_2 for several days, SO_2:Mn ratios of 0.3, 0.24, and 0.56 were found for X=Cl, Br, and I, respectively. The uptake of SO_2 is higher in the solid state (no figures given), and gas absorption is usually complete after ~30 min [3].

References:

[1] Booth, B. L.; Mu-guang, L.; McAuliffe, C. A. (J. Chem. Soc. Dalton Trans. **1987** 1415/7).
[2] Hosseiny, A.; Macky, A. G.; McAuliffe, C. A.; Minten, K. (Inorg. Chim. Acta **49** [1981] 99/105).
[3] Booth, B. L.; Li, Ke-Si; McAuliffe, C. A. (J. Chem. Soc. Dalton **1987** 2959/61).

39.2 Complexes with Bis(phosphanes)

Survey. The bis(phosphanes) are chelating ligands containing two functional PR_2 groups. The most important type of complex is the *trans*-octahedral $Mn^{II}L_2X_2$ species with four P atoms in equatorial and two halide or pseudohalide atoms in axial positions. But depending on the steric and electronic requirements of R and X, the tetrahedral $Mn^{II}LX_2$ type also occurs. The $Mn^{II}L_2X_2$ complexes can be oxidized to give the $[Mn^{III}L_2X_2]^+$ and $[Mn^{IV}L_2X_2]^{2+}$ species for L=1,2-phenylene-bis(dimethylphosphane). In addition, several interesting Mn^I complexes are reported (see below, p. 76, and p. 79).

39.2.1 With 1,2-Ethanediylbis(dialkylphosphanes)

ligand 1 $(CH_3)_2PCH_2CH_2P(CH_3)_2$ $(=C_6H_{16}P_2)$
ligand 2 $(C_2H_5)_2PCH_2CH_2P(C_2H_5)_2$ $(=C_{10}H_{24}P_2)$
ligand 3 $(i\text{-}C_3H_7)_2PCH_2CH_2P(C_3H_7\text{-}i)_2$ $(=C_{14}H_{32}P_2)$

$\mathbf{[Mn^I(C_6H_{16}P_2)_3]PF_6}$. A solution of NH_4PF_6 in tetrahydrofuran was added dropwise to an equimolar amount of Mn(syn-η^3-C_5H_7)$(C_6H_{16}P_2)_2$ in tetrahydrofuran maintained at −78°C. The solution was then allowed to warm to room temperature and stirred for 12 h. The resulting yellow precipitate was collected and washed with pentane. The product was purified by extraction with methanol and recrystallization from methanol-pentane at −30°C. The yield was 46.1%. It could be raised to 73.7% by the addition of excess ligand 1 (twofold) to the reaction solution in the preparation step. All manipulations were carried out under N_2. The reaction is believed to proceed by initial protonation at the metal center, followed by migration of the proton to the η^3-pentadienyl ligand. The resulting η^2-pentadiene ligand is then displaced by one bis(phosphane) ligand. The $[Mn^I(C_6H_{16}P_2)_3]^+$ cation containing six Mn–P bonds represents the first homoleptic manganese phosphane complex.

NMR measurements were carried out on the complex in CD_2Cl_2 at 22°C and the following chemical shifts (in ppm; s = singlet) observed: ^1H NMR (100 MHz), δ = 1.40 (s, br); $^{13}C\{^1H\}$ NMR (25.14 MHz), δ = 23.32 (s, methyl), 24.84 (s, methyl), 33.38 (s, methylene); $^{31}P\{^1H\}$ NMR (40.25 MHz, referenced to $P(CH_3)_3$), δ = 124.65 (6 lines, ligand, J_{P-Mn} = 242 Hz), −81.97 (binominal heptet, anion, J_{P-F} = 706 Hz). The NMR spectra are consistent with a D_3 point group symmetry for the $Mn(C_6H_{16}P_2)_3^+$ cation. Bands in the IR spectrum of the complex (in CH_2Cl_2) were assigned as follows: 1425, 1305, 930, 850 to δ(CH) and ν(PC) vibrations; 850 cm^{-1} to ν(P–F) vibrations [1].

$[Mn^I(AlH_4)(C_6H_{16}P_2)_2]_2$. To a solution of $Mn(C_6H_{16}P_2)_2Br_2$ (~1.5 mmol) in toluene at −78°C was added a suspension of excess $LiAlH_4$ (~10.5 mmol) in toluene at −78°C. The solution was allowed to warm to room temperature and stirred for 12 h. Then the solvent was removed and the residue extracted with light petroleum. The filtered extract was cooled to −20°C to give yellow prisms of the complex. The yield was 44%. The complex may also be recrystallized from toluene. It melts at 200°C with decomposition [2].

An X-ray crystal structure determination at 273 K reveals a monoclinic lattice, space group $P2_1/n-C_{2h}^5$, No. 14 (standard setting $P2_1/c$) with the lattice parameters a = 9.313(2), b = 13.515(3), c = 16.982(3) Å, β = 97.01(2)°; Z = 2. The structure was solved to R = 4.15%. The calculated density is D = 1.21 g/cm³. In the solid state, the compound exists as a loosely bound centrosymmetric dimer, due to the formation of a $(\mu-H)_2AlH(\mu-H)_2AlH(\mu-H)_2$ bridge (see **Fig. 19**). The geometry about the Mn atoms is *cis*-octahedral. The Al atoms adopt a distorted trigonal-bipyramidal geometry with the terminal H atoms occupying equatorial positions. The central $Al(\mu-H)_2Al$ unit is asymmetric (short and long Al–H distances). In the $Mn(\mu-H)_2Al$ units the Mn–H distances are equal and shorter than the two Al–H distances by 0.2 Å, indicating a strong affinity of the H atoms for the Mn atoms. Selected bond lengths (in Å) and angles (in °) are:

Mn(1)–H(1)	1.610(26)	Mn(1)–P(1)	2.206(3)	Mn(1)–P(3)	2.218(3)
Mn(1)–H(1′)	1.631(28)	Mn(1)–P(2)	2.218(3)	Mn(1)–P(4)	2.215(3)
Mn(1)–Al(1)	2.393(4)				

H(1)–Mn(1)–H(1′)	97.9(14)	P(4)–Mn(1)–P(3)	83.1(2)	P(1)–Mn(1)–P(3)	95.8(2)
H(1′)–Mn(1)–P(1)	78.5(10)	P(3)–Mn(1)–H(1′)	91.4(10)	H(1)–Mn(1)–P(1)	173.6(8)
P(1)–Mn(1)–P(4)	100.0(2)	P(3)–Mn(1)–H(1)	89.6(10)	P(3)–Mn(1)–P(2)	177.3(1)
P(2)–Mn(1)–P(4)	94.3(2)	P(2)–Mn(1)–P(1)	84.0(2)	H(1′)–Mn(1)–P(4)	174.1(10)

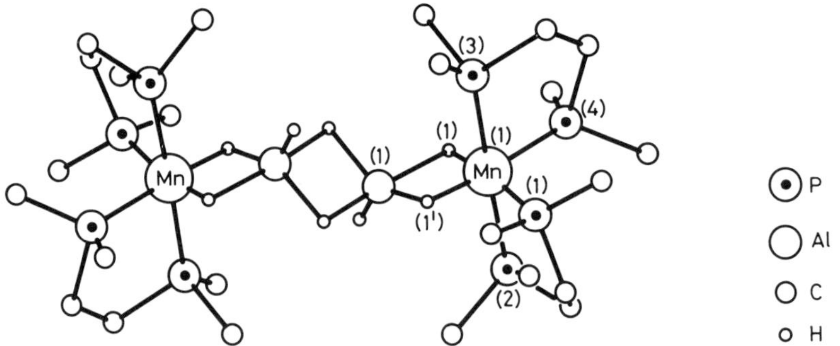

Fig. 19. Molecular structure of $[Mn(AlH_4)(C_6H_{16}P_2)_2]_2$ (hydrogen atoms omitted except for those in the μ-AlH_4 groups) [2].

NMR measurements were performed on the complex in C_6D_6 (positive chemical shifts, δ, in ppm downfield of $Si(CH_3)_4$; s = singlet): ^1H (298 K), δ = −15.02 (s, Mn–H–Al, br), 5.05 (s, Al–H);

^1H (183 K), $\delta = -15.02$ (s, Mn–H–Al, br), 5.34 (s, Al–H–Al), 4.76 (s, Al–H). The splitting of the peak due to the Al–H groups at low temperatures ($T_C = 210$ K) shows that there is a fluxional process involving the $HAl(\mu\text{-}H)_2AlH$ unit. The activation energy of the process, $\Delta G = 40$ kJ/mol, was calculated from the NMR data. The ^1H and ^{13}C NMR resonances of the coordinated bis(phosphane) ligand (also presented) confirm the *cis*-octahedral arrangement around the Mn atom. The NMR data suggest a monomeric structure for the complex in solution. IR bands at 1740, 1610, and 970 cm^{-1} may be attributed to Al–H and Mn–H–Al stretching vibrations. Other IR bands in the range between 2800 and 250 cm^{-1} were not assigned [2], preliminary communication [3].

$Mn^{II}(C_6H_{16}P_2)_2X_2$ (X = Br, I) and **$Mn^{II}(C_6H_{16}P_2)Cl_2$**. To a suspension of $MnBr_2$ or MnI_2 in dichloromethane was added the bis(phosphane) ligand (1:2 mole ratio). After 24 h (for X = Br) or 30 min (for X = I) the Mn salts had dissolved to give a pale yellow (for X = Br) or pale orange solution (for X = I). The solution was filtered, concentrated, and cooled to −20°C. The colorless prisms which formed were recrystallized from 2:1 toluene-dichloromethane (for X = Br) or toluene (for X = I) to give transparent plates. All attempts to prepare $Mn(C_6H_{16}P_2)_2Cl_2$ by an analogous method were unsuccessful. Instead, a colorless insoluble complex of composition $Mn(C_6H_{16}P_2)Cl_2$ formed. The compound, suggested to contain Cl bridges, was not investigated further. All operations were carried out in vacuum or under purified argon using standard techniques. $Mn(C_6H_{16}P_2)_2Br_2$ melts at 155 to 159°C, $Mn(C_6H_{16}P_2)_2I_2$ at 210 to 215°C [4].

An X-ray crystal structure determination was performed for $Mn(C_6H_{16}P_2)_2Br_2$ at 295 K. The compound crystallizes in the tetragonal system with a = b = 14.014(2), c = 12.220(7) Å, $\alpha = 90°$; space group $P4_2/ncm–D_{4h}^{16}$ (No. 138); Z = 4. The structure was solved to R = 0.072. The calculated density is D = 1.424 g/cm³. The four P atoms of the two ligand molecules are arranged in equatorial positions and the two Br atoms in axial positions, resulting in a *trans*-octahedral geometry. The bis(phosphane) ligands show disorder. While the P atoms can be represented by a single position, the methyl and methylene carbon atoms are each disordered over two sites. (Fractional atomic coordinates involving six C positions are presented in the paper.) Selected bond lengths (in Å) and angles are: Mn–Br = 2.666(3), Mn–P = 2.655(4); P–Mn–P = 78.8(1)°; Mn–P–C(CH_2) = 102(1)°; Mn–P–C(CH_3) = 118(3)° to 123(1)° [4]. The magnetic moment of both compounds in toluene is $\mu_{eff} = 5.9\ \mu_B$ indicating a high-spin d^5 configuration. X-band ESR spectra of the complexes were measured in toluene glass at −196 K. The parameters D = 0.63 cm^{-1}, $\lambda = 0.03$, and A(^{55}Mn) = 0.0079 cm^{-1} for X = Br, and D = 0.85 cm^{-1}, $\lambda = 0.03$, and A(^{55}Mn) = 0.0070 cm^{-1} for X = I have been evaluated. The ESR spectrum of $Mn(C_6H_{16}P_2)_2Br_2$ is identical to that of manganocene in an $Mg(\eta^5\text{-}C_5H_5)_2$ matrix, indicating the close electronic similarity of the two compounds. The ^1H NMR spectra of the complexes in C_6D_6 show two broad resonances (chemical shifts, δ, in ppm): 51.0 (s, CH_3), 26.0 (s, CH_2) for X = Br and 53.5 (s, CH_3), 29.5 (s, CH_2) for X = I. It is suggested that the contact shift contribution to the chemical shifts dominates, as is the case for manganocene [4].

The complexes are air-sensitive, especially in solution. They are soluble in toluene, tetrahydrofuran, dichloromethane, and acetonitrile, sparingly soluble in diethyl ether, and essentially insoluble in light petroleum [4]. The reaction of $Mn(C_6H_{16}P_2)_2Br_2$ with excess $LiAlH_4$ in toluene gives $[Mn^I(AlH_4)(C_6H_{16}P_2)_2]_2$ (see p. 76) [3]. Alkylation of $Mn(C_6H_{16}P_2)_2Br_2$ with $Mg(CH_3)_2$ in diethyl ether gives $Mn(CH_3)_2(C_6H_{16}P_2)_2$ [3, 4] (defined and characterized in [4], but not described in the "Manganese" D series). Alkylation with $Mg(C_2H_5)_2$ gives $MnH(CH_2{=}CH_2)$-$(C_6H_{16}P_2)_2$ (reported in [3]).

$Mn^{II}(C_{10}H_{24}P_2)_2Br_2$ and **$Mn^{II}(C_{10}H_{24}P_2)_2(NCS)_2$**. The compounds were prepared by the reaction of $MnBr_2$ or $Mn(NCS)_2$ with ligand 2 in a solvent. Details of the preparation method used are as described on p. 38 for Mn^{II} complexes with monophosphanes. An X-ray diffraction study of the bromo complex revealed a monomeric *trans*-octahedral structure, as was described for

$Mn^{II}(C_6H_{16}P_2)_2Br_2$ (see above). The Mn–ligand distances are within the range expected for high-spin d^5 six-coordinated Mn^{II} compounds: Mn–P 2.697(1) and 2.720(1) Å; Mn–Br 2.649(1) Å. The P–Mn–P angle is 77.5(1)° [5].

The compounds, either in the solid state or in tetrahydrofuran or toluene, react with dioxygen reversibly at room temperature and at about atmospheric partial pressure of dioxygen. Isotherm data for binding of dioxygen have been obtained using a volume reduction method. (A plot of dioxygen reacted versus partial pressure of dioxygen at 4.4 and 21.1°C is presented in the paper.) The following equilibrium data were calculated from the isotherm data by use of the Hill equation (see p. 45): log $K_{O_2}=-2.225$, n = 0.95 at 4.4°C and log $K_{O_2}=-1.486$, n = 1.11 at 21.1°C for $Mn(C_{10}H_{24}P_2)_2Br_2$ and log $K_{O_2}=-2.434$, n = 0.81 at 4.4°C and log $K_{O_2}=-2.086$, n = 1.40 at 21.1°C for $Mn(C_{10}H_{24}P_2)_2(NCS)_2$ (K_{O_2} in $Torr^{-1}$). For $Mn(C_{10}H_{24}P_2)_2(NCS)_2$ there exists a peak temperature (about 18.3°C) at which the amount of dioxygen bound is maximum. At higher temperatures dioxygenation of the complex decreases and becomes zero at about 74°C. Kinetic studies with $Mn(C_{10}H_{24}P_2)_2(NCS)_2$ also show that the rate of reaction with dioxygen reaches a maximum near 20°C. Equilibrium studies with $Mn(C_{10}H_{24}P_2)_2(NCS)_2$ in tetrahydrofuran solution show results similar to those with the solid compound, except that the rate of reaction with dioxygen is higher for the solution [5].

$\mathbf{Mn^{II}(C_{14}H_{32}P_2)X_2}$ (X = Cl, Br) and $\mathbf{Mn^{II}(C_{14}H_{32}P_2)I_2 \cdot C_6H_5CH_3}$. The chloro complex was prepared by combining $MnCl_2$ and ligand 3 (1:~1.5 mole ratio) and heating the mixture to ~70°C for 12 h. The resulting tan solid was extracted with hot toluene, and pale green prisms were obtained after filtering the solution and cooling to −20°C. The yield was 23%. The bromo and iodo complexes were prepared by the reaction of $MnBr_2$ or MnI_2 with ligand 3 (1:1 mole ratio) in toluene. After being stirred for 24 h, the suspension was warmed to 60°C (for X = Br) and to 75°C (for X = I), and the solution was filtered while hot. Cooling to −20°C resulted in the formation of pale green crystals (for X = Br), while pale orange needles of the toluene solvate crystallized upon cooling to room temperature (for X = I). The yields were 73 and 86%, respectively. The complexes melt at 150°C (X = Cl), 128 to 129°C (X = Br), and 84 to 86°C (X = I).

The magnetic moment, determined by a modification of the Evans method, is $\mu_{eff}=5.7\ \mu_B$ for all the compounds. The X-band ESR spectra, taken from frozen glasses in toluene-dichloromethane, show resonances near 160 mT that are characteristic of rhombically-distorted tetrahedral $S=5/2$ species (zero-field splitting parameter $\lambda \sim 0.3$). A weak low-field transition at ~70 mT shows hyperfine coupling of 0.0072 cm^{-1} to the ^{55}Mn nucleus. No hyperfine splittings due to ^{31}P or ^1H nuclei are detected, indicating localization of the spin density on the manganese atom [6].

$\mathbf{Mn^{III}H_3(C_6H_{16}P_2)_2}$ was prepared by adding distilled water to a solution of $[Mn^I(AlH_4)(C_6H_{16}P_2)_2]_2$ (see p. 76) in ether until gas evolution ceased. The solution was filtered, evaporated, and the residue extracted with light petroleum. The filtered extract was concentrated and cooled to −20°C to give yellow prisms of the complex, melting point 144°C (with decomposition), in a 75% yield. The complex is readily volatile in high vacuum at 50°C. The IR spectrum (Nujol mull) shows a single ν(Mn–H) band at 1680 cm^{-1}, while the ^1H NMR spectrum has a binomial quintet at $\delta=-12.67$ ppm due to the H atoms, J(P–H) = 28.6 Hz. The spectrum is unchanged on cooling to −60°C, as is the $^{31}P\{^1H\}$ NMR spectrum which shows a broad singlet at $\delta=88.0$ ppm. On the basis of these data and by analogy to the analogous rhenium complex, $ReH_3(C_6H_{16}P_2)_2$, a pentagonal-bipyramidal structure with the H atoms in equatorial positions was suggested. $MnH_3(C_6H_{16}P_2)_2$ reacts with carbon monoxide in toluene to give mixtures of various hydridocarbonylphosphane species. Photolysis in the presence of ethylene gives $MnH(C_2H_4)(C_6H_{16}P_2)_2$ in high yield [2].

References:

[1] Bleeke, J. R.; Kotyk, J. J. (Organometallics **4** [1985] 194/7).
[2] Girolami, G. S.; Howard, C. G.; Wilkinson, G.; Dawes, H. M.; Thornton-Pett, M.; Motevalli, M.; Hursthouse, M. B. (J. Chem. Soc. Dalton Trans. **1985** 921/9).
[3] Girolami, G. S.; Wilkinson, G.; Thornton-Pett, M.; Hursthouse, M. B. (J. Am. Chem. Soc. **105** [1983] 6752/3).
[4] Girolami, G. S.; Wilkinson, G.; Galas, A. M. R.; Thornton-Pett, M.; Hursthouse, M. B. (J. Chem. Soc. Dalton Trans. **1985** 1339/48).
[5] Kulkarni, V. B.; Govind, R. (Inorg. Chim. Acta **150** [1988] 11/2).
[6] Hermes, A. R.; Girolami, G. S. (Inorg. Chem. **27** [1988] 1775/81).

39.2.2 With 1,2-Ethanediylbis(diphenylphosphane) $(C_6H_5)_2PCH_2CH_2P(C_6H_5)_2$ $(=C_{26}H_{24}P_2)$

$Mn^I(C_{26}H_{24}P_2)_2(N_2)Cl$. The buff-colored complex was prepared by the reduction of $Mn^{II}(C_{26}H_{24}P_2)Cl_2$ in benzene-alcohol solution with $NaBH_4$ in the presence of molecular hydrogen and then passing nitrogen through the solution. It is suggested that $Mn(C_{26}H_{24}P_2)_2(H)Cl$ might have been formed as an intermediate and that the nitrogen complex is formed by displacement of the coordinated hydride. The presence of coordinated dinitrogen in the complex was confirmed by an IR band at 2100 cm^{-1} [1]. MO calculations show that end-on coordination of the dinitrogen is more favorable than side-on coordination. The high value of the ν(N–N) frequency suggests that σ donation is more important for the formation of the $Mn–N_2$ bond than π back-donation which would mainly contribute to the weakening of the N–N bond [2]. The compound is moderately stable in dry air and decomposes without melting. It is sparingly soluble in benzene, alcohol, and chloroform [1].

$Mn^{II}(C_{26}H_{24}P_2)_2Cl_2$ is reported as a starting material for the preparation of $Mn^I(C_{26}H_{24}P_2)_2$-$(N_2)Cl$ [1].

$(NO)_3Mn(C_{26}H_{24}P_2)Mn(NO)_3$. A solution of $Mn(NO)_3 \cdot THF$ in THF (prepared by the photo-induced nitrosylation of Mn_2CO_{10}) was treated with the bis(phosphane) ligand in a 2:1 mole ratio and the solution stirred overnight at room temperature. After removal of the solvent the residue was redissolved in benzene and chromatographed on silica gel/benzene. The benzene solution obtained by elution of the dark green zone was reduced in volume and the concentrated solution treated with hexane to give dark green crystals of the product. All manipulations were carried out under purified nitrogen. A molecular weight determination shows that the complex is dinuclear (710, calculated 688). The complex starts decolorizing at 135°C, it melts (with decomposition) at ~240°C. The IR spectrum of the complex (in THF) shows the $\nu_{NO}(A_1)$ band at 1776 and the split $\nu_{NO}(E)$ bands at 1685 and 1679 cm^{-1}. The splitting of the $\nu_{NO}(E)$ band reflects the remarkable disturbance of the C_{3v} symmetry at the Mn center by the bis(phosphane) ligand [3].

References:

[1] Taqui Khan, M. M.; Anwaruddin, Q. (unpublished results from Taqui Khan, M. M.; Martell, A. E.; Homogeneous Catalysis by Metal Complexes, Vol. I, Academic, New York 1974, p. 240).
[2] Yamabe, T.; Hori, K.; Minato, T.; Fukui, K. (Inorg. Chem. **19** [1980] 2154/9).
[3] Herberhold, M.; Razavi, A. (J. Organometal. Chem. **67** [1974] 81/6).

39.2.3 With 1,2-Phenylenebis(dimethylphosphane)

$C_6H_4(P(CH_3)_2)_2$ $(=C_{10}H_{16}P_2)$

General. The ligand is able to stabilize high oxidation states. The $[Mn^{III}X_2(C_{10}H_{16}P_2)_2]^+$ cations were obtained by oxidation of the $[Mn^{II}X_2(C_{10}H_{16}P_2)_2]$ compounds with trityl hexafluorophosphate. Nitric acid oxidation of the chloro Mn^{III} complex generated the $[Mn^{IV}Cl_2(C_{10}H_{16}P_2)]^{2+}$ ion. The cations were isolated as the perchlorates. The complexes undergo electrochemical one-electron-transfer reactions. The following reduction potentials (V vs. Ag|$AgClO_4$ in acetonitrile, 0.1M tetraethylammonium perchlorate) were observed:

$$[Mn^{II}Cl_2(C_{10}H_{16}P_2)_2] \xleftarrow{\sim -0.6} [Mn^{III}Cl_2(C_{10}H_{16}P_2)_2]^+ \xleftrightarrow{0.78} [Mn^{IV}Cl_2(C_{10}H_{16}P_2)_2]^{2+}$$

$$[MnBr_2(C_{10}H_{16}P_2)_2] \xleftarrow{\sim 0.4} [Mn^{III}Br_2(C_{10}H_{16}P)_2]^+ \xleftrightarrow{0.72} [Mn^{IV}Br_2(C_{10}H_{16}P_2)_2]^{2+}$$

The $Mn^{III} \rightarrow Mn^{II}$ couple is irreversible because the Mn^{II} complexes decompose in the electrolyte medium [1].

X-ray crystal structure determinations of the compounds listed below (no details reported) indicate that the complexes have a *trans*-octahedral structure. The following Mn–Cl and Mn–P distances (in Å) have been found [2]:

complex	Mn–Cl	Mn–P (average)
$[Mn^{II}Cl_2(C_{10}H_{16}P_2)_2]$	2.502(1)	2.625
$[Mn^{III}Cl_2(C_{10}H_{16}P_2)_2]ClO_4$	2.239(2)	2.344
$[Mn^{IV}Cl_2(C_{10}H_{16}P_2)_2][H(NO_3)_2]_2$	2.195(1)	2.428

The decrease in the Mn–Cl bond lengths with increasing oxidation state suggests that the metal-halogen interaction is an important feature in the stabilization of the higher oxidation states. In the Mn^{IV} complex the close approach of the axial Cl atoms to the metal results in repulsion of the equatorial phosphorus ligands to a greater distance than found in the Mn^{III} complex [1].

$[Mn^{II}X_2(C_{10}H_{16}P_2)_2]$ (X = Cl, Br). To a solution of anhydrous $MnCl_2$ or $MnBr_2$ in deoxygenated isopropyl alcohol an excess of the bis(phosphane) ligand was added under nitrogen and the solution stirred for a few hours. Then the mixture was cooled in an ice bath. The precipitated yellow solid was filtered under nitrogen using a Schlenk apparatus and washed successively with cold deoxygenated isopropyl alcohol and toluene to obtain the complex in a 45% yield. The unusually long Mn–Cl and Mn–P bond lengths in the chloro compound (see above) are in accord with the ionic character expected for a stable high-spin d^5 electronic configuration (μ_{eff} = 6.04 μ_B at 296°C). The IR spectrum of this compound (Nujol mulls) shows the ν(Mn–Cl) band at 319 cm^{-1}. The UV-visible spectra of the complexes (in Nujol) show band maxima at 404 nm for X = Cl and 437 nm for X = Br. The complexes are stable in dry air, but they dissociate in solution, as judged by the odor of free bis(phosphane) and in some cases (e.g., water) by the disappearance of the yellow color. The bromo complex is less soluble than the chloro complex. The Mn^{II} species can be oxidized to form $[Mn^{III}X_2(C_{10}H_{16}P_2)]^+$ ions (see below) [1].

$[Mn^{III}X_2(C_{10}H_{16}P_2)_2]ClO_4$ (X = Cl, Br) and **$[Mn^{III}Cl_2(C_{10}H_{16}P_2)_2]PF_6$**. Solid $[Mn^{II}Cl_2(C_{10}H_{16}P_2)_2]$ was added to a solution of $[(C_6H_5)_3C]PF_6$ (~1:1 mole ratio) in dichloromethane with stirring, and the resulting orange suspension was stirred for another 20 h under nitrogen. Filtering and washing with toluene yielded orange crystalline $[Mn^{III}Cl_2(C_{10}H_{16}P_2)_2]PF_6$. Purification was effected by dissolving this material in the minimum amount of acetonitrile and treating the solution with excess solid anhydrous $LiClO_4$. The bromo compound was prepared similarly from

$[Mn^{II}Br_2(C_{10}H_{16}P_2)_2]$ and $[(C_6H_5)_3C]PF_6$, but the resulting deep orange mixture was evaporated to dryness in vacuum at room temperature and the residue extracted with a small quantity of dry acetonitrile. The filtered solution was then treated with excess solid anhydrous $LiClO_4$. The orange crystalline solids were filtered and washed with 1-propanol. The yield was 42% for the bromo compound. For structural data, see p. 80. Magnetic measurements at 296°C yielded $\mu_{eff}=3.10\ \mu_B$ (for X = Cl) and 3.14 μ_B (for X = Br), indicating low-spin manganese(III) d^4 complexes. The IR spectrum of the chloro compound (in Nujol) shows the ν(Mn–Cl) band at 374 cm^{-1} (hexafluorophosphate at 381 cm^{-1}). The UV-visible spectra of the complexes in acetonitrile show band maxima at 519 nm (for X = Cl) and at 548 nm (for X = Br). Solutions of the complexes in protic solvents decompose fairly rapidly. Acetonitrile was found to be one of the best solvents. The complexes can be oxidized to give the manganese(IV) species, $[Mn^{IV}X_2(C_{10}H_{16}P_2)_2]^{2+}$ [1].

$[Mn^{IV}Cl_2(C_{10}H_{16}P_2)_2](ClO_4)_2$ was prepared by dissolving $[Mn^{III}Cl_2(C_{10}H_{16}P_2)_2]ClO_4$ in the minimum quantity of cold (0°C) concentrated nitric acid and adding dropwise cold dilute $HClO_4$ until the orange-brown crystalline complex precipitated. The complex was washed with water and dried in air. The magnetic moment is $\mu_{eff}=3.99\ \mu_B$, indicating a high-spin d^3 complex. The UV-visible spectrum shows bands at 504(sh), 471, and 415 nm [1].

References:

[1] Warren, L. F.; Bennett, M. A. (Inorg. Chem. **15** [1976] 3126/40).

[2] Rosalky, J. M. (Diss. Australian National Univ. 1975 from Warren, L. F.; Bennett, M. A.; Inorg. Chem. **15** [1976] 3126/40, 3139).

39.3 Complexes with Phosphanylacetic Acids or Phosphanetriyltris(acetic Acid)

No.	ligand	formula
1	$(C_6H_5)_2P(CH_2COOH)$	$C_{14}H_{13}O_2P$
2	$(C_6H_5)P(CH_2COOH)_2$	$C_{10}H_{11}O_4P$
3	$P(CH_2COOH)_3$	$C_6H_9O_6P$

$(HOOCH_2C)_2PCH_2CH_2P(CH_2COOH)_2$

ligand 4 ($=C_{10}H_{16}O_8P_2$)

Complexes in Solution. Mn^{2+} ions form 1:1 complexes with ligands 1 to 3 in aqueous solution and aqueous dioxane (50:50 v/v). Stability constants, log K_1, determined by the pH method under nitrogen at 25°C, I = 0.1 mol/L (Na)ClO_4, are <1 and 1.9(2) for $Mn(C_{14}H_{12}O_2P)^+$, <1 and 2.69(2) for $Mn(C_{10}H_9O_4P)$, and 2.04(3) and 3.99(2) for $Mn(C_6H_6O_6P)^-$ in aqueous solution and aqueous dioxane, respectively. The gradual increase of the stability of the complexes going from the unidentate O-donor ligand $(C_6H_5)_2P(CH_2COO^-)$ to the terdentate O-donor ligand $P(CH_2COO^-)_3$ suggests coordination by the carboxyl oxygen atoms only and not by the phosphorus atoms [1].

The formation of the $Mn(C_{10}H_{12}O_8P_2)^{2-}$ complex (ligand 4) and of the protonated $Mn(C_{10}H_{13}O_8P_2)^-$ and $Mn(C_{10}H_{14}O_8P_2)$ complexes in aqueous solution was revealed by potentiometric titrations at 25°C and I = 0.1 ($NaClO_4$). Stability constants are: log K_1 = 2.92(2), determined by the pH method, log β = 8.47(4) for $\beta=[MnHL^-]/[Mn^{2+}][H^+][L^{4-}]$, and log β′ = 12.46(6) for $\beta'=[MnH_2L]/[Mn^{2+}][H^+]^2[L^{4-}]$. The values of β and β′ were obtained by a combination of pH and competition titrations, using Zn as an auxiliary metal [2].

Measurements of the magnetic susceptibility on solutions with the composition 0.05 M $MnCl_2$ and 0.2 M $Na_4(C_{10}H_{12}O_8P_2)$ by the modified Evans method (using $NiCl_2$ solutions for calibration) yielded an effective magnetic moment of 5.95 μ_B. Solutions with ligand concentrations of 0.1 mol/L for IR studies (in CaF_2 cells) were prepared in a dry inert atmosphere from $MnCl_2$ and $Na_4(C_{10}H_{12}O_8P_2)$ in D_2O. The IR spectrum of a solution of pD 3.99 (pD = pH-meter reading + 0.41) shows bands at the following frequencies (relative intensities): 1688 (33%), 1607 (40%), and 1561 (27%), which were assigned to the $PCH_2COOH + HP^+CH_2COOH$, $HP^+CH_2COO^-$, and PCH_2COO^- groups, respectively. The bands are not shifted compared to the free ligand bands. However, an increase in the intensity of the band at 1607 cm^{-1} from the free ligand value (35%) can be attributed to a competition of an Mn^{2+} ion and a proton on the carboxyl group forcing partial shift of the protons to the P atom. The IR spectral data therefore indicate the presence of Mn–carboxylate bonds (and no Mn–P bonds) [3], as had already been assumed on the basis of titration studies [2]. A solution of pD 8.38 shows only the band of the PCH_2COO^- group at 1562 cm^{-1} (100%), which had not shifted from the free ligand spectrum. The UV-visible spectrum of an aqueous complex solution shows an absorption band at >50000 cm^{-1} with the molar extinction coefficient $\varepsilon > 8000$ $L \cdot mol^{-1} \cdot cm^{-1}$ [3].

$Mn_3[Ni(C_{10}H_{12}O_8P)_2] \cdot 16H_2O$. The compound was prepared from $MnBr_2$, $NiBr_2 \cdot 3H_2O$, and $C_{10}H_{16}O_8P_2 \cdot 2HBr$ in aqueous solution in the presence of CH_3COONa and NaOH [4]. An X-ray structure analysis shows that in the compound the Ni^{2+} ion is coordinated by two P,P-chelating ligand anions in a distorted square-planar arrangement, whereas the three Mn^{2+} ions are octahedrally surrounded by 8 O atoms from five carboxylate groups and the O atoms from 10 H_2O molecules [5]. A detailed description of this compound will be included with the Ni^{II} complexes treated later.

References:

[1] Podlahová, J. (Collection Czech. Chem. Commun. **44** [1979] 2460/4).

[2] Podlahová, J.; Podlaha, J. (Collection Czech. Chem. Commun. **47** [1982] 1078/85).

[3] Podlahová, J.; Podlaha, J. (Collection Czech. Chem. Commun. **50** [1985] 445/53).

[4] Podlahová, J.; Kavan, L.; Šilha, J.; Podlaha, J. (Polyhedron **3** [1984] 963/8).

[5] Podlahová, J.; Kratochvíl, B.; Langer, V.; Podlaha, J. (Polyhedron **5** [1986] 799/803).

39.4 Cyclohexyl- or Phenylphosphidomanganate(II) Complexes

$Li[Mn^{II}\{P(C_6H_{11}\text{-}c)_2\}_3]$ was prepared by the reaction of $MnBr_2$ (10.2 mmol) with $LiP(C_6H_{11}\text{-}c)_2$ (35.7 mmol) in THF. The reaction mixture was stirred for 3 to 4 h at room temperature. The solution was evaporated to ~1/5 of the initial volume, then cooled with CO_2 snow. The red-brown crystals which separated were washed with cold THF until bromide-free, then dried in vacuum at 50 to 60°C, the color of the crystals turning to ochre. Addition of petroleum ether to the filtrate yielded additional product. $Li[Mn^{II}\{P(C_6H_{11}\text{-}c)_2\}_3]$ melts at 253 to 255°C. It is pyrophoric and extremely sensitive to hydrolysis. It dissolves readily in THF, only sparingly in dioxane, and is insoluble in benzene, diethyl ether, and petroleum ether [1]. According to [2] the compound is possibly not monomeric.

$Li(C_4H_{10}O_2)[Mn_2\{P(C_6H_{11}\text{-}c)_2\}_5]$ containing 1,2-dimethoxyethane (= $C_4H_{10}O_2$) was prepared by the reaction of $Mn(thf)_2Br_2$ with $LiP(C_6H_{11}\text{-}c)_2$ in tetrahydrofuran at −80°C. It was separated from LiBr by fractional crystallization from 1,2-dimethoxyethane at −30°C. The complex is paramagnetic ($\mu_{eff} \approx 2.0$ μ_B per Mn atom). It is insoluble in hexane or benzene. The properties suggest an oligomeric structure [2].

$K[Mn\{P(C_6H_5)_2\}_3] \cdot 2thf$ was prepared by the reaction of $MnBr_2$ (~16.3 mmol) with ~46.4 mmol of $KP(C_6H_5)_2 \cdot 2C_4H_8O_2$ ($C_4H_8O_2$ = dioxane) in tetrahydrofuran at 40 to 50°C. The solution was filtered and reduced to ~1/4 of its initial volume. Yellow-orange crystals of the product were precipitated with petroleum ether. They were washed with petroleum ether and dried in vacuum at room temperature. The compound is sensitive to air. On heating, the tetrahydrofuran is split off at temperatures above 110°C. At 195 to 200°C decomposition occurs. The solubility properties are the same as described for $Li[Mn\{P(C_6H_{11}\text{-}c)_2\}_3]$ (see on p. 82). $K[Mn\{P(C_6H_5)_2\}_3]\cdot 2thf$ decomposes in water and ethanol [1].

$K_2[Mn\{P(C_6H_5)_2\}_4]\cdot 2C_4H_8O_2$ was obtained by the reaction of $MnBr_2$ (~16.7 mmol) with $KP(C_6H_5)_2\cdot 2C_4H_8O_2$ (~50 mmol) in tetrahydrofuran. The mixture was filtered and the solution evaporated to dryness at reduced pressure. The red oily residue was treated with dioxane and the dioxane distilled off. The residue was refluxed with dioxane and the hot solution filtered, then cooled. Orange-red crystals of the product separated and were recrystallized from dioxane. $K_2[Mn\{P(C_6H_5)_2\}_4]\cdot 2C_4H_8O_2$ is sensitive to air and moisture. The dioxane is lost at temperatures over 145°C. At 217 to 221°C decomposition occurs. The complex dissolves in tetrahydrofuran, is sparingly soluble in dioxane, but is insoluble in benzene, ether, and petroleum ether [1].

References:

[1] Issleib, K.; Wenschuh, E. (Z. Naturforsch. **17b** [1962] 778/9).
[2] Baker, R. T.; Krusic, P. J.; Tulip, T. H.; Calabrese, J. C.; Wreford, S. S. (J. Am. Chem. Soc. **105** [1983] 6763/5).

39.5 Complexes with Phosphane Oxides

General Aspects. Manganese(II) complexes with phosphane oxides are prepared mainly by reaction of MnX_2 salts with the phosphane oxide in organic solvents under anhydrous conditions. They are obtained also by refluxing MnX_2 salts with phosphanes, e.g., triphenylphosphane in tetrahydrofuran, or by oxidation of mixed ligand complexes of Mn^{II} containing phosphanes and tetrahydrofuran. Another method for preparing several Mn^{II} phosphane oxide complexes is the reaction of $MnCl_3$ in ether solution with the corresponding phosphane.

The steric properties of the phosphane oxides largely determine the number of phosphoryl ligands (L) attached to the Mn^{2+} ion. An $[MnL_6]^{2+}$ complex is formed only with phosphoryltrismethanol. Triphenylphosphane oxide, one of the most extensively studied monodentate ligands, forms different complexes, depending on the coordinated anion; e.g., $[MnL_4X_2]$ where Mn^{II} is coordinated octahedrally, $[MnL_5]^{2+}$ and $[MnL_4X]^+$ with pentacoordinated Mn^{II} in a square-pyramidal surrounding, or $[MnL_2X_2]$ with a tetrahedral geometry. A trigonal-bipyramidal structure was found for the $[MnL_4(NCS)Cl]_2$ complex with triphenylphosphane oxide. Formation of an anionic $[MnLBr_3]^-$ complex was observed with dimethylphenylphosphane oxide.

The phosphane oxides are usually bonded to Mn by the oxygen atom. The infrared spectra show that complex formation causes the P–O stretching frequencies to shift ~50 cm^{-1} to lower values. Only in the spectrum of the complex with tris(2-pyridinyl)phosphane oxide an increase in frequency of $\nu(PO)$ was observed as manganese is coordinated by the nitrogen atom of the pyridine rings.

Phosphane oxides are used to extract Mn^{2+} ions from aqueous solution into organic solvents. Mixed ligand complexes form in the presence of other ligands. An interesting compound of this type is *trans*-$[MnL_4(SO_2)_2]I_2$ with $L=(C_6H_5)_3PO$. The X-ray crystal structure shows it to contain O-bonded sulfur dioxide groups; one of these can be lost to form $[MnL_4(SO_2)]I_2$. The reaction was found to be reversible.

Only a few manganese(III) complexes are known; e.g., MnL_3Cl_3 and MnL_2X_3 compounds with L = triphenylphosphane oxide (X = NO_3, Cl) or $[MnL(C_6Cl_4O_2)_2]^-$ complexes containing tetrahalocatecholato anions (X = Cl, Br).

39.5.1 With Dimethylphosphane Oxide $(CH_3)_2HPO$ (= C_2H_7OP)

$Mn^{II}(C_2H_7OP)_2Cl_2\cdot0.5H_2O$ was obtained from manganese(II) chloride and the ligand in butanol at 50°C. The light pink oil which separated on cooling the mixture was triturated with ether at −10°C to yield crystals [1]. The IR spectrum of the complex was studied in the 4000 to 200 cm^{-1} range together with those of analogous Cd or Ni^{II} complexes. The experimental results and the analysis of a normal vibration model for the Cd complex were used to assign the principal vibration modes. Characteristic bands of the Mn^{II} complex are the vibration modes: ν(OH) at 3285, ν(PH) at 2424, 2408, 2395, 2383, δ(HOH) at 1635, 1610, ν(PO) at 1168, 1130, 1095, 1080, ν(Mn–O) at 397, and ν(Mn–Cl) at 306 and 325 cm^{-1}. The increase of ν(PH) and the decrease of ν(PO) on complexation indicate bonding of the ligand through the phosphoryl oxygen atom [2]. This is confirmed by the similar ^1H NMR spectra of the free ligand and the complex [1]. A comparative estimate of the metal-oxygen bond strength indicates a slight decrease in the sequence Cd → Ni → Mn [2].

References:

[1] Popova, I. A.; Sarukhanov, M. A.; Kharitanov, Yu. Ya. (Koord. Khim. **10** [1984] 858/9; C.A. **101** [1984] No. 162544).

[2] Sarukhanov, M. A.; Popova, I. A.; Kharitanov, Yu. Ya. (Zh. Neorgan. Khim. **32** [1987] 87/92; Russ. J. Inorg. Chem. **32** [1987] 47/50).

39.5.2 With Trimethylphosphane Oxide $(CH_3)_3PO$ (= C_3H_9OP)

$[Mn^{II}(C_3H_9OP)_5]X_2$ (X = ClO_4, BF_4) and **$[Mn(C_3H_9OP)_5(H_2O)](ClO_4)_2$**. The $[Mn(C_3H_9OP)_5]X_2$ complexes were prepared by reaction of the hydrated Mn^{II} salt with a slight excess of ligand in hot acetone [1, 2] or methanol, or ethanol [2] containing triethyl orthoformate as dehydrating agent. The white complexes precipitated upon cooling the mixture [1] or reducing the volume by evaporation in vacuum [2]. The compounds were washed with dry ether; the perchlorate was dried in high vacuum [1], the fluoroborate in a desiccator over P_4O_{10}. The complex melts at 267 to 274°C [2]. If exposed to atmospheric moisture, $[Mn(C_3H_9OP)_5](ClO_4)_2$ is changed into $[Mn(C_3H_9OP)_5(H_2O)](ClO_4)_2$. An octahedral coordination at Mn^{II} was proposed for this complex on the basis of the reflectance spectra of the corresponding isomorphous Ni^{II} and Co^{II} compounds [1]. X-ray powder photographs show that the $[Mn(C_3H_9OP)_5]X_2$ complexes are isomorphous with the corresponding Ni^{II} and Co^{II} compounds [1, 2].

The magnetic moment, $\mu_{eff} = 6.05\ \mu_B$, was determined for $[Mn(C_3H_9OP)_5](ClO_4)_2$ by susceptibility measurements at room temperature. The complex shows IR bands of ν(PO) at 1145 and 1135 cm^{-1}, the free ligand only one band at 1166 cm^{-1}. Bands at 404 and 400 cm^{-1} were assigned to ν(Mn–O) vibrations; further bands were observed at 363, 320, and 255 cm^{-1} [1]. $[Mn(C_3H_9OP)_5](BF_4)_2$ shows a ν(Mn–O) band at 393 cm^{-1} [2]. For the $[Mn(C_3H_9OP)_5]X_2$ complexes no bands of coordinated anion groups were observed. A square-pyramidal geometry at Mn was proposed by comparison with the electronic spectra of the corresponding Ni^{II} and Co^{II} complexes. Preparation and handling of the complexes require a dry-box because the compounds are hygroscopic [1, 2]. The perchlorate decomposes in acetone solution [1].

$Mn^{II}(C_3H_9OP)X_2$ (X = Cl, Br). The white bromo complex was obtained by exposing an $MnBr_2$ film to trimethylphosphane oxide vapor for 13 h at room temperature, followed by evacuation for 2 h [3]. Formation of the compounds with X = Cl or Br was assumed [3, 4] after exposing the corresponding phosphane complexes, $Mn\{P(CH_3)_3\}X_2$, as thin films on KBr windows to dry dioxygen at 100 Torr under strictly anhydrous conditions. Based on the observed color change of white → blue → off-white and the IR bands, the formation of a dioxygen adduct, $Mn\{P(CH_3)_3\}Br_2 \cdot O_2$, as an intermediate was discussed [3, 4]. More recent data [5, 6], however, have shown that $Mn\{P(CH_3)_3\}X_2$ complexes (X = Cl, Br, I) may be oxidized with dry dioxygen to highly colored $Mn^{III}\{P(CH_3)_3\}X_3$ complexes; see p. 43.

References:

[1] Brodie, A. M.; Hunter, S. H.; Rodley, G. A.; Wilkins, C. J. (Inorg. Chim. Acta **2** [1968] 195/8).
[2] de Bolster, M. W. G.; Boutkan, C.; van der Knaap, T. A.; van Zweeden, L.; Kortram, I. E.; Groeneveld, W. L. (Z. Anorg. Allgem. Chem. **443** [1978] 269/78).
[3] Burkett, H. D.; Newberry, V. F.; Hill, W. E.; Worley, S. D. (J. Am. Chem. Soc. **105** [1983] 4097/9).
[4] Newberry, V. F.; Burkett, H. D.; Worley, S. D.; Hill, W. E. (Inorg. Chem. **23** [1984] 3911/7).
[5] Beagley, B.; McAuliffe, C. A.; Minten, K.; Pritchard, R. G. (J. Chem. Soc. Chem. Commun. **1984** 658/9).
[6] Beagley, B.; McAuliffe, C. A.; Minten, K.; Pritchard, R. G. (J. Chem. Soc. Dalton Trans. **1987** 1999/2003).

39.5.3 With Phosphoryltris(methanol) $(HOCH_2)_3PO$ (= $C_3H_9O_4P$)

$[Mn^{II}(C_3H_9O_4P)_6](ClO_4)_2$. Hydrated Mn^{II} perchlorate and the ligand were dissolved separately in the minimum amounts of triethyl orthoformate, stirred for 1 h at 60°C, and then combined in an Mn : ligand ratio of 1:6.2. After stirring for 30 min at 60°C, the solvent was evaporated slowly at 30 to 40°C. The viscous liquid was treated with ether or butanol to yield a soft solid or viscous liquid of pale yellow color, which was stored in an evacuated desiccator over $CaCl_2$. A magnetic moment, $\mu_{eff} = 6.08\ \mu_B$, was determined by susceptibility measurements at 297 K (Faraday method). The IR spectrum of the complex in Nujol shows bands of ν(OH) at 3390 and ν(PO) at 1100 cm^{-1}; these were observed for the free ligand at 3300 and 1130 cm^{-1}, respectively. A band at 624 cm^{-1} was assigned to ionic perchlorate groups. The ligand is coordinated to Mn^{II} through the phosphoryl oxygen atom, but not through the hydroxymethyl oxygen atoms. This is supported by the positive shift of the ν(OH) vibration upon complexation. In the electronic spectrum of the complex in Nujol the d–d transitions are masked by charge-transfer bands. The complex is very hygroscopic, soluble in nitromethane or polar solvents, sparingly soluble in alcohol, and insoluble in ether. The molar conductivity, $\Lambda = 125\ cm^2 \cdot \Omega^{-1} \cdot mol^{-1}$, of a 10^{-3} M nitromethane solution at 25°C is that of a 1:2 electrolyte [1].

$[Mn^{II}(C_3H_9O_4P)_4Cl_2]$ was prepared similarly to $[Mn(C_3H_9OP)_5](BF_4)_2$, see p. 84, using butanol as solvent. The white complex melts at 99 to 103°C. In the IR spectrum of the complex in Nujol the ν(OH) and ν(PO) bands were observed at 3300 and 1130 cm^{-1}, compared with bands at 3250 and 1145 cm^{-1} for the free ligand. This indicates that the phosphoryl oxygen of the ligand bonds to Mn and hydrogen bonding is present. Bands at 462 and 440 cm^{-1} were assigned to ν(Mn–O), a band at 240 cm^{-1} to ν(Mn–Cl). Corresponding Raman bands were observed at 470, 440, and at 227 cm^{-1}. The complex is very hygroscopic. It behaves as a nonelectrolyte in acetonitrile [2].

References:

[1] Mikulski, C. M.; Skryantz, J. S.; Karayannis, N. M.; Pytlewski, L. L.; Gelfand, L. S. (Inorg. Chim. Acta **27** [1978] 69/73).

[2] de Bolster, M. W. G.; Boutkan, C.; van der Knaap, T. A.; van Zweeden, L.; Kortram, I. E.; Groenefeld, W. L. (Z. Anorg. Allgem. Chem. **443** [1978] 269/78).

39.5.4 With Other Trialkylphosphane Oxides R_3PO

ligand	1	2	3	4
R	C_2H_5	C_3H_7	C_4H_9	C_8H_{17}
formula	$C_6H_{15}OP$	$C_9H_{21}OP$	$C_{12}H_{27}OP$	$C_{24}H_{51}OP$

Complexes in Solution. Mn^{2+} ions are extracted from aqueous solution into an organic phase by tributyl- or trioctylphosphane oxide. $Mn(C_{24}H_{51}OP)_3(NO_3)_2$ and $Mn(C_{24}H_{51}OP)_2Cl_2$ are formed on extraction of manganese(II) nitrate or chloride with a 0.7M solution of ligand 4 in carbon tetrachloride [1]. Extraction of Mn^{II} perchlorate into a solution of ligand 4 in hexane results in the formation of $Mn(C_{24}H_{51}OP)_4(ClO_4)_2$. The distribution ratio and the extraction constant at 25°C have been determined [2]. A mixed ligand complex, $Mn(C_{24}H_{51}OP)(HL)_2L_2$, is formed on extraction of Mn^{2+} from sulfate solution with bis(2-ethylhexyl) hydrogen phosphate, $(C_4H_9CH(C_2H_5)CH_2O)_2P(O)OH$ (= HL), see p. 155, into kerosine in the presence of trioctylphosphane oxide according to: $Mn(HL)_2L_2 + C_{24}H_{51}OP \rightarrow Mn(C_{24}H_{51}OP)(HL)_2L_2$. Its formation constant, log K = 3.24, was determined at 25°C [3]. Mixed ligand complexes are also formed by the synergistic extraction of Mn^{II} into cyclohexane with HL = $C_6H_5COCH_2COCF_3$ (= $C_{10}H_7F_3O_2$) or the thienyl compound $C_4H_3SCOCH_2COCF_3$ (= $C_8H_5F_3O_2S$), and trioctylphosphane oxide [4 to 6]. The adduct formation constants, K for the reaction $MnL_2 + C_{24}H_{51}OP \rightarrow Mn(C_{24}H_{51}OP)L_2$ and K′ for the reaction $MnL_2 + 2C_{24}H_{51}OP \rightarrow Mn(C_{24}H_{51}OP)_2L_2$, were determined by distribution measurements at I = 0.1M ($NaClO_4$) and room temperature: log K = 7.2 and log K′ = 12.0 for the complexes with HL = $C_{10}H_7F_3O_2$ [5]; log K = 5.7 [5] and log K′ = 10.8 [4, 5], 13.90 [6], and 10.65 (in benzene [6]) for the complexes with HL = $C_8H_5F_3O_2S$. The extraction was enhanced by trioctylphosphane oxide, which is more effective than tributyl phosphate [1, 3, 6].

$[Mn^{II}(C_{12}H_{27}OP)_4](ClO_4)_2$ precipitates after 5 to 15 min from a warmed (40 to 50°C) mixture containing stoichiometric amounts of manganese(II) perchlorate and ligand 3 dissolved in triethyl orthoformate. The white complex was washed with dry ether and kept in an evacuated desiccator over $Mg(ClO_4)_2$. It melts at 149.5 to 150°C. The X-ray powder diffraction pattern is similar to that of the corresponding Ni^{II} compound, for which a distorted tetrahedral geometry was proposed from its electronic spectrum. Susceptibility measurements at 297 K (Faraday method) show a magnetic moment of $\mu_{eff} = 5.83\ \mu_B$. The IR spectrum shows ν(PO) and ν(Mn–O) bands at 1130 and 454 cm^{-1}, respectively (ν(PO) of the free ligand at 1157 cm^{-1}). Bands observed at 1078 and 623 cm^{-1} are indicative of ionic perchlorate groups. Pure T_d symmetry is proposed for the $Mn-O_4$ moiety and the somewhat lower D_{2d} symmetry for the complex. The electronic spectrum of the solid complex in Nujol exhibits a broad absorption maximum between 500 and 415 nm and another band at 304 nm. A shoulder at 561 nm and maxima at 427 and 415 nm with ε = 0.6 and 0.8 $L \cdot mol^{-1} \cdot cm^{-1}$, respectively, were observed in the spectrum of a 10^{-3}M nitromethane solution. The molar conductivity of this solution, Λ = 185 $cm^2 \cdot \Omega^{-1} \cdot mol^{-1}$, is characteristic of a 1:2 electrolyte. The complex is hygroscopic [7].

$Mn^{II}(C_6H_{15}OP)X_2$ complexes with X = Cl or Br were reported to form (together with irreversible SO_2 adducts of $Mn\{P(C_2H_5)_3\}X_2$ complexes, see p. 57) on exposing $Mn\{P(C_2H_5)_3\}(O_2)X_2$ complexes to SO_2 at high temperature, but were characterized only by IR spectroscopy [8, 9].

[$Mn^{II}(C_9H_{21}OP)(NCS)_2$] and **[$Mn^{II}(C_{12}H_{27}OP)X_2$]** (X = I, NCS). The iodo complex was formed on keeping a tetrahydrofuran solution of $Mn\{P(C_4H_9)_3\}(O_2)I_2$, see p. 55, under a dioxygen pressure of 100 Torr at 60°C for 24 h. After removing the solvent a bluish oil remained. It was washed with dry toluene to yield the white-yellow solid complex. The isothiocyanato complexes were obtained from the Mn^{II} salt and ligand 2 or 3 by a procedure described in detail for the phosphane complex $Mn\{P(C_6H_5)_3\}(NCS)_2$, see p. 38. The magnetic moment at room temperature, $\mu_{eff} = 5.2\ \mu_B$ for both complexes, suggests some spin pairing. The IR spectra show the ν(PO) bands at 1150 cm^{-1} (X = I) or 1145 cm^{-1} (X = NCS), that of the free ligand 3 at 1175 cm^{-1}. The anion ν(CN) vibrations were observed at 2140 and 2100 cm^{-1}. A band at 296 cm^{-1} is tentatively assigned to ν(Mn–O) of the iodo complex. The complexes are nonelectrolytes in tetrahydrofuran or toluene. On heating [$Mn(C_{12}H_{27}OP)I_2$] in vacuum at 180°C a waxy sublimate of ligand 3 was obtained [10].

Mixed Ligand Complexes. $Mn^{II}(C_{12}H_{27}OP)_2(C_5HF_6O_2)_2$ was prepared by extracting a 0.1N acidic aqueous Mn^{II} solution with cyclohexane containing stoichiometric amounts of tributylphosphane oxide and $CF_3COCH_2COCF_3$ (= $C_5H_2F_6O_2$). The organic layer was dried with $MgSO_4$, the solvent was evaporated in vacuum at 40 to 50°C, and the residue was sublimed at 20 Torr. The complex melts at 37 to 38°C, yield 84%. Formation of the volatile complex is useful for the gas-chromatographic determination of trace Mn [11]. A similar procedure using trioctylphosphane oxide is reported in [12].

References:

[1] Rozen, A. M.; Skotnikov, A. S.; Martinova, M. K. (Zh. Neorgan. Khim. **29** [1984] 1798/802; Russ. J. Inorg. Chem. **29** [1984] 1031/3).
[2] Kusakabe, S.; Niitsu, M.; Sekine, T. (Proc. Symp. Solvent Extr. Metals **1981** 33/6; C.A. **99** [1983] No. 219653).
[3] Shen, J.; Jiang, D.; Sun, S. (Gaodeng Xuexiao Huaxue Xuebao **5** No. 1 [1984] 7/14; C.A. **100** [1984] No. 127567).
[4] Honjo, T.; Horiuchi, M.; Kiba, T. (Bull. Chem. Soc. Japan **47** [1974] 1176/80).
[5] Ueda, K.; Gokoh, M.; Yamamoto, Y. (Mem. Fac. Technol. Kanazawa Univ. **13** [1980] 83/91; C.A. **93** [1980] No. 102125).
[6] Nakamura, S.; Imura, H.; Suzuki, N. (J. Radioanal. Nucl. Chem. **81** No. 1 [1984] 33/44).
[7] Karayannis, N. M.; Mikulski, C. M.; Pytlewski, L. L.; Labes, M. M. (Inorg. Chem. **9** [1970] 582/7).
[8] Newberry, V. F.; Hill, W. E.; Worley, S. D. (Phosphorus Sulfur **27** [1986] 253/60).
[9] Newberry, V. F.; Burkett, H. D.; Worley, S. D.; Hill, W. E. (Inorg. Chem. **23** [1984] 3911/7).
[10] McAuliffe, C. A.; Al Khateeb, H. F.; Barratt, D. S.; Briggs, J. C.; Challita, A.; Hosseiny, A.; Little, M. G.; Mackie, A. G.; Minten, K. (J. Chem. Soc. Dalton Trans. **1983** 2147/53).

[11] Hellmuth, K. H.; Mirzai, H. (Z. Anal. Chem. **321** [1985] 49/55).
[12] Hellmuth, K. H. (Z. Anal. Chem. **321** [1985] 121/5).

39.5.5 With Triphenylphosphane Oxide $(C_6H_5)_3PO$ (= $C_{18}H_{15}OP$)

39.5.5.1 Manganese(II) Compounds

$Mn(C_{18}H_{15}OP)_4I_2$ and **[$Mn(C_{18}H_{15}OP)_4I$]I_3.** Formation of $Mn(C_{18}H_{15}OP)_4I_2$ was observed on keeping MnI_2 and an excess of the ligand in acetone overnight at 0°C. Some green crystals of $Mn(C_{18}H_{15}OP)_2I_2$ (see p. 90) were filtered off and the filtrate stored at 0°C for 3 d to yield plate-like colorless crystals of $Mn(C_{18}H_{15}OP)_4I_2$ [1]. On refluxing MnI_2 (2 mmol) and the ligand (12 mmol) in 50 mL of ethanol for 6 h the yellow-green $Mn(C_{18}H_{15}OP)_2I_2$ (p. 90) was formed in 43% yield. It was filtered off and the filtrate evaporated in vacuum. The dry residue was treated with 50 mL of toluene and the ivory white complex separated (yield 35%). $Mn(C_{18}H_{15}OP)_4I_2$ melts at 175°C. The complex has a magnetic moment of $\mu_{eff} = 6.3\,\mu_B$ at 298 K. The IR spectrum of the solid in Nujol shows the ν(PO) bands at 1185 and 1165 cm^{-1} [2]. ESR X-band measurements gave ambiguous structural information, showing only one line near $g_{eff} = 2$ [1]. The complex is soluble in ethanol, tetrahydrofuran, acetone, dichloromethane, slightly soluble in benzene, toluene, insoluble in ligroin. The conductivity of a 10^{-3} M nitrobenzene solution at 20°C is $\Lambda = 7.2\ cm^2 \cdot \Omega^{-1} \cdot mol^{-1}$. The complex reacts with $[Mn(CO)_5I]$ in tetrahydrofuran at 50°C according to $Mn(C_{18}H_{15}OP)_4I_2 + [Mn(CO)_5I] \rightleftharpoons [Mn(C_{18}H_{15}OP)_4I][Mn(CO)_4I_2] + CO$. The equilibrium can be reversed by bubbling CO through a tetrahydrofuran solution of $[Mn(C_8H_{15}OP)_4I][Mn(CO)_4I_2]$, see p. 89 [2].

$[Mn(C_{18}H_{15}OP)_4I]I_3$ is cited in [3]. Details of the preparation are not given, but the average bond lengths are Mn–I = 2.745(4) and Mn–O = 2.123 Å in the cation, which contains a pentacoordinated Mn atom [3]. These distances result from unpublished crystallographic data [4]. An X-ray structure determination of $[Mn(C_{18}H_{15}OP)_4I][Mn(CO)_4I_2]$, mentioned on p. 89, shows a square-pyramidal C_{4v} environment for Mn^{II} in the $[Mn(C_{18}H_{15}OP)_4I]^+$. A perspective view of the cation with the most relevant bond angles and distances in shown is **Fig. 20**. The Mn atom lies at 0.47(2) and 0.94(1) Å over the O_4 and P_4 planes, respectively. The longer-than-expected Mn–I distance suggests a highly ionic character for this bond [5].

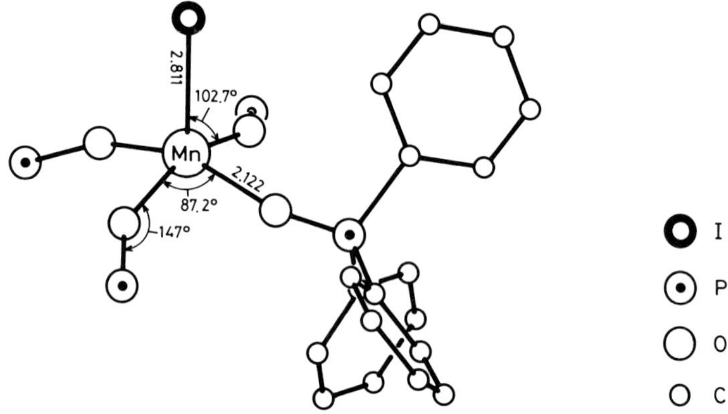

Fig. 20. A perspective view of the $[Mn(C_{18}H_{15}OP)_4I]^+$ ion. Only one of the triphenylphosphane oxide molecules has been drawn completely; hydrogen atoms are omitted [5].

$[Mn(C_{18}H_{15}OP)_4(ClO_4)]ClO_4$ was prepared by the reaction of Mn^{II} perchlorate and a slight excess of triphenylphosphane oxide in ethanol or acetone containing triethyl orthoformate. The white precipitate was washed with solvent and ether and dried in vacuum [6]. The complex was also obtained by evaporating methanol or ethanol solutions of $Mn(ClO_4)_2 \cdot 6H_2O$ (1 mmol) and the ligand (4.2 mmol) [7]. The magnetic moment, $\mu_{eff} = 5.97\ \mu_B$, was determined by susceptibility measurements at room temperature. X-ray powder photographs show that $[Mn(C_{18}H_{15}OP)_4(ClO_4)]ClO_4$ is isomorphous with the corresponding Fe^{II}, Co^{II}, and Ni^{II} complexes [6, 7] (for the Ni^{II} complex a square-pyramidal geometry with an axial monodentate ClO_4 group was made obvious by IR data [6], in contrast to a tetracoordinated species previously assumed by [7]).

The IR spectrum of the complex in Nujol shows a ν(PO) band at 1151 cm^{-1}, which is negatively shifted (Δν = −44 cm^{-1}) in comparison to the free ligand [1]. Similar results were observed by [6, 8]. The anion ν_1 band at 925 cm^{-1} and a splitting of the ν_3 and ν_4 bands (ν_3 observed at 1099 and 1044 cm^{-1}, ν_4 at 621 and 624 cm^{-1}) are in accord with the deductions from the ESR spectrum: At least one perchlorate ion is bonded to manganese, although no IR bands definitely attributable to metal anion bands could be identified down to 100 cm^{-1}. The ESR X-band spectrum is centered on g_{eff} = 2, but there are numerous resolved bands throughout the 0 to 9000 G range. A good correlation was found with D = 0.0875 cm^{-1} and λ = 0.067 [1]. The combination of the IR data with stereochemical considerations favors a binuclear oxide-bridged structure [9]. The electronic reflectance spectrum of the solid compound reveals absorption maxima at ~19600(sh), ~24500, and ~28400(sh) cm^{-1}. Absorption maxima of the dichloromethane solution observed at ~24500, 28400, 29400, and 30200 cm^{-1} with extinctions of ε = 1.2, 2.2, 2.7, or 2.9 L·mol^{-1}·cm^{-1}, respectively, were assigned to transitions calculated for a square-pyramidal complex with C_{4v} symmetry [10].

[Mn($C_{18}H_{15}OP$)$_4$(NCS)]NCS and **[Mn($C_{18}H_{15}OP$)$_4$(BF_4)]BF_4.** The isothiocyanate was prepared by mixing calculated quantities of the Mn^{II} salt and the ligand in ethanol [1]. A slight excess of the ligand in methanol containing trimethyl orthoformate was used for the preparation of the tetrafluoroborate [11]. The white precipitates were washed with the solvent, then with ether, and dried in vacuum [1, 11]. The fluoroborate complex is isomorphous with the corresponding Cd and Ni compounds. The ligand field spectral data of the Ni compound agree with a square-pyramidal arrangement of the ligands around the five-coordinated metal atom. The complex melts at 276 to 280°C [11]. The IR spectrum of [Mn($C_{18}H_{15}OP$)$_4$(NCS)]NCS in Nujol shows bands of ν(PO) at 1171, ν(C–N) at 2079 sh and 2067, ν(C–S) at 781, and ν(Mn–N) at 202 cm^{-1}. Bands at 401 and 316 cm^{-1} are considered to have some ν(Mn–O) character. Coordination of a monodentate N-bonded thiocyanato group was suggested. The ESR X-band spectrum of the complex does not allow making a distinction between five- and six-coordination. A strong band with a series of shoulders was observed in the g = 2 region. By comparison with calculated spectra the parameters D = 0.0375 cm^{-1} and λ = 0.067 have been deduced [1].

[Mn($C_{18}H_{15}OP$)$_4$X] [*cis*-Mn^I(CO)$_4X_2$] complexes with X = Cl, Br, or I are formed on reaction of [Mn^I(CO)$_5$X] compounds with triphenylphosphane oxide. The bromo complex has been obtained in good yields by reacting the components in carbon tetrachloride at room temperature. In the case of the iodo complex, [Mn(CO)$_5$I] and triphenylphosphane oxide were allowed to react at 60°C in tetrahydrofuran, benzene, or n-heptane for 8 h. Yellow-orange crystals separated on cooling the solution. The iodo complex was also formed by the reaction of [Mn(CO)$_5$I] with Mn($C_{18}H_{15}OP$)$_4I_2$ (see p. 88). The chloro complex has not been isolated, but the IR spectrum of the solution indicates that the reaction proceeds as in the case of the bromo and iodo derivatives [2].

The compounds will be treated in detail in the volumes describing organometallic compounds together with other dihalotetracarbonyl manganates(I). A perspective view of the [Mn($C_{18}H_{15}OP$)$_4$I]$^+$ cation is shown in Fig. 20.

[Mn($C_{18}H_{15}OP$)$_4$(NCSe)$_2$] was prepared under strictly anhydrous conditions. It was precipitated on addition of a hot solution of the ligand (1.79 g) in 20 mL dry ethanol to a solution of Mn(NCSe)$_2$·C_2H_5OH (0.5 g) in 50 mL of the same solvent. The IR spectrum of the complex in Nujol shows anion vibrations of ν(CN) at 2087sh, 2079, 2065sh and of ν(CSe) at 620 cm^{-1}; other bands are ν(Mn–O) at 406, 322, 302 and ν(Mn–N) at 255 cm^{-1}. The IR data are consistent with the presence of *cis* and *trans* isomers of the octahedral complex. An ESR X-band spectrum of powdered [Mn($C_{18}H_{15}OP$)$_4$(NCSe)$_2$] shows a large number of absorptions at room temperature in the 600 to 5050 G region, indicative of a small D value (<0.3 cm^{-1}). The complex must be handled with exclusion of moisture [12].

 References on pp. 96/7

[Mn($C_{18}H_{15}OP$)$_2$X$_2$] (X = NO_3, Cl, Br, I). To prepare the nitrato complex, a slight excess of the ligand dissolved in absolute ethanol was added to a solution of $Mn(NO_3)_2$ in aqueous ethanol (~50%). The mixture was heated to boiling, then cooled and placed in an evacuated desiccator over sulfuric acid for 3 d. The pink precipitate was recrystallized from chlorobenzene, washed with benzene, and dried in vacuum. The halo complexes precipitated from solutions of the appropriate Mn^{II} salt and the ligand (1:2.2 mole ratio) in hot absolute ethanol, were washed with cold absolute ethanol, and were dried in vacuum [13]. The bromo complex, synthesized in ethanol, crystallized nicely on slow evaporation of a methanol solution [14]. Preparation of the bromo complex by heating stoichiometric amounts of $MnBr_2$ and ligand dissolved in a small volume of glacial acetic acid was reported in [15]. [Mn($C_{18}H_{15}OP$)$_2$X$_2$] complexes with X = Cl, Br, or I were also obtained by stirring the MnX_2 salts (dehydrated at 130°C in vacuum) in dry tetrahydrofuran at 40°C together with the stoichiometric amount of triphenylphosphane, $C_{18}H_{15}P$, under nitrogen for 3 d. The hot mixtures were filtered and some ether was added. Crystals of the phosphane oxide complexes separated on standing and were dried in vacuum [16, 17]. [Mn($C_{18}H_{15}OP$)$_2$Cl$_2$] was also formed by fusing anhydrous $MnCl_2$ and triphenylphosphane [17], or by reaction of an excess of $MnCl_3$ dissolved in ether with a solution of triphenylphosphane in tetrahydrofuran. The mixture became colorless overnight. The solution was rotary-evaporated to dryness. The complex was extracted from the residue with hot dichloromethane [18]. A yellow-green [Mn($C_{18}H_{15}OP$)$_2$I$_2$], was obtained as by-product during the preparation of Mn($C_{18}H_{15}OP$)$_4$I$_2$, see p. 87 [1, 2].

X-ray structure determinations reveal the following data (a, b, c in Å, α, β, γ in °):

complex	[Mn($C_{18}H_{15}OP$)$_2$Cl$_2$]	[Mn($C_{18}H_{15}OP$)$_2$Br$_2$]	[Mn($C_{18}H_{15}OP$)$_2$I$_2$]	
lattice	orthorhombic	monoclinic	triclinic	
space group	Fdd2–D_{2v}^{19} (No. 43)	—	P1–C_i^1 (No. 1)	
a	20.990(3)	24.33	10.105(4)	α = 114.77(3)
b	33.028(6)	14.33 β = 93	10.471(4)	β = 116.29(3)
c	9.756(8)	13.18	10.703(5)	γ = 90.14(3)
D_{calc} in g/cm^3	1.340(1)*)	1.51	1.602	
Z	8	7	1	
Ref.	[19]	[20]	[3]	

*) Measured density D = 1.36(1) g/cm^3.

Atomic coordinates and equivalent isotropic thermal parameters for [Mn($C_{18}H_{15}OP$)$_2$Cl$_2$] are given in [19]. The structure was refined to R = 0.046. As shown in **Fig. 21** the Mn atom lies on a twofold axis and is tetrahedrally coordinated with distances Mn–O = 2.069(6), Mn–Cl = 2.294(3) Å and angles O–Mn–Cl = 112.1(2)°, Mn–O–P = 156.0(2)° [19]. [Mn($C_{18}H_{15}OP$)$_2$Br$_2$] was preliminarily investigated by [20].

The structure of [Mn($C_{18}H_{15}OP$)$_2$I$_2$] was refined to R = 0.029. Fractional atomic coordinates and vibrational parameters for nonhydrogen atoms are given in [3]. The structure consists of isolated asymmetrical molecules, shown in **Fig. 22**, related only by the translational symmetry afforded by the P1 triclinic space group. Selected bond lengths (in Å) and angles (in °) are shown below:

Mn(1)–I(1)	2.670(1)	I(1)–Mn(1)–I(2)	113.8(1)	O(1)–Mn(1)–O(2)	103.9(2)
Mn(1)–O(1)	2.014(5)	I(1)–Mn(1)–O(1)	113.1(1)	I(2)–Mn(1)–O(1)	110.0(1)
Mn(1)–I(2)	2.666(1)	I(1)–Mn(1)–O(2)	104.5(1)	I(2)–Mn(1)–O(2)	110.9(2)
Mn(1)–O(2)	2.027(5)	Mn(1)–O(1)–P(1)	157.2(3)	Mn(1)–O(2)–P(2)	148.7(4)

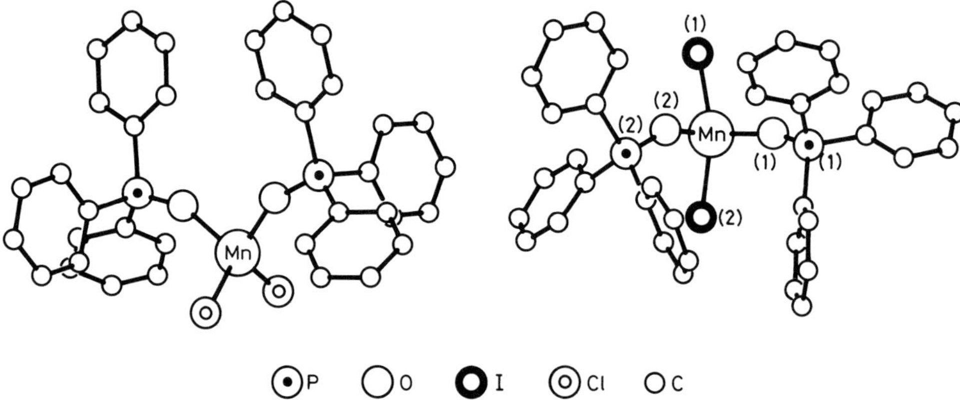

Fig. 21. Molecular structure of $[Mn(C_{18}H_{15}OP)_2Cl_2]$. Hydrogen atoms are omitted [19].

Fig. 22. Molecular structure of $[Mn(C_{18}H_{15}OP)_2I_2]$. Hydrogen atoms are omitted [3].

The mean Mn–I distance of 2.668(1) Å is in the range of values covered by other Mn–I bonds in tetrahedral environments. The mean Mn–O distance, 2.021(4) Å, is shorter than in an octahedral environment of Mn (e.g., 2.177 Å in $MnI_2 \cdot 4H_2O$ or 2.24 Å in MnI_2 when dissolved in tetrahydrofuran). The variability of the O–P–C–C torsion angles (see the paper) demonstrate the asymmetry of the molecule [3]. X-ray powder diffraction patterns of $[Mn(C_{18}H_{15}OP)_2Cl_2]$ and $[Mn(C_{18}H_{15}OP)_2Br_2]$ show them to be isomorphous with their Co^{II} and Ni^{II} analogs, respectively [13]. Melting points (in °C) and magnetic moments at room temperature (in μ_B) of the pale yellow $[Mn(C_{18}H_{15}OP)_2X_2]$ complexes are tabulated below, together with absorption bands of the ν(PO) vibration mode; ν(PO) of the free ligand was observed at 1195 cm^{-1}.

X	m.p.	μ_{eff}	ν(PO) in cm^{-1}	Ref.	X	m.p.	μ_{eff}	ν(PO) in cm^{-1}	Ref.
NO_3	236	6.10	1189	[13]	Br	243	6.02	1163sh, 1155	[13]
						265	–	1160sh, 1150	[17]
Cl	244	5.29	1155	[13]		243	–	1163sh, 1152	[14]
	234	–	1152	[17]		243	6.0	1163, 1150	[2]
	239	6.0	1170, 1153	[2]					
	–	–	1152	[18]	I	230	6.02	1161, 1148	[13]
						227	–	1160, 1148	[17]
						229	5.9	1160, 1145	[2]

Coordination of monodentate nitrato groups is indicated by bands observed at 1499, 1292, 1031, and 821 cm^{-1} [13]. Bands of the ν(Mn–X) vibrations are observed at 310 and 290 [17], at 327, 317, and 292 cm^{-1} [21] for X = Cl; at 240 and 225 [17], at 248 and 229 cm^{-1} [21] for X = Br.

The electronic spectra of the complexes with X = Cl, Br, or I provide evidence for a tetrahedral geometry around Mn, which, however, is not confirmed for the nitrato complex. The reflectance spectrum of solid $[Mn(C_{18}H_{15}OP)_2(NO_3)_2]$ shows maxima (not assigned) at 24390(sh), 28010(sh), 29400, and 31250 cm^{-1} [13]. Bands observed in the reflectance spectra of the solid halogeno complexes, in the absorption spectra of their acetonitrile solutions, and in the excitation spectra are summarized in table on p. 92.

References on pp. 96/7

Electronic Spectral Data of $[Mn(C_{18}H_{15}OP)_2Cl_2]$ Complexes.
Transitions from $^6A_1(S)$; ν_{max} in cm^{-1}.

complex	spectra	$^4T_{1g}(G)$	$^4T_{2g}(G)$	$^4A_{1g}, ^4E_g(G)$	$^4T_{2g}(D)$	$^4E_g(D)$	$^4A_{2g}(F)$	Ref.
$[Mn(C_{18}H_{15}OP)_2Cl_2]$	I[a]	21280 sh	22570	23260 sh	26670 sh[e]	28570[e]	30300[e]	[13]
$[Mn(C_{18}H_{15}OP)_2Br_2]$	I[a]	21500	22470	23260 sh	26670[e]	28410[e]	30300 sh[e]	[13]
	I[a]	21500	22600	23500				[22]
	II[b]	21834 (0.66)	22727 (0.74)	23584 (0.45)	26738 (0.68)	28325 (2.15)		[15, 16]
	III[c]	20880, 21300	22620	22950, 23310	26600, 27170	28010		[22]
$[Mn(C_{18}H_{15}OP)_2I_2]$	I[a]	21050 sh			26670[e]		30300 sh[e]	[13]
	I[a]	21400	22100	22800				[22]
	II[b]		22500		27300			[22]
	III[d]	20320, 20980	21950	22900	26450	27030	32950, 33400	[22]

[a] Reflectance spectrum of the solid complex. – [b] Absorption spectrum of the acetonitrile solution with extinction coefficients in parentheses. – [c] Excitation spectrum monitored at 19420 cm^{-1}. – [d] Excitation spectrum monitored at 19230 cm^{-1}. – [e] Assignments made by [22].

$[Mn(C_{18}H_{15}OP)_2I_2]$ shows additional bands at 34100 and 36100 cm^{-1} in the acetonitrile solution spectrum [22] and at 34900 and 35500 cm^{-1} in the excitation spectrum [22], which were assigned to the transitions $^6A_1(S) \rightarrow {}^4T_{1g}(F)$ or $\rightarrow {}^4T_{2g}(F)$, respectively.

Due to the tetrahedral geometry at Mn^{II} the solid $[Mn(C_{18}H_{15}OP)_2X_2]$ complexes with X = Br or I are photo- and triboluminescent at room temperature, whereas the complex with X = Cl is not photoluminescent. It shows triboluminescence only on cooling to 80 K [12]. The triboluminescence is centered at 520 nm (= 19230 cm^{-1}) for X = Cl, at 500 nm (= 20000 cm^{-1}) for X = Br [13]. The triboluminescence spectrum of $[Mn(C_{18}H_{15}OP)_2Br_2]$ is superimposed on the photoluminescence spectrum assigned to the $^4T_1 \rightarrow {}^6A_1$ transition [23, 24]. The photoluminescence is centered at 19400 cm^{-1} (X = Br), at 19200 cm^{-1} (X = I), revealing a Stoke's shift, Φ = 1480 and 1120 cm^{-1}, respectively ($\Phi = \nu_{abs} - \nu_{lum}$), which is in the range of tetrahedral complexes [22]. The emission lifetime of $[Mn(C_{18}H_{15}OP)_2Br_2]$ is 602 μs [23]. Photo- and triboluminescence spectra of complexes with X = Cl or Br are plotted in [23 to 25]. The effect of pressure on the energy of the photoluminescence emission maximum of $[Mn(C_{18}H_{15}OP)_2Br_2]$, plotted in [23, 25], shows the maximum shifted from 19400 to 18300 cm^{-1} at 36 kbar. The effect of temperature on photo- and triboluminescence was investigated. The triboluminescence of the chloro and bromo complexes disappear at 100 and 200°C, respectively, much lower than the melting point temperature (see above) [14, 25 to 27]. A decrease in spectral efficiency with temperature is caused by lower quantum yield due to increased radiationless transition and by changing noncentrosymmetric crystal sites into centrosymmetric sites. Electroluminescence spectra of complexes with X = Cl or Br are almost the same as the triboluminescence spectra, showing a slight shift towards shorter wavelengths. The intensity increases with the increase of both frequency (1 to 5000 Hz) and applied electric field (100 to 1000 V) [25, 27].

ESR X- and Q-band spectra of powdered $[Mn(C_{18}H_{15}OP)_2X_2]$ with X = Cl, Br, I [28] or of Mn^{II}-doped single crystals of $[Zn(C_{18}H_{15}OP)_2X_2]$ with X = Cl [29] or Br [30] show band positions and a hyperfine structure depending mainly on the X group. The spectra were not altered by dilution, i.e., by incorporation of Mn^{II} in the corresponding Zn complexes [28]. Spin-Hamiltonian (zero-field splitting) parameters D, E (in cm^{-1}), λ (E/D) and g values are tabulated below. They gave good agreement between observed and calculated transitions of both X-band (~9 GHz) and Q-band (~35 GHz) spectra:

complex	g_{eff} [a)]	g_{eff} [b)]	D	λ	Ref.
$[(Mn)(C_{18}H_{15}OP)_2Cl_2]$	4.3	2	0.29	0.244	[28]
$[(Mn)(C_{18}H_{15}OP)_2Br_2]$	4.3	4.3	0.425	0.267	[28]
$[(Mn)(C_{18}H_{15}OP)_2I_2]$	4.3, 2	4.3	0.81	0.222	[28]

a) From X-band spectrum – b) From Q-band spectrum.

complex	g_x	g_y	g_z	D	E	Ref.
$[Zn(Mn)(C_{18}H_{15}OP)_2Cl_2]$	2.03	2.02	2.02	0.172	0.047	[29]
$[Zn(Mn)(C_{18}H_{15}OP)_2Br_2]$	2.020	2.019	2.017	0.523	0.133	[30]

The λ values are close to 1/3, as predicted for tetrahedral C_{2v} complexes. D increases in the order of size of the X group, which is the order of increasing electron release from X [28]. A more detailed analysis is given by ESR X- and Q-band studies at 298 K of Mn^{II} doped into $[Zn(C_{18}H_{15}OP)_2X_2]$ single crystals with an orthorhombic (X = Cl) [29] or triclinic (X = Br) [30] lattice. A calculation of sign and magnitude of intrinsic second-order zero-field splitting parameters ($\bar{\beta}_2$), which are quite sensitive with respect to small structural changes, were calculated for $[Mn(C_{18}H_{15}OP)_2X_2]$ complexes with X = Cl or Br from Zn^{II}-host crystal structure data and compared with experimental ESR data [31]. The ^{31}P NMR spectrum of $[Mn(C_{18}H_{15}OP)_2Br_2]$ was measured in acetone and relaxation times T_1 and T_2 were determined [32].

References on pp. 96/7

The $[Mn(C_{18}H_{15}OP)_2X_2]$ complexes with X = NO_3, Cl, Br, or I are nonelectrolytes in nitromethane [13, 17] or nitrobenzene [13, 16]. The nitrato complex is soluble in alcohols, acetone, dioxane, chloroform, acetonitrile, and hot chlorobenzene, but insoluble in ethyl acetate, benzene, cyclohexane, ligroin, or carbon tetrachloride. The complexes with X = Cl, Br, and I are soluble in acetone or acetonitrile but only very slightly soluble in other organic solvents [13].

$[Mn(C_{18}H_{15}OP)_2(NCS)Cl]_2$. An X-ray structure determination of the colorless complex (the preparation is not given) reveals an orthorhombic lattice, Pbac–D_{2h}^{15} (No. 61), with a = 10.992(3), b = 22.331(2), c = 27.722(4) Å; Z = 8; The structure was refined to R = 0.102. Atomic coordinates and equivalent isotropic thermal parameters are tabulated in the paper. Selected bond lengths (in Å) and angles (in °) are:

Mn–O(1)	2.020(9)	Cl(1)–Mn–O(1)	133.2(3)	O(2)–Mn–N	96.5(4)
Mn–O(2)	2.110(8)	Cl(1)–Mn–N	119.1(3)	Cl(1)–Mn–Cl(1′)	82.0(1)
Mn–N	2.08(1)	O(1)–Mn–N	107.3(4)	Cl(1′)–Mn–O(1)	88.5(2)
Mn–Cl(1)	2.468(3)	Cl(1)–Mn–O(2)	90.0(2)	Cl(1′)–Mn–N	95.2(3)
Mn–Cl(1′)	2.600(4)	O(1)–Mn–O(2)	90.6(3)	Cl(1′)–Mn–O(2)	168.0(2)

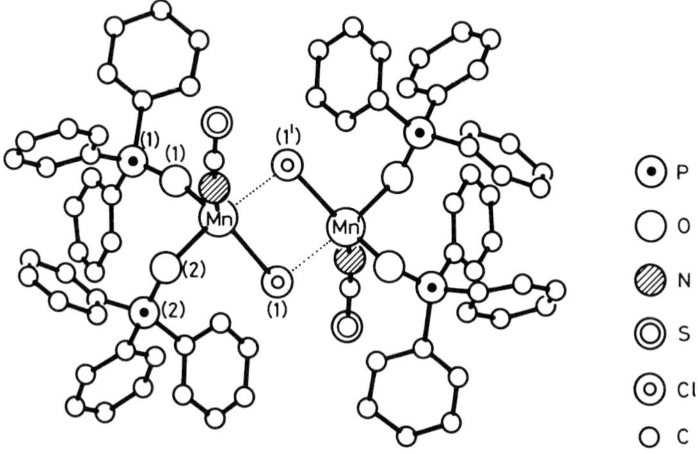

Fig. 23. Molecular structure of the dimeric unit, $[Mn(C_{18}H_{15}OP)_2(NCS)Cl]_2$. Hydrogen atoms are omitted [33].

Pairs of molecules are linked by chloro bridges to form dimeric units, see **Fig. 23.** Because of this dimerization, the Mn atom assumes a distorted trigonal-bipyramidal geometry. The Mn atom is 0.08 Å out of the equatorial plane formed by the O(1), N, and Cl(1) atoms. Pairs of the bipyramids share Cl(1)–Cl(1′) edges. The density D = 1.37(1) g/cm^3 was determined by flotation in a CH_3OH–$CHBr_3$ mixture; D_{calc} = 1.376(5) g/cm^3 [33].

$Mn(C_{18}H_{15}OP)(NCS)_2$ was prepared from $Mn(NCS)_2$ and triphenylphosphane oxide by the procedure referred to for the corresponding triphenylphosphane complex, see p. 38. The pale yellow compound melts at 226°C. The susceptibility, independent of temperature, reveals a magnetic moment of μ_{eff} = 5.5 μ_B. The IR spectrum of the complex in Nujol shows bands at 2100 and 2075 cm^{-1}, giving evidence of S- and N-bonded, but not of bridging anion groups. Further bands are assigned as follows: ν(PO) at 1140, δ(NCS) at 420, ν(Mn–S) at 300, and ν(Mn–N) at 254 cm^{-1}. A three-coordinated structure is proposed with the ligand bonded through the phosphoryl oxygen atom. The diffuse reflectance spectrum shows maxima at 434 and 360 nm. The compound is a nonelectrolyte in dichloromethane solution [34].

Mixed Ligand Complexes. The $[Mn(C_{18}H_{15}OP)_4(SO_2)_2]I_2$ complex was prepared by reaction of stoichiometric amounts of anhydrous MnI_2 and triphenylphosphane oxide, stirred in propyl acetate for 7d. The beige suspension was then saturated with gaseous SO_2 and the orange suspension stirred for 2d. The solid was filtered off and recrystallized from toluene. $[Mn(C_{18}H_{15}OP)_4(SO_2)]I_2$ was obtained by keeping $[Mn(C_{18}H_{15}OP)_4(SO_2)_2]I_2$ under vacuum for 2 to 3 h or by heating to 129°C. $[Mn(C_{18}H_{15}OP)_4(SO_2)]I_2$ melts at 148 to 150°C without further loss of sulfur dioxide. Upon exposure of $[Mn(C_{18}H_{15}OP)_4(SO_2)]I_2$ to SO_2 in the solid state or in propyl acetate solution, $[Mn(C_{18}H_{15}OP)_4(SO_2)_2]I_2$ is reformed. The reaction is reversible.

An X-ray structure determination of $[Mn(C_{18}H_{15}OP)_4(SO_2)_2]I_2$ shows a triclinic lattice, space group $P\bar{1}-C_i^1$ (No. 2), with lattice constants a = 13.426(7), b = 13.022(8), c = 13.863(11) Å, α = 125.5(1)°, β = 120.1(1)°, γ = 83.2(1)°; Z = 1. The structure was refined to R = 0.050. Bond lengths (in Å) and angles (in °) are:

Mn–O(1)	2.084(3)	P(1)–O(1)	1.498(3)	Mn–O(3)–S	146.9(3)
Mn–O(2)	2.147(3)	P(2)–O(2)	1.485(3)	O(3)–S–O(4)	116.2(4)
Mn–O(3)	2.282(4)	S–O(3)	1.448(4)		
Mn–S	3.59	S–O(4)	1.410(6)		

The Mn atom is at a center of symmetry with monodentate oxygen-bonded sulfur dioxide groups at the apical positions of an octahedron, see **Fig. 24.** Distances involving the I^- anion are: I–O(3) = 3.39, I–O(4) = 3.345, I–S = 2.859(2) Å. The short cation-anion interaction between sulfur and iodine atoms is less than the sum of their van der Waals radii of 4.0 Å. The complex has a calculated density of D = 1.55 g/cm³. The IR spectra of the complexes show bands assignable to the ν(SO) vibration at 1275 and 1234 cm^{-1} for $[Mn(C_{18}H_{15}OP)_4(SO_2)_2]I_2$ and at 1275 and 1260 cm^{-1} for $[Mn(C_{18}H_{15}OP)_4(SO_2)]I_2$ [35].

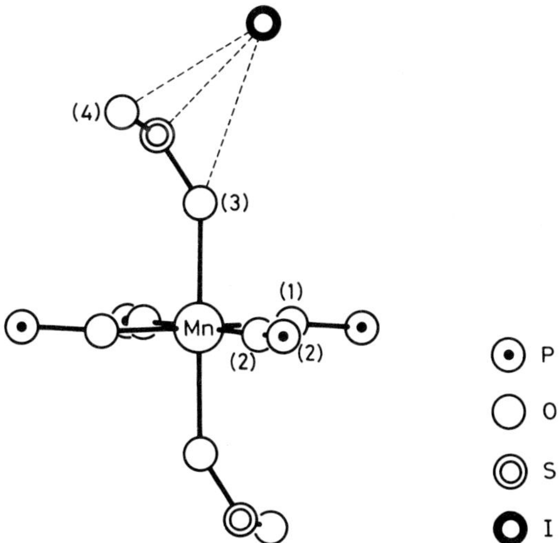

Fig. 24. Molecular structure of $[Mn(C_{18}H_{15}OP)_4(SO_2)_2]I_2$. The phenyl groups have been omitted and the position of only one iodine anion is shown [35].

References on pp. 96/7

References:

[1] Goodgame, D. M. L.; Goodgame, M.; Hayward, P. J. (J. Chem. Soc. A **1970** 1352/6).
[2] Sartorelli, V.; Canziani, F.; Garlaschelli, L. (Gazz. Chim. Ital. **104** [1974] 567/80).
[3] Beagley, B.; McAuliffe, C.A.; Pritchard, R. G.; White, E. W. (Acta Chem. Scand. A **42** [1988] 544/53).
[4] Al-Farhan, K.; Beagley, B.; McAuliffe, C.A.; Pritchard, R. G. (unpublished data, cited in [3]).
[5] Ciani, G.; Manassero, M.; Sansoni, M. (J. Inorg. Nucl. Chem. **34** [1972] 1763/7).
[6] Hunter, S. H.; Nyholm, R. S.; Rodley, G. A. (Inorg. Chim. Acta **3** [1969] 631/4).
[7] Bannister, E.; Cotton, F. A. (J. Chem. Soc. **1960** 1878/82).
[8] Cotton, F. A.; Barnes, R. D.; Bannister, E. (J. Chem. Soc. **1960** 2199/203).
[9] Karayannis, N. M.; Mikulski, C. M.; Strocko, M. J.; Pytlewski, L. L.; Labes, M. M. (J. Inorg. Nucl. Chem. **33** [1971] 2691/3).
[10] Ciampolini, M.; Mengozzi, C. (Gazz. Chim. Ital. **104** [1974] 1059/66).

[11] de Bolster, M. W. G.; Kortram, I. E.; Groeneveld, W. L. (J. Inorg. Nucl. Chem. **34** [1972] 575/80).
[12] Benson, C. G.; Little, M. G.; McAuliffe, C. A. (Inorg. Chim. Acta **87** [1984] 169/75).
[13] Goodgame, D. M. L.; Cotton, F. A. (J. Chem. Soc. **1961** 3735/41).
[14] Chandra, B. P.; Deshmukh, N. G.; Jaiswal, A. K. (Mol. Cryst. Liquid Cryst. **142** [1987] 157/72).
[15] Negoiu, D.; Furlani, C. (Atti Accad. Nazl. Lincei Classe Sci. Fis. Mat. Nat. Rend. [8] **35** [1963] 58/67).
[16] Negoiu, D. (Analele Univ. Bucuresti Ser. Stiint. Nat. Chim. **14** No. 2 [1965] 145/50; C.A. **67** [1967] No. 60439).
[17] Casey, S.; Levason, W.; McAuliffe, C.A. (J. Chem. Soc. Dalton Trans. **1974** 886/9).
[18] Levason, W.; McAuliffe, C.A. (J. Inorg. Nucl. Chem. **37** [1975] 340/2).
[19] Tomita, K. (Acta Cryst. C **41** [1985] 1832/3).
[20] Negoiu, D.; Ciomârtan, D. (Rev. Roumaine Chim. **11** [1966] 349/55).

[21] Allan, J. R.; Brown, D. H.; Nuttall, R. R.; Sharp, D. W. A. (J. Inorg. Nucl. Chem. **27** [1965] 1305/9).
[22] Furlani, C.; Cervone, E.; Cancellieri, P. (Atti Accad. Nazl. Lincei Classe Sci. Fis. Mat. Nat. Rend. [8] **37** [1964] 446/56).
[23] Hardy, G. E.; Zink, J. I. (Inorg. Chem. **15** [1976] 3061/5).
[24] Hardy, G. E.; Baldwin, J. C.; Zink, J. I.; Kaska, W. C.; Liu, P. H.; Dubois, L. (J. Am. Chem. Soc. **99** [1977] 3552/8).
[25] Chandra, B. P.; Kaza, B. R. (J. Lumin. **27** [1982] 101/7).
[26] Chandra, B. P.; Khokhar, M. S. K.; Gupta, R. S.; Majumdar, B. (Pramana **29** [1987] 399/407; C.A. **107** [1987] No. 245703).
[27] Chandra, B. P.; Jaiswal, A. K.; Chandraker, T. R.; Kaza, B. R. (Phys. Status Solidi A **68** [1981] K207/K211).
[28] Dowsing, R. D.; Gibson, J. F.; Goodgame, D. M. L.; Goodgame, M.; Hayward, D. J. (J. Chem. Soc. A **1969** 1242/8).
[29] Vivien, D.; Gibson, J. F. (J. Chem. Soc. Faraday Trans. II **71** [1975] 1640/53).
[30] Kosky, C. A.; Gayda, J. P.; Gibson, J. F.; Jones, S. F.; Williams, D. J. (J. Inorg. Chem. **21** [1982] 3173/9).

[31] Heming, M.; Remme, S.; Lehmann, G. (J. Magn. Resonance **69** [1986] 134/43).
[32] Bertini, I.; Luchinat, C.; Scozzafava, A. (Inorg. Nucl. Chem. Letters **15** [1979] 89/91).

[33] Tomita, K. (Acta Cryst. C **43** [1987] 1628/30).
[34] Hosseiny, A.; Mackie, A. G.; McAuliffe, C. A.; Minten, K. (Inorg. Chim. Acta **49** [1981] 99/105).
[35] Gott, G. A.; Fawcett, J.; McAuliffe, C. A.; Russell, D. R. (J. Chem. Soc. Chem. Commun. **1984** 1283/4).

39.5.5.2 Manganese(III) Compounds

$[Mn(C_{18}H_{15}OP)_3Cl_3]$ was reported to form on addition of a cold solution of triphenylphosphane oxide dissolved in absolute ethanol to a solution of $MnCl_3$ in ether, free of HCl and kept below −30°C under dry conditions. The blue precipitate turns white at 142°C due to decomposition and reduction to Mn^{II}. The white solid melts at 243°C [1, 2]. The magnetic susceptibilities, measured with the Faraday method at five different temperatures between 70 and 300 K, follow the Curie-Weiss law with $\Theta = -2.63$ K. A magnetic moment of $\mu_{eff} = 5.37$ μ_B has been calculated [2]. The IR spectra of the complex in Nujol show the bands of ν(PO) at 1167 and 1155 [1] or 1160 and 1145 cm^{-1} [2]. Bands at 450, 425 [1] or 410 cm^{-1} [2] were assigned to ν(Mn–O), at 340 cm^{-1} to ν(Mn–Cl) vibrations [2]. The electronic spectrum of the complex in $CHCl_3$ shows a maximum at 17240 cm^{-1}, assigned to the transition $^5B_{1g} \rightarrow {}^5B_{2g}$ ($\varepsilon = 560$ L·mol^{-1}·cm^{-1}), or at 23810 cm^{-1} in pyridine solution. A distorted octahedral environment at Mn is proposed. A molecular weight of 972.9 was determined in benzene solution (theoretical value 996.2). The complex behaves as a nonelectrolyte in nitromethane, acetone, or pyridine. It is soluble in most organic solvents, slightly soluble in hexane, cyclohexane, ether, or carbon tetrachloride. The complex hydrolyzes in water or alkaline solution to form MnO_2; solutions in acidic aqueous media change color from blue to pale pink [1].

$Mn(C_{18}H_{15}OP)_2(NO_3)_3$ was precipitated in the form of dark red crystals on addition of triphenylphosphane oxide to the dark brown solution obtained as a by-product of the synthesis of $K_2[Mn(NO_3)_4]$ by reaction of $KMnO_4$ with N_2O_4 in nitromethane. The structure of the complex is not known. From the molar susceptibility of $\chi_{corr} = 10150 \times 10^{-6}$ cm³/mol, determined by the Faraday method at 22°C, a magnetic moment of $\mu_{eff} = 4.9 \pm 0.1$ μ_B, was calculated. The IR spectrum reveals a ν(PO) band at 1150 cm^{-1}. Bands at 1552sh, 1520, 1265, 968, 953, 798, and 787 cm^{-1} may be assigned to nonequivalently coordinated nitrate groups. The complex dissolves in benzene or acetonitrile to give yellow solutions and is insoluble in carbon tetrachloride, nitromethane, and water [3].

$[Mn(C_{18}H_{15}OP)_2Cl_3]$ was obtained by mixing the ligand and the dioxane solvate, $MnCl_3 \cdot 2C_4H_8O_2$, both dissolved in dry ethanol. The reaction was carried out at −40°C under nitrogen with exclusion of daylight and moisture. The precipitate formed after stirring for 4 h was washed with ether at −40°C and dried in vacuum. The deep blue complex melts at 159°C with decomposition [4]. Magnetic susceptibilities determined by the Faraday method between 300 and 65 K reveal magnetic moments of $\mu_{eff} = 4.96$ to 5.00 μ_B and the Weiss constant, $\Theta = -1.37$ K [4].

The IR spectrum shows ν(PO) vibrations at 1170 and 1160 cm^{-1}; bands at 430, 365, 325, 295, and 275 cm^{-1} are attributed to ν(Mn–O) and ν(Mn–Cl) vibrations, but they are not useful for structural assignment. The UV spectra of the complex in the solid state as well as in solution were investigated. The diffuse reflectance spectrum shows maxima at 15600 (d–d transition), 24400, and 29900 cm^{-1} (charge-transfer transition); a solution in dichloromethane shows maxima at 16200, 27800, and 30300sh cm^{-1} with $\varepsilon = 500$, 2860, and 2540 L·mol^{-1}·cm^{-1}, respectively. The complex with pentacoordinated Mn is stable at room temperature. It is soluble in nitrobenzene and behaves as a nonelectrolyte in nitrobenzene. A molecular weight of 704 was determined cryoscopically in nitrobenzene (theoretical value 714). The complex decomposes when exposed to daylight [4].

Mixed ligand complexes of composition **K[Mn($C_{18}H_{15}$OP)($C_6X_4O_2$)$_2$]·(CH_3)$_2$CO·H_2O,** containing tetrahalocatecholato anions, $C_6X_4O_2^{2-}$, were prepared as follows: For X = Br, 1 mmol of tetrabromo-1,2-benzoquinone in 15 mL water, 2 mmol of triphenylphosphane in 10 mL acetone, and 2 mmol of KOH in 5 mL water were added to a solution of $MnCl_2$·4 H_2O (1 mmol) in 5 mL water. The mixture was refluxed for 15 min and reduced in volume. The yellow-green precipitate (yield 35%) was recrystallized from aqueous acetone to give the solvated complex. The complex was also formed by combining a suspension of [Mn($C_{18}H_{15}$OP)$_2$$Cl_2$] (1 mmol), see p. 90, in 50 mL acetone, tetrabromocatechol (= $C_6H_2Br_4O_2$) (2 mmol) in 20 mL ethanol, and KOH (4 mmol) in 6 mL water. The mixture was heated to reflux and reduced in volume to separate crystals (yield 73%). The complex with X = Cl was formed by both the above reactions, using $C_6Cl_4O_2$ or $C_6H_2Cl_4O_2$, the complexes with X = Cl or Br have the same structure, according to X-ray analysis, magnetic and IR data [5].

An X-ray structure determination of K[Mn($C_{18}H_{15}$OP)($C_6Br_4O_2$)$_2$]·(CH_3)$_2$CO·H_2O reveals the monoclinic space group P2$_1$/n–C_{2h}^5 (No. 14) with a = 10.256(2), b = 26.434(3), c = 14.965(2) Å, β = 102.23(1)°; Z = 4. The structure was refined to R = 0.059. Selected bond distances (in Å) and angles (in °) are:

Mn–O(1)	1.913 (0.013)	O(2)–Mn–O(1)	85.6 (0.6)	O(3)–Mn–O(1)	174.6 (0.6)
Mn–O(3)	1.897 (0.013)	O(3)–Mn–O(2)	94.6 (0.6)	O(4)–Mn–O(1)	93.4 (0.6)
Mn–O(5)	2.135 (0.015)	O(4)–Mn–O(2)	168.4 (0.6)	O(4)–Mn–O(3)	85.3 (0.6)
Mn–O(2)	1.888 (0.013)	O(5)–Mn–O(1)	93.5 (0.6)	O(5)–Mn–O(2)	98.0 (0.6)
Mn–O(4)	1.909 (0.012)	O(5)–Mn–O(3)	91.8 (0.6)	O(5)–Mn–O(4)	93.5 (0.6)
Mn–O(4′)	2.723 (0.013)				
Mn–Mn′	3.533 (0.008)				

The coordination geometry about each Mn^{3+} ion is distorted octahedral with catecholate ligands chelated to the metal in a planar arrangement and the phosphane oxide bonded in an apical site. The sixth coordination site at Mn is occupied by the catecholato oxygen of the adjacent complex anion. The result is a dimeric structure with the [Mn($C_{18}H_{15}$OP)($C_6Br_4O_2$)$_2$]$^-$ ions in a face-to-face arrangement, see **Fig. 25** a. The Mn–O length is short for Mn^{III}. The separation between planes of catecholato ligands in the dimeric unit is approximately 3.3 Å, a value commonly associated with charge-transfer interactions between planar unsaturated organic molecules. The bridging O(4) atom lies 2.71 Å off the plane defined by the atoms of the catecholate containing O(1) and O(2). Potassium ions interact with the catecholate oxygen atoms, forming a polymeric crystal structure consisting of cation and anion pairs, see Fig. 25 b. Atomic positional and isotropic thermal parameters are tabulated in the paper. The calculated and measured densities of the complex are 2.171 and 2.16(2) g/cm^3 [5].

A magnetic moment, μ_{eff} = 4.89 μ_B, was determined for both complexes from susceptibility measurements at room temperature (Faraday method). IR spectra of the complexes in KBr pellets show bands at 1360, 1240, and 1120 cm^{-1} associated with the catecholate anions, at 1425 and 1170 cm^{-1} with the phosphane oxide, and at 1780 cm^{-1} with the acetone molecule. The electronic spectra of the complexes in acetonitrile or dichloromethane solution are identical with maxima at 640, 340, 294, and 248 nm with ε = 110, 8500, 20000, or 36000 L·mol^{-1}·cm^{-1}, respectively. K[Mn($C_{18}H_{15}$OP)($C_6Cl_4O_2$)$_2$]·(CH_3)$_2$CO·H_2O dissolved in acetonitrile or dichloromethane shows an irreversible one-electron oxidation at +0.30 V (vs. ferrocene/ferrocene$^+$), which is coupled with a reduction at −0.18 V [5].

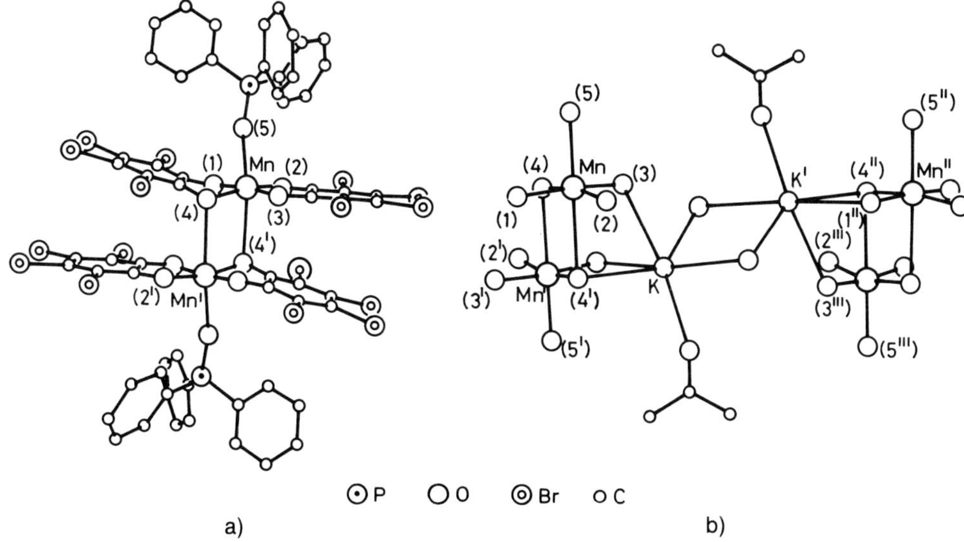

Fig. 25. a) View of the dimeric interaction between $[Mn(C_{18}H_{15}OP)(C_6Br_4O_2)_2]^-$ units. b) View of the interaction between K^+ ions, the interaction between K^+ ions and complex anions, and the repeating unit defining the crystal structure of $K[Mn(C_{18}H_{15}OP)(C_6Br_4O_2)_2]\cdot(CH_3)_2CO\cdot H_2O$ [5].

References:

[1] Contreras, E.; Riera, V.; Usón, R. (Rev. Acad. Cienc. Exact. Fis. Quim. Nat. Zaragoza [2] **28** [1973] 43/66; C.A. **80** [1974] No. 43582).
[2] Contreras, E.; Riera, V.; Usón, R. (Inorg. Nucl. Chem. Letters **8** [1972] 287/91).
[3] Johnson, D. W.; Sutton, D. (Can. J. Chem. **50** [1972] 3326/31).
[4] Usón, R.; Riera, V.; Ciriano, M. A.; Valderrama, M. (Transition Metal. Chem. [Weinheim] **1** [1976] 122/6).
[5] Larsen, S. K.; Pierpont, C. G.; de Munno, G.; Dolcetti, G. (Inorg. Chem. **25** [1986] 4828/31).

39.5.6 With Tritolyl- or Tribenzylphosphane Oxide

$(CH_3C_6H_4)_3PO$ — ligand 1 ($=C_{21}H_{21}OP$)

$(C_6H_5CH_2)_3PO$ — ligand 2 ($=C_{21}H_{21}OP$)

$Mn^{II}(C_{21}H_{21}OP)_2X_2$. The bromo complex with ligand 1 was prepared by the reaction of $MnBr_2$ with tritolylphosphane oxide in glacial acetic acid [1]. The preparation of $Mn(C_{21}H_{21}OP)_2Cl_2$ and $Mn(C_{21}H_{21}OP)_2Br_2$ complexes with ligand 2 is reported in [2]. Single crystals were obtained on cooling saturated solutions of the compounds in ethanol or benzene, respectively. The saturated solutions were prepared at ~50°C. Crystallographic data of the complexes with ligand 2 are shown below (lattice constants in Å):

complex	lattice	space group	a	b	c	Z
$Mn(C_{21}H_{21}OP)_2Cl_2$	orthorhombic	$P2_12_12_1-D_2^4$ (No. 19)	15.68(6)	16.52(7)	15.16(6)	4
$Mn(C_{21}H_{21}OP)_2Br_2$	orthorhombic	$P2_12_12_1-D_2^4$ (No. 19)	15.84(6)	16.79(7)	15.43(6)	4

The molecular structure does not present an internal symmetry. The calculated densities of the complexes are D = 1.296(5) and 1.386(6) g/cm³ for X = Cl and Br, respectively. The experimental values are 1.28(1) and 1.37(1) g/cm³ [3]. The electronic spectrum of the bromo complex with ligand 1 dissolved in acetonitrile shows maxima assigned as follows (extinction values ε in parentheses): $^6A_{1g} \rightarrow {}^4T_{1g}(G)$ at 21584 (0.70), $\rightarrow {}^4T_{2g}(G)$ at 22950 (1.50), $\rightarrow {}^4A_{1g}, {}^4E_g(G)$ at 24357 (1.24), $\rightarrow {}^4T_{2g}(D)$ at 29520 (2.25). The spectrum is in agreement with a tetrahedral geometry at Mn^{II} [1].

References:

[1] Negoiu, D. (Analele Univ. Bucuresti Ser. Stiint. Nat. Chim. **16** No. 1 [1967] 25/33; C.A. **70** [1969] No. 110242).
[2] Davolos, R. (Diss. Univ. Sao Paulo 1978).
[3] Periotto, D.; Antonio, S. A.; Tomita, K. (Ecletica Quim. **8** [1983] 11/2; C.A. **102** [1985] No. 70588).

39.5.7 With Tris(2-pyridinyl)phosphane Oxide

$(2\text{-}C_5H_4N)_3PO$ (= $C_{15}H_{12}N_3OP$)

$[Mn^{II}(C_{15}H_{12}N_3OP)_2](ClO_4)_2$ was obtained from $Mn(ClO_4)_2 \cdot 6\,H_2O$ and the ligand by the procedure used for the analogous 2-pyridinylphosphane complex, see p. 71. The IR spectrum of the complex in mineral oil or hexachlorobutadiene mulls shows bands at 1580, 1565sh, 1450, and 1420 cm^{-1}, assigned to the vibrations of nearly equivalent 2-pyridinyl rings that are coordinated to Mn through the nitrogen atom. An interesting feature is the increase in the frequency of the ν(PO) vibration on complexation: from 1215 (free ligand) to 1231 cm^{-1}. Coordination of the positive metal ion has apparently increased the electronegativity of the pyridinyl ring and consequently imposed a higher electronegativity on the phosphorus atom, lessening the contributions of the $P^+–O^-$ form and increasing the P=O frequency. No splitting observed for perchlorate $ν_3$ and $ν_4$ bands indicates ionic ClO_4 groups. The electrical conductivity, $Λ = 298$ cm$^2 \cdot Ω^{-1} \cdot$ mol^{-1}, and the equivalent conductivity at infinite dilution, $Λ_0 = 181.6$ cm$^2 \cdot Ω^{-1} \cdot$ mol^{-1}, determined in 10^{-3} M acetonitrile solution at room temperature are consistent with a 1:2 electrolyte.

Reference:

Boggess, R. K.; Zatko, D. A. (J. Coord. Chem. **4** [1975] 217/24).

39.5.8 With Phosphane Oxides of the $R_2R'PO$ Type

ligand	1	2	3
R	CH_3	C_6H_5	$C_6H_5CH_2$
R'	C_6H_5	CH_3	C_6H_5
formula	$C_8H_{11}OP$	$C_{13}H_{13}OP$	$C_{20}H_{19}OP$

$[Mn^{II}(C_{13}H_{13}OP)_4(ClO_4)]ClO_4$ was prepared by the reaction of Mn^{II} perchlorate and a slight excess of ligand 2 in acetone or ethanol containing triethyl orthoformate as dehydrating agent. The precipitate was washed with solvent and ether and dried in high vacuum. The complex is isomorphous with the corresponding Zn^{II}, Ni^{II}, Co^{II}, Fe^{II}, Cu^{II} complexes, according to X-ray powder photographs. IR bands of the compound in Nujol are ν(PO) at 1145 cm^{-1} (free ligand at 1175 cm^{-1}), $\nu(ClO_4)$ vibrations of coordinated (at 1143 and 1077 cm^{-1}) and ionic (at 1089 cm^{-1}) ClO_4 groups, $\nu(Mn\text{–}O_{ligand})$ at 417 and 324 cm^{-1}. A pentacoordinate square-pyramidal structure with distortion toward a square-planar arrangement is suggested from IR data [1]. A binuclear oxygen-bridged structure with two bridging and three terminal ligand molecules was suggested from split $\nu_3(ClO_4)$ and ν(Mn–O) vibrations [2]. The complex is sensitive to moisture [1].

$Mn^{II}(C_{13}H_{13}OP)_4I_2 \cdot H_2O$ was prepared by stirring MnI_2 (incompletely dehydrated by heating in vacuum) and methyldiphenylphosphane, $C_{13}H_{13}P$, (1:2 mole ratio) in dry tetrahydrofuran at 40°C in a sealed evacuated tube for 3d. The mixture was filtered and some ether added. Crystals separated on standing (yield ~15%), were filtered off under nitrogen, and dried in vacuum. The compound melts at 160°C. In the IR spectrum of the complex the presence of water is shown, as well as the ν(PO) vibration at 1145 cm^{-1}. The mechanism of phosphane oxidation is unknown, but residual water is proposed to be the oxygen source. The electrical conductivity of the complex in nitromethane, $\Lambda = 46\ cm^2 \cdot \Omega^{-1} \cdot mol^{-1}$, is below the value of a 1:1 electrolyte [3].

$Mn^{II}(C_{20}H_{19}OP)_2Cl_2$. For preparation, see [4]. Single crystals were obtained on cooling a solution of the complex in acetone saturated at ~50°C. X-ray diffraction data reveal a monoclinic lattice; space group B2–C_2^3 (No. 5) or Bm–C_s^3 (No. 8) or B2/m–C_{2h}^3 (No. 12), with lattice parameters a = 20.45(9), b = 15.19(8), c = 11.47(6) Å, β = 93.0(5)°; Z = 4. The calculated density, D_{calc} = 1.321(7) g/cm^3, agrees with the measured value D_{exp} = 1.32(1) g/cm^3 [5].

$Mn^{II}(C_{13}H_{13}OP)_2Cl_2 \cdot 2H_2O$ was prepared analogously to $Mn(C_{13}H_{13}OP)_4I_2 \cdot H_2O$, see above [3]. The complex is also formed on the reaction of $MnCl_3$ with methyldiphenylphosphane in ether, as cited for $[Mn(C_{18}H_{15}OP)_2Cl_2]$; see p. 90 [6]. The compound melts at 110°C [3]. The IR spectrum shows the presence of water by the ν(OH) bands at 3530 and δ(HOH) at 1635 cm^{-1} [6]; further bands at 1151 [3] or 1152 cm^{-1} [6] were assigned to ν(PO); at 273 cm^{-1} to ν(Mn–Cl) [3]. The complex is a nonelectrolyte in nitromethane [3].

$Mn^{II}(C_8H_{11}OP)X_2$ (X = Br, I, NCS). The bromo complex was prepared by stirring anhydrous $MnBr_2$ (3.3 mmol) and dimethylphenylphosphane oxide (7.8 mmol) in 40 mL of dry toluene. The oil separating after 3 h was washed with toluene and the excess solvent removed in vacuum to yield a yellow oil. This was also obtained on exposing a mixture of $MnBr_2 \cdot 2\,thf$ and dimethylphenylphosphane in tetrahydrofuran to an excess of dioxygen [7]. Formation of $Mn(C_8H_{11}OP)Br_2$ was observed when thin films of the corresponding phosphane complex on infrared windows were exposed to dioxygen [8], see also p. 43. $Mn(C_8H_{11}OP)X_2$ complexes with X = Br or I were obtained as decomposition products on reaction of the reversibly-formed SO_2 adducts (p. 58) of the corresponding phosphane complexes (p. 38) with dioxygen [9].

$[Mn(C_8H_{14}OP)(NCS)_2]$ was prepared from $Mn(NCS)_2$ and the phosphane oxide according to a procedure recommended in [10] for the corresponding complex with triphenylphosphane, see p. 38. A magnetic moment, $\mu_{eff}=5.3$ μ_B, determined at room temperature suggests a considerable degree of spin pairing [10]. The IR spectrum of $Mn(C_{18}H_{11}(^{18}O)P)Br_2$ is reported in [8]. The spectrum of $[Mn(C_8H_{11}OP)(NCS)_2]$ shows the bands of the ν(NCS) vibrations at 2130, 2100, 2080; of the ν(PO) at 1160 cm^{-1}. The complex behaves as a nonelectrolyte in 10^{-3} M toluene or tetrahydrofuran solution [11].

$[(c\text{-}C_6H_{11})P(C_6H_5)_3][Mn^{II}(C_8H_{11}OP)Br_3]$. The complex was prepared by addition of dimethylphenylphosphane oxide (0.19 mmol) dissolved in 10 mL of ethanol to an ethanolic solution (20 mL) containing stoichiometric amounts of the corresponding tetrabromomanganate, $[(c\text{-}C_6H_{11})P(C_6H_5)_3]MnBr_4$. The solvent was allowed to evaporate slowly at 35°C and the resulting yellow-green crystals dried in the air. The ESR X-band spectrum of the compound in ethanol glass at 93 K suggests that coordination of two ethanol molecules occurs in solution to yield a hexacoordinated Mn in the anion moiety [12].

$[Mn^{II}(C_8H_{11}P)_2(t\text{-}C_4H_9NC)_4][Mn^{II}(C_8H_{11}OP)Br_3]_2$. The mixed ligand complex containing dimethylphenylphosphane ($=C_8H_{11}P$) and *tert*-butyl isocyanide in the cation and dimethylphenylphosphane oxide in the anion is formed, if the red ethanolic solution of $[Mn(C_8H_{11}P)_2$-$(t\text{-}C_4H_9NC)_4]MnBr_4$ is allowed to stand in air, see also p. 46. The structure is reported. In the cation, Mn^{II} is octahedrally coordinated and lies on a crystallographic inversion center. The latter is true of each pair of anions where Mn^{II} is tetrahedrally coordinated. Details are given in the volume describing organometallic compounds together with other mononuclear isonitrile complexes of manganese. The IR spectrum shows the bands of ν(PO) at 1148 cm^{-1} and ν(Mn–Br) at 226 cm^{-1} [12].

References:

[1] Hunter, S. H.; Nyholm, R. S.; Rodley, G. A. (Inorg. Chim. Acta **3** [1969] 631/4).
[2] Karayannis, N. M.; Mikulski, C. M.; Strocko, M. J.; Pytlewski, L. L.; Labes, M. M. (J. Inorg. Nucl. Chem. **33** [1971] 2691/3).
[3] Casey, S.; Levason, W.; McAuliffe, C. A. (J. Chem. Soc. Dalton Trans. **1974** 886/9).
[4] Davalos, R. (Diss. Univ. Sao Paulo 1978).
[5] Periotto, D.; Antonio, S. A.; Tomita, K. (Ecletica Quim. **8** [1983] 11/2; C.A. **102** [1985] No. 70588).
[6] Levason, W.; McAuliffe, C. A. (J. Inorg. Nucl. Chem. **37** [1975] 340/2).
[7] Brown, R. M.; Bull, R. E.; Green, M. L. H.; Grebenik, P. D.; Martin-Polo, J. J.; Mingos, D. M. P. (J. Organometal. Chem. **201** [1980] 437/46).
[8] Newberry, V. F.; Burkett, H. D.; Worley, S. D.; Hill, W. E. (Inorg. Chem. **23** [1984] 3911/7).
[9] Newberry, V. F.; Hill, W. E.; Worley, S. D. (Phosphorus Sulfur **27** [1986] 253/60).
[10] Hosseiny, A.; Mackie, A. G.; McAuliffe, C. A.; Minten, K. (Inorg. Chim. Acta **3** [1969] 631/4).
[11] McAuliffe, C. A.; Al-Khateeb, H. F.; Barratt, D. S.; Briggs, J. C.; Challita, A.; Hosseiny, A.; Little, M. G.; Mackie, A. G.; Minten, K. (J. Chem. Soc. Dalton Trans. **1983** 2147/51).
[12] Beagley, B.; Benson, C. G.; Gott, G. A.; McAuliffe, C. A.; Britchard, R. G.; Tanner, S. P. (J. Chem. Soc. Dalton Trans. **1988** 2261/6).

39.5.9 With Alkanediyl- or Ethylenebis(diphenylphosphane Oxides)

$$(C_6H_5)_2\underset{\underset{O}{\|}}{P}-R-\underset{\underset{O}{\|}}{P}(C_6H_5)_2 \qquad (=L)$$

ligand	1	2	3	4
R	CH_2	C_2H_4	C_3H_6	CH=CH
formula	$C_{25}H_{22}O_2P_2$	$C_{26}H_{24}O_2P_2$	$C_{27}H_{26}O_2P_2$	$C_{26}H_{22}O_2P_2$

$[Mn^{II}(C_{25}H_{22}O_2P_2)_2]Br_2$ was obtained on heating dehydrated $MnBr_2$ and the corresponding bis(phosphane), $C_{25}H_{22}P_2$, in tetrahydrofuran. The complex precipitated on cooling the mixture. IR bands at 1165 and 1158sh cm^{-1} were assigned to the ν(PO) vibration, indicating that the phosphoryl oxygen atoms are coordinated to Mn. The complex behaves as a 1:2 electrolyte in 10^{-3} M nitromethane solution; the conductivity is $\Lambda = 126\ cm^2 \cdot \Omega^{-1} \cdot mol^{-1}$ [1].

$Mn^{II}LX_2$ Compounds. The $Mn(C_{26}H_{24}O_2P_2)Cl_2$ and $Mn(C_{26}H_{22}O_2P_2)Cl_2$ complexes were obtained by refluxing equimolar amounts of $MnCl_2$ (dehydrated by heating in vacuum) and ligand 2 or 4 in tetrahydrofuran for 30 min. The compounds precipitated on cooling, were filtered off, and recrystallized from dichloromethane. The yield is 60 to 70%. $MnLX_2$ complexes with L = ligand 1, 2, or 4 were obtained in ~15% yield by heating dehydrated MnX_2 salts and the corresponding bis(phosphane) in tetrahydrofuran for 3 to 4 d. The precipitates were then filtered off and separated from excess MnX_2 and unreacted bis(phosphane) [1]. The chloro complexes $MnLCl_2$ with ligand 1, 2, or 4 and the hydrate $Mn(C_{27}H_{26}O_2P_2)Cl_2 \cdot H_2O$ were obtained by adding an excess of $MnCl_3$ in ether solution to a tetrahydrofuran solution of the corresponding bis(phosphane). The mixtures became colorless overnight. The insoluble solid (largely $MnCl_2$) was separated and the solution was rotary evaporated to dryness. Yellow compounds were obtained by extracting the residues with hot dichloromethane and reducing the volume of the extract [2]. The table lists melting points (in °C) and IR band positions (in cm^{-1}) of the complexes in Nujol or hexachlorobutadiene mulls:

	$MnLCl_2$				$MnLBr_2$				$MnLI_2$	
ligand	m.p.	ν(PO)	ν(Mn–Cl)	ligand	m.p.	ν(PO)	ν(Mn–Br)	ligand	m.p.	ν(PO)
1	302	1160	270					1	220	1163, 1155sh
2	280	1150	280	2	305	1150	225	2	290	1155
4	252	1160	273	4	295	1155	225	4	230	1165

The ν(PO) vibrations of free ligands 2 and 4 were observed at 1184 and 1180 cm^{-1}, respectively. The ν(PO) band of $Mn(C_{27}H_{26}O_2P_2)Cl_2 \cdot H_2O$ is shown at 1150 cm^{-1}, the ν(OH) and δ(HOH) vibrations are observed at 3500 and 1630 cm^{-1}, respectively. The complexes with ligand 1, 2, or 4 are nonelectrolytes in nitromethane [1].

References:

[1] Casey, S.; Levason, W.; McAuliffe, C. A. (J. Chem. Soc. Dalton Trans. **1974** 886/9).
[2] Levason, W.; McAuliffe, C. A. (J. Inorg. Nucl. Chem. **37** [1975] 340/2).

39.5.10 With Polymer- or Silica-Supported Phosphane Oxides

Polymer-supported complexes, $\mathbf{Mn^{II}X_2\{(C_6H_5)_2(pol\text{-}C_6H_4)PO\}_n}$ with n = 1 or 3 for X = Cl, n = 1.5 or 3 for X = Br, and n = 1.5 or 5 for X = I, containing a phosphane oxide ligand as a styrene-divinylbenzene polymer (= pol-C_6H_4) were obtained by stirring anhydrous manganese(II) salt and a 1:2 or 1:4 molar quantity of polymer ligand in a dry, degassed 1:1 pentane-tetrahydrofuran mixture under strictly anhydrous conditions. Stirring was continued for 5 d at room temperature before filtration under argon, followed by drying the powders in vacuum. The complexes absorb SO_2 in the solid state or when suspended in a mixture of toluene and dichloromethane (60:40 v/v). In both cases the SO_2-absorption efficiencies are found to be in the order I > Br ~ Cl. The sulfur dioxide binding in the adducts is partially reversible by application of vacuum at 0.01 mm Hg [1].

The synthesis of a silica-supported phosphane oxide complex is reported in [2]: Oxidation of $(C_6H_5)_2PCH_2CH_2Si(OCH_3)_3$ gives $[(C_6H_5)_2P(O)CH_2CH_2Si(O^-)_3]_n$, from which the polymeric complex $\mathbf{[MnI_2\{(C_6H_5)_2P(O)CH_2CH_2Si(O^-)_3\}_{1.5}]_n}$ can be formed by reaction with powdered anhydrous MnI_2 in dry pentane under nitrogen. The ESR spectra of the solid complex at room temperature and at −170°C show a broad single-line signal with g = 2.0178 and line widths of 850 and 1025 G, respectively. When exposed to dry sulfur dioxide the finely ground complex becomes deep red. A nonreversible SO_2 uptake occurred until an Mn:SO_2 ratio of 1:0.85 was reached after 20 min [2].

References:

[1] Booth, B. L.; Li, K. S.; McAuliffe, C. A. (J. Chem. Soc. Dalton Trans. **1987** 2959/61).
[2] Booth, B. L.; Mu-guang, L.; McAuliffe, C. A. (J. Chem. Soc. Dalton Trans. **1987** 1415/7).

39.6 Complexes with Diorganylphosphinous Acids and a Derivative

ligand 1	$(CH_3)_2POH$	(= C_2H_7OP)
ligand 2	$(C_6H_5)_2POH$	(= $C_{12}H_{11}OP$)
ligand 3	$(CH_3)_2POCH_3$	(= C_3H_9OP)

Remark. A complex of composition $Mn^{II}(C_2H_7OP)_2Cl_2 \cdot 0.5H_2O$, prepared by reaction of $MnCl_2$ with dimethylphosphinous acid in butanol, contains dimethylphosphane oxide, the tautomeric form of ligand 1. The complex is therefore described in Section 39.5.1, p. 84. Secondary phosphane oxides, R_2HPO (R = CH_3, C_6H_5), can be transformed into the complex-stabilized phosphinic acids, R_2POH, via the pentacarbonylmanganese halides, $Mn(CO)_5X$. These P-isomeric manganese compounds, $Mn(CO)_4(R_2POH)X$, first prepared by [1], will be treated in the volumes describing organometallic compounds of manganese.

$\mathbf{Mn^{II}(C_{12}H_{10}OP)_2}$ was obtained from anhydrous $MnCl_2$ or $MnBr_2$ salts (dehydrated in vacuum at 110 to 130°C for 8 to 10 h) when they reacted with diphenylphosphane (1:2 mole ratio) in THF solution. The mixture was heated to 40°C in evacuated sealed tubes for 3 to 4 d. The resulting yellow solution was filtered and dry diethyl ether was added to the filtrate. After several days, small pale yellow crystals separated; these were filtered under nitrogen and dried in vacuum. The compound prepared from $MnCl_2$ melts at 160 to 163°C, that from $MnBr_2$ at 164°C. The products show identical IR spectra, which exhibit no ν(PH) or ν(OH) bands [2]. $Mn(C_{12}H_{10}OP)_2$ was also obtained on addition of excess $MnCl_3$ dissolved in diethyl ether to a solution of $(C_6H_5)_2PH$ in THF. Standing overnight, the mixture became colorless and a white solid deposited. After evaporation to dryness, the residue was extracted with hot CH_2Cl_2. The insoluble solid (largely $MnCl_2$) was filtered off, and the filtrate evaporated to a small volume. After standing for 2 to 3 d, yellow crystals appeared and they were recrystallized from CH_2Cl_2 [3].

$Mn^{II}(C_3H_9OP)Br_2$. The formation of the complex with ligand 3 as an intermediate species on decomposition of $Mn\{P(CH_3)_3\}Br_2 \cdot O_2$, see p. 42, was suggested by [4]. The IR band at 1060 cm^{-1} was attributed to a complex species containing the methyl ester of dimethylphosphinous acid which isomerizes to the phosphane oxide. $Mn(C_3H_9OP)Br_2$ is the final decomposition product, see p. 85.

References:

[1] Lindner, E.; Schilling, B. (Chem. Ber. **110** [1977] 3266/71).
[2] Casey, S.; Levason, W.; McAuliffe, C. A. (J. Chem. Soc. Dalton Trans. **1974** 886/9).
[3] Levason, W.; McAuliffe, C. A. (J. Inorg. Nucl. Chem. **37** [1975] 340/2).
[4] Newberry, V. F.; Burkett, H. D.; Worley, S. D.; Hill, W. E. (Inorg. Chem. **23** [1984] 3911/7).

39.7 Complexes with Phosphinic Acids

General. Metal(II) bis(phosphinates) have received attention because of their polymeric structures. A number of single-crystal X-ray studies show different coordination modes and polymer configurations of the complexes. The structures of the complexes $Mn(R_2PO_2)_2$ with straight-chain R substituents include polymers with both octrahedrally and tetrahedrally coordinated metal centers. Complexes with bulky alkyl R substituents all appear to have the same structure, which contains coordinated metal centers and symmetrically bridging O,O′ phosphinato groups. In the complex with dimethylphosphinic acid, $[Mn(C_2H_6O_2P)_2(H_2O)_2]$, chains are formed by magnanese atoms doubly bridged by phosphinato groups. Four oxygen atoms from four different phosphinates form a square-planar array around the metal atoms and two coordinated water ligands complete the octahedral geometry around the manganese atoms in the polymer. The diphenylphosphinato complex, $[Mn(C_{12}H_{10}O_2P)_2dmf] \cdot dmf$, contains pentacoordinated manganese atoms which are doubly bridged by diphenylphosphinato groups and coordinated to one dimethylformamide molecule, completing a slightly distorted trigonal-bipyramidal geometry. Methylphenylphosphinic acid reacts with Mn^{2+} in formamide forming a polymeric mixed ligand complex of composition $[Mn(C_7H_8O_2P)_2(CH_3NO)] \cdot CH_3NO$. The structure consists of one-dimensional polymeric chains in which two phosphinato groups form double bridges and a formamide oxygen atom forms a μ-O bridge between the manganese atoms. The results indicate that conformations of the phosphinato polymers can be determined by several factors such as the size and the nature of R in the $R_2PO_2^-$ ligands.

39.7.1 With Dialkyl- or Diphenylphosphinic Acids $R_2P(O)OH$ (=HL)

ligand	R	formula
1	CH_3	$C_2H_7O_2P$
2	C_2H_5	$C_4H_{11}O_2P$
3	C_4H_9	$C_8H_{19}O_2P$
4	$(CH_3)_3C$	$C_8H_{19}O_2P$
5	$C_3H_7(CH_3)CH$	$C_{10}H_{23}O_2P$
6	C_8H_{17}	$C_{16}H_{35}O_2P$
7	C_6H_5	$C_{12}H_{11}O_2P$

Remark. A manganese(II) complex with a "cyclic phosphinic acid" is described in the Section "Complexes with Heterocycles Containing Phosphorus" p. 182.

39.7.1.1 Manganese(II) Compounds

MnL_2 compounds were prepared by different methods: The complex with ligand 1 was obtained by dehydrating the hydrates $[Mn(C_2H_6O_2P)_2(H_2O)_2]$ or $Mn(C_2H_6O_2P)_2 \cdot 0.5H_2O$ (see p. 108) at 70°C in vacuum [1]. For preparation of the complex with ligand 2, 3, 6, or 7 the phosphinic acid and manganese(II) acetate (1:2 mole ratio) were reacted in boiling benzene or ethanol [2 to 5]. The compound with ligand 4, 5 [8], or 6 [4] was obtained by reaction of the corresponding potassium phosphinate with the stoichiometric amount of Mn^{II} sulfate in aqueous solution. The complex with dioctylphosphinic acid, $Mn(C_{16}H_{34}O_2P)_2$, was recrystallized from CCl_4 [4]. Complexes with ligands 3 and 7 were also formed on refluxing $Mn_2(CO)_{10}$ with the corresponding phosphinic acid dissolved in benzene or THF, respectively. The mixture was irradiated with UV light under nitrogen atmosphere; the dibutylphosphinate, $Mn(C_8H_{18}O_2P)_2$, was purified by pouring the hot benzene solution into absolute ethanol [3]. The yellow diphenylphosphinate, $Mn(C_{12}H_{10}O_2P)_2$, precipitated from THF solution, and was washed with petroleum ether and ethanol [5]. The compound was also prepared by adding a solution of $MnCl_2 \cdot 4H_2O$ to a solution containing ligand 7 and triethylamine. The precipitate was washed with ether and acetone [7]. Formation of the complex was also observed upon refluxing Mn^{III} acetylacetone with ligand 7 in xylene. The complex was purified by extraction with ethanol [6].

The complexes are mostly white and polymeric. The octahedrally coordinated metal ions are linked in a linear chain-like fashion by diorganylphosphinato bridges. As shown by IR data, the structures of the complexes with straight-chain alkyl groups ($R = C_4H_9$, C_8H_{17}) contain unsymmetrically bridging O,O′ phosphinato groups, whereas the compounds with bulky alkyls ($R = (CH_3)_3C$ or $C_3H_7(CH_3)CH$) reveal symmetrically bridging O,O′ phosphinato groups [4, 8]. Attempts to obtain crystals of the anhydrous dimethylphosphinato complex, $Mn(C_2H_6O_2P)_2$, suitable for X-ray studies have not been successful. However, it seems likely that the removal of the two water molecules on dehydration of the dihydrate (see p. 108) does not significantly alter its doubly phosphinato-bridged backbone structure. A square-planar environment about manganese is assumed for the pale pink anhydrous compound [1]. Tetrahedral structures are assumed for the complexes with $R = C_2H_5$ [2], $(CH_3)_3C$ [8], and C_6H_5 [5]. The X-ray powder patterns show the complex with ligand 4, $Mn(C_8H_{18}O_2P)_2$, to be isomorphous with the di-*tert*-butylphosphinates of Zn^{II}, Ni^{II}, and Co^{II} and, therefore, it should also have a distorted tetrahedral structure [8].

The different coordination modes of the phosphinato groups are indicated by IR spectral data. The IR spectra of the dimethylphosphinates of various metals have been recorded. The vibrations have been calculated in [9] for various models of the salt. It has been shown that the anion of dimethylphosphinic acid corresponds to a symmetrical structure with equivalent PO bonds. The spectra of nonalkali metal salts show in the range of PO_2 vibrations significant changes, depending on the nature of the cation. The observed displacement towards lower frequencies, relative to the spectrum of the free ion, indicates the formation of coordinative bonds in these compounds. The manganese complex with ligand 1, $Mn(C_2H_6O_2P)_2$, is described by a chain model with unsymmetrical PO_2 coordination to the metal [9]. The spectra of the compounds with ligands 3 and 6 ($R = C_4H_9$ and C_8H_{17}) reveal three strong PO_2 stretching bands, which are characteristic for complexes with unsymmetrically coordinated phosphinato groups. The separation of the symmetric and antisymmetric PO_2 stretching bands was found to be considerably greater in comparison to complexes with bulky alkyls ($R = (CH_3)_3C$ or $C_3H_7(CH_3)CH$). These compounds contain symmetrically bridging phosphinato groups. Their spectra reveal only two strong PO_2 stretching bands. Wavenumbers of PO_2 stretching frequencies (in cm^{-1}, Nujol or hexachlorobutadiene mulls) of MnL_2 complexes with various ligands are shown on p. 107:

HL	complex	$\nu_{as}(PO_2)$	$\nu_s(PO_2)$	Ref.
1	$Mn(C_2H_6O_2P)_2$	1123	1031	[1]
3	$Mn(C_8H_{18}O_2P)_2$	1140	1065, 1010	[8]
4	$Mn(C_8H_{18}O_2P)_2$	1128	1035	[8]
5	$Mn(C_{10}H_{22}O_2P)_2$	1135	1065	[8]
6	$Mn(C_{16}H_{34}O_2P)_2$	1140	1067, 1015	[4]
7	$Mn(C_{12}H_{10}O_2P)_2$	1140, 1100	1045, 1020	[5]

The IR spectum of the dioctylphosphinate, $Mn(C_{16}H_{34}O_2P)_2$, in the molten state (>112°C) shows only two strong bands: at 1130 and 1050 cm^{-1}. In CCl_4 solution bands were observed at 1135 and 1065 cm^{-1}, quite similar to those found for the corresponding complexes of Co^{II} or Zn^{II}. This suggests symmetrically coordinated phosphinato groups for the $Mn(C_{16}H_{34}O_2P)_2$ complex in the melt and in solution [4]. A graph of the IR spectrum of the diphenylphosphinate, $Mn(C_{12}H_{10}O_2P)_2$, in the 1300 to 400 cm^{-1} range is reported in [2].

The compounds show antiferromagnetic behavior. Magnetic susceptibilities (in cm^3/mol) of the dimethylphosphinate, $Mn(C_2H_6O_2P)_2$, recorded in [1] are shown below:

T in K	$10^6\ \chi_m$	T in K	$10^6\ \chi_m$	T in K	$10^6\ \chi_m$	T in K	$10^6\ \chi_m$
303.4	12410	128.5	23700	36.3	36900	19.7	35000
276.6	13340	109.4	26530	34.5	37000	17.9	34400
253.9	14090	92.9	28700	32.8	37000	15.9	33700
234.8	14990	82.4	30300	31.0	36900	13.4	32800
214.9	16210	70.8	32300	29.1	36800	9.82	31400
184.6	18350	61.3	34000	27.4	36600	7.02	30300
177.0	19210	52.9	35300	25.2	36300	6.16	29900
156.7	20890	44.5	36300	23.2	35900	5.32	29700
139.6	22340	39.7	36800	21.5	35500	4.22	29400

The magnetic moment varies from just above 5.4 μ_B at room temperature to about 1.0 μ_B at 4.2 K, see **Fig. 26**, p. 108. A susceptibility vs. temperature plot shows a broad maximum at 34 K. A one-dimensional antiferromagnetic exchange is suggested. The data were analyzed according to the scaling model of Wagner and Friedberg [12] and the interpolation scheme developed by Weng [13]. The scaling model gives $J = -2.94\ cm^{-1}$ and $g = 2.02$, the interpolation scheme gives $J = -2.69\ cm^{-1}$ and $g = 2.01$. The magnitude of the exchange coupling is compared to that observed in related manganese compounds. Possible reasons for the observed damping of the exchange on hydration are discussed, see the dihydrate on p. 108 [1]. Magnetic moments of the other MnL_2 compounds and g values resulting from susceptibility measurements (Faraday method) or from ESR data (in parentheses) are given below:

ligand	2	3	4	4	5	6	6	7	7
μ_B	—	—	5.4	5.58	5.5	5.6	5.38	—	—
g	—	—	—	1.89 ± 0.04	—	—	1.82 ± 0.04	—	—
	(2.010)	(2.019)	—	(2.00 ± 0.01)	—	—	(2.00 ± 0.01)	(2.006)	(2.020)
Ref.	[2]	[2]	[8]	[12]	[8]	[8]	[12]	[7]	[2]

Most of the compounds melt with decomposition. As shown by DSC studies, the dimethylphosphinate, $Mn(C_2H_6O_2P)_2$, exhibits an endothermic peak at 200°C ($\Delta H = 41$ J/g) and undergoes oxidative decomposition above 310°C [1]. The dibutylphosphinate, $Mn(C_8H_{18}O_2P)_2$, begins to decompose in air at approximately 220°C (heating rate 4.5°C/min) [3]. The dioctylphosphinate, $Mn(C_{16}H_{34}O_2P)_2$, melts at 112°C (ΔH = 20.5 cal/g) [4]. DTA data for the diphenylphosphinate, $Mn(C_{12}H_{10}O_2P)_2$, indicate that endotherms occurred at 240 and 350°C; the weight loss was 5% at 525°C and 10% at 540°C [6]. Cryoscopic measurements in benzene have shown the polymeric state of the diethyl-, dibutyl-, and diphenylphosphinates in solution [2]. The polymeric state of the dibutylphosphinate in various solutions is indicated by different values of the reduced viscosity. Depending on the preparation conditions (in ethanol at 20°C or in benzene at 80°C) different values of η were observed in 1% toluene solutions [3]. The molecular weight of the compound with ligand 5 in carbon tetrachloride is >35000 [8].

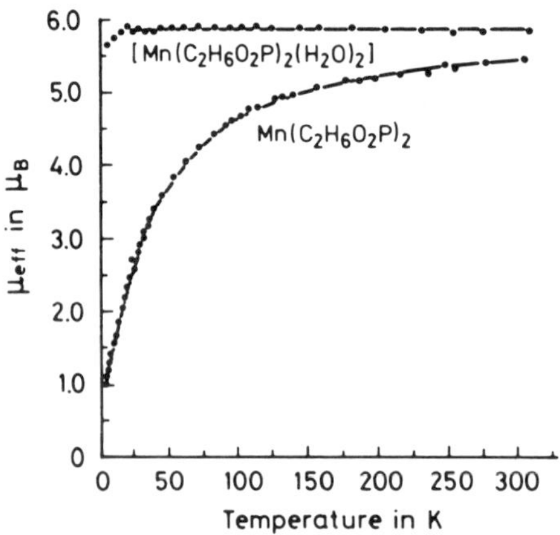

Fig. 26. Magnetic moment versus temperature plots for the dimethylphosphinato complexes, $Mn(C_2H_6O_2P)_2$ and $[Mn(C_2H_6O_2P)_2(H_2O)_2]$. Solid lines are the best fit to the Wagner-Friedberg model [1].

$Mn(C_2H_6O_2P)_2 \cdot xH_2O$ (x = 2, 1, 0.5). In order to obtain pure crystals of $[Mn(C_2H_6O_2P)_2(H_2O)_2]$ it is necessary to crystallize it slowly from dilute solutions: A solution of 5 mmol $MnCl_2 \cdot 4H_2O$ in 400 mL ethanol was added dropwise to a solution of dimethylphosphinic acid (10 mmol) in 600 mL ethanol. The mixture was stirred for 2 d, then the volume was reduced to about 300 mL by flash evaporation. Over the next 5 d small pale pink crystals, suitable for X-ray diffraction studies, were obtained. They were washed with cold ethanol and air-dried. More crystals were obtained by reducing the filtrate to about half the volume and allowing the solution to stand for about a week. More rapid crystallization yields noncrystalline materials of intermediate degrees of hydration: e.g., $Mn(C_2H_6O_2P)_2 \cdot 0.5H_2O$ precipitates immediately on mixing 7.5 mmol $Mn(CH_3COO)_2 \cdot 4H_2O$ in 250 mL ethanol with a solution of 15.1 mmol dimethylphosphinic acid in 125 mL ethanol [1]. $Mn[(CH_3)_2PO_2]_2 \cdot H_2O$ was reported to form on reaction of $MnCO_3$ with dimethylphosphinic acid [11].

Single-crystal X-ray diffraction studies revealed monoclinic symmetry for $[Mn(C_2H_6O_2P)_2(H_2O)_2]$, space group C2/c–$C^6_{2h}$ (No. 15) with lattice parameters a = 20.722(3), b = 4.8652(2), c = 11.0689(14) Å, β = 102.209(7)°; Z = 4. The structure was refined to R = 0.030. The structure consists of infinite centrosymmetric chains of Mn atoms linked by double phosphinato bridges and extending along the crystallographic b axis, see **Fig. 27**. The water molecules are involved in both interchain and bifurcated intrachain hydrogen bonding. The coordination about Mn is slightly distorted octahedral. Four oxygens from four different phosphinato groups form a roughly square-planar array around each manganese. Water molecules coordinated above and below this plane complete an MnO_6 chromophore. The libration-corrected bond lengths are: Mn–O(phosphinato) = 2.156(2) and 2.212(2) Å; Mn–O(H_2O) = 2.247(2) Å. Bond distances and angles involving the $(CH_3)_2PO_2$ moieties are comparable to those found in polymeric Pb or Cu phosphinates. The intrachain distance between the manganese atoms is 4.8652(2) Å and the minimum interchain distance is 6.046(1) Å. Intra- and interchain hydrogen bond parameters are given in the paper. The shorter interchain distances suggest that hydrogen bonding is stronger between chains than within chains. A view showing hydrogen-bonding interactions connecting the chains in the c direction is given in the paper. The calculated density of the diaqua complex is 1.687 g/cm^3 [1].

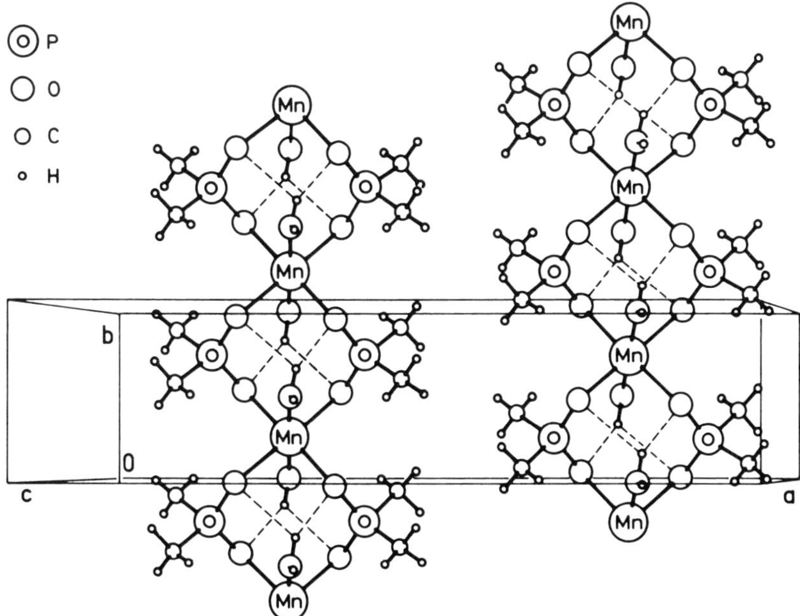

Fig. 27. View of the dimethylphosphinato complex, $[Mn(C_2H_6O_2P)_2(H_2O)_2]$, showing two chains extending along the b axis. Broken lines indicate hydrogen bonds [1].

Magnetic susceptibility studies from 300 to 4.2 K reveal a magnetic moment of ~5.9 μ_B for $[Mn(C_2H_6O_2P)_2(H_2O)_2]$ over most of the range and give no evidence for significant magnetic exchange (see Fig. 26), contrary to the anhydrous compound. Obviously, on hydration the water molecules, through their coordination to manganese, will effectively decrease the metal's acidity towards the bridging phosphinato groups and, through hydrogen bonding to the phosphinato oxygens, they will effectively decrease the phosphinate basicity towards manganese. The net effect is weaker phosphinate–manganese bonds, poorer orbital overlap, and weaker interchain exchange. Another explanation is that antiferromagnetic intrachain exchange effects are also present in the dihydrate, but are approximately canceled by interchain ferromagnetic exchange via the water molecules [1].

The IR spectrum of $[Mn(C_2H_6O_2P)_2(H_2O)_2]$ in Nujol or hexachlorobutadiene mulls is recorded in the 4000 to 250 cm^{-1} range [1], that of $Mn(C_2H_6O_2P)_2 \cdot H_2O$ (in KBr) between 1500 and 430 cm^{-1} [9]. DSC studies exhibit an endothermic peak at 107°C ($\Delta H = 343$ J/g) due to loss of water, followed by a second endothermic event at 208°C ($\Delta H = 40$ J/g) [1].

$[Mn(C_{12}H_{10}O_2P)_2dmf] \cdot dmf$ and **$[Mn(C_{12}H_{10}O_2P)_2(C_{12}H_{11}O_2P)_2(dmf)_2]$** complexes were formed by reaction of diphenylphosphinic acid (ligand 7) with Mn^{II} salts in the presence of dimethylformamide. A colorless precipitate containing a few long needles of $[Mn(C_{12}H_{10}O_2P)_2dmf] \cdot dmf$

and large prisms of composition $[Mn(C_{12}H_{10}O_2P)_2(C_{12}H_{11}O_2P)_2(dmf)_2]$ was obtained from a solution of 0.18 g of $Mn(ClO_4)_2 \cdot 6H_2O$ and 0.055 g of diphenylphosphinic acid in a mixture of 2 mL of dimethyl sulfoxide, 5 mL of dimethylformamide, 5 mL of acetone, and 0.014 g of CH_3ONa. Pure samples of $[Mn(C_{12}H_{10}O_2P)_2(C_{12}H_{11}O_2P)_2(dmf)_2]$ were obtained if 0.045 g of $Mn(ClO_4)_2 \cdot 6H_2O$ and 0.11 g ligand 7 were used [13]. $[Mn(C_{12}H_{10}O_2P)_2dmf] \cdot dmf$ was prepared by adding a solution of 1 g $MnCl_2 \cdot 4H_2O$ in dimethylformamide to a similar solution containing 2.18 g of ligand 7 and 1 g of triethylamine. The precipitate was washed with ethanol and acetone and air-dried [7].

The single-crystal X-ray diffraction data of $[Mn(C_{12}H_{10}O_2P)_2dmf] \cdot dmf$ reveal triclinic symmetry, space group $P\bar{1}-C_i^1$ (No. 2) with lattice constants a = 13.174(2), b = 13.364(2), c = 9.995(1) Å, α = 96.05(3)°, β = 110.23(3)°, γ = 104.54(2)°; Z = 2. The refinement to R = 0.099 (Mn, P, O anisotropic; N, C isotropic) shows polymeric chains of fused centrosymmetric eight-membered rings formed by two Mn atoms bridged by two phosphinato groups. Mn is coordinated to four oxygen atoms of different phosphinato groups with the average Mn–O distance of 2.08(1) Å and to the oxygen atom of dimethylformamide with a distance $Mn–O_{dmf}$ = 2.29(2) Å. The trigonal-bipyramidal geometry around the manganese atom is slightly distorted, see **Fig. 28**.

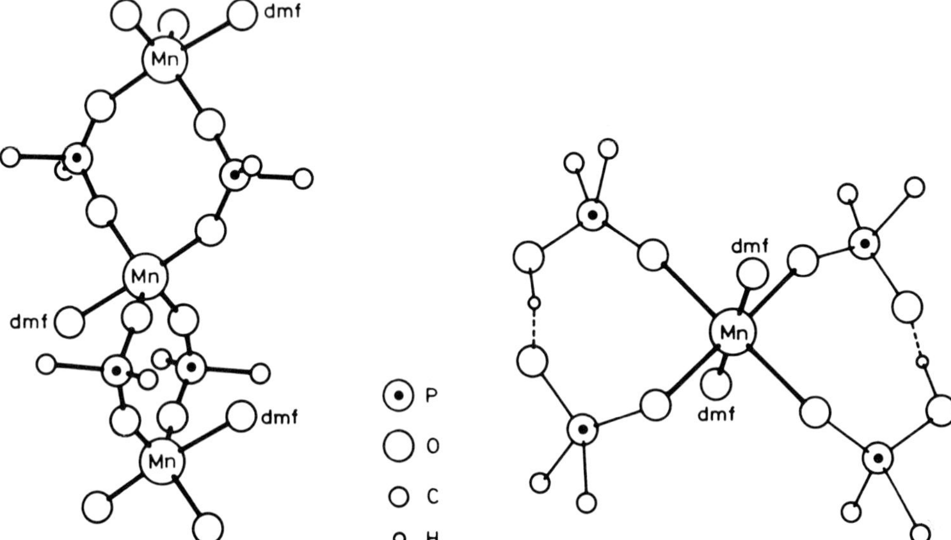

Fig. 28. A section of $[Mn(C_{12}H_{10}O_2P)_2dmf]$.

Fig. 29. Molecular structure of $[Mn(C_{12}H_{10}O_2P)_2(C_{12}H_{11}O_2P)_2(dmf)_2]$.

Only the α-carbon atoms of the phenyl rings and the oxygen atoms of the coordinated dmf molecules are shown in Fig. 28 and Fig. 29 [13].

The Mn–Mn separation is 4.583 Å. Positional parameters, bond lengths and angles are given in the paper. The axial-metal-axial angle is 173.3(7)°, while the three equatorial angles are 123.7(6)°, 116.8(5)°, and 117.4(6)°. The six axial-metal-equatorial angles range from 83° to 97°. The calculated density of the complex is 1.354 g/cm³ [13]. The ESR spectrum of $[Mn(C_{12}H_{10}O_2P)_2dmf] \cdot dmf$, 0.2% doped into the corresponding Cd^{II} complex, is poorly resolved at room temperature. The values of g, A, D, and E parameters are similar to those of the O,O-diphenyl phosphato complex, $Mn(C_{12}H_{10}O_4P)_2 \cdot 2dmf$, see p. 157 [7].

The X-ray diffraction data for $[Mn(C_{12}H_{10}O_2P)_2(C_{12}H_{11}O_2P)_2(dmf)_2]$ also reveal triclinic symmetry, space group $P\bar{1}-C_i^1$ (No. 2) with the lattice constants a = 12.019(1), b = 12.390(1), c = 9.479(1) Å, α = 94.90(3)°, β = 101.71(2)°, γ = 102.65(3)°; Z = 1. The structure was refined to R = 0.040 using anisotropic thermal parameters for all non-hydrogen atoms and isotropic ones for the OH hydrogen atom; ring hydrogen atoms were introduced but unrefined. Each Mn on the symmetry center has octahedral coordination, see **Fig. 29**, with square-planar arrangement of two phosphinato and two phosphinic acid groups with Mn–O distances of 2.129(2) and 2.150(2) Å (two times each). The Mn–O separation for two dimethylformamide oxygens above and below the plane is 2.279 Å. The O–O separation in the very strong intramolecular hydrogen bond is 2.444(3) Å. The hydrogen bonds formed complete the two eight-membered rings of the equatorial plane which may stabilize the monomeric structure of this compound. Bond lengths and angles are given in the paper. The calculated density of the complex is 1.332 [13].

References:

[1] Cicha, W. V.; Haynes, J. S.; Oliver, K. W.; Rettig, S. J.; Thompson, R. C.; Trotter, J. (Can. J. Chem. **63** [1985] 1055/62).
[2] Rosca, I. (Bul. Inst. Politeh. Iasi [3/4] **17** [1971] 13/25; C.A. **77** [1972] No. 48835).
[3] Vinogradova, S. V.; Korshak, V. V.; Vinogradova, O. V.; Polyakova, A. M.; Anisimov, K. N.; Kolobova, N. E. (Vysokomol. Soedin. A **15** [1973] 516/20; Polym. Sci. [USSR] A **15** [1973] 581/7).
[4] Gillman, H. D. (Inorg. Chem. **13** [1974] 1921/4).
[5] Korshak, V. V.; Polyakova, A. M.; Vinogradova, O. V.; Anisimov, K. N.; Kolobova, N. E.; Kotova, M. N. (Izv. Akad. Nauk SSSR Ser. Khim. **1968** 642/5; Bull. Acad. Sci. USSR Div. Chem. Sci. **1968** 618/21).
[6] Krüger, P. C. (U.S. 3681265 [1972]; C.A. **77** [1972] No. 115353).
[7] Cookson, D. J.; Wakefield, R. F.; Smith, T. D.; Boas, J. F.; Pilbrow, J. R.; Hicks, P. R.; Lamble, D. H. (J. Chem. Soc. Faraday Trans. II **75** [1979] 500/8).
[8] Gillman, H. D.; Eichelberger, J. L. (Inorg. Chem. **15** [1976] 840/3).
[9] Matrosov, E. I.; Fisher, B. (Izv. Akad. Nauk SSSR Neorgan. Materialy **3** [1967] 545/50; Inorg. Materials [USSR] **3** [1967] 485/9).
[10] Wagner, G. R.; Friedberg, S. A. (Phys. Letters **9** [1964] 11/3).

[11] Weng, C. H. (Diss. Carnegie-Mellon Univ., Pittsburgh 1968).
[12] Scott, J. C.; Garito, A. F.; Heeger, A. J.; Nannelli, P.; Gillman, H. D. (Phys. Rev. [3] B **12** [1975] 356/61).
[13] Betz, P.; Bino, A. (Inorg. Chim. Acta **147** [1988] 109/13).

39.7.1.2 Manganese(III) Compound

An insoluble and infusible lilac-colored polymeric compound $\mathbf{Mn(C_8H_{18}O_2P)_3}$ was obtained on reaction of dibutylphosphinic acid with manganese(III) acetate in ethanol. It has the same thermal stability as the corresponding manganese(II) compound, see p. 106.

Reference:

Vinogradova, S. V.; Korshak, V. V.; Vinogradova, O. V.; Polyakova, A. M.; Anisimov, K. N.; Kolobova, N. E. (Vysokomol. Soedin. A **15** [1973] 516/20; Polym. Sci. [USSR] A **15** [1973] 581/7).

39.7.2 With Distyrylphosphinic Acid $(C_6H_5CH{=}CH)_2P(O)OH$ $(=C_{16}H_{15}O_2P)$

The polymeric complex $Mn^{II}(C_{16}H_{14}O_2P)_2$ was precipitated from an ethanolic solution of the phosphinic acid on addition of the appropriate amount of Mn^{II} acetate dissolved in ethanol. The mixture was refluxed for 30 min, and the precipitate filtered off, washed with hot distilled water and dried at 80°C. The heat resistance of several polymeric distyrylphosphinates was studied and found to decrease in an inert atmosphere in the following order: $Zn^{II} > Co^{II} > Mn^{II} \gtrsim Cd^{II} > Ni^{II} > Cu^{II}$. The light gray manganese compound begins to decompose at 265°C.

Reference:

Korshak, V. V.; Krukovskii, S. P.; Knyazeva, Ye. K.; Danilov, V. G. (Vysokomol. Soedin. A **11** [1969] 3/6; Polym. Sci. [USSR] A **11** [1969] 1/4).

39.7.3 With Methylphenylphosphinic Acid $CH_3(C_6H_5)P(O)OH$ $(=C_7H_9O_2P)$

$\mathbf{Mn^{II}(C_7H_8O_2P)_2}$ was prepared by heating the suspension of manganese(II) chloride in a large excess of the methyl ester, $CH_3(C_6H_5)P(O)OCH_3$, at a rate of 1 to 2°C/min. The salt dissolved and heating was continued until the complex precipitated. It was washed with dry ether and stored in vacuum over $CaSO_4$. A more rapid increase in preparation temperature produced a rubber-like material. A highly cross-linked polymeric structure is proposed for the pink complex with bidentate bridging ligands. The Mn atom is obviously hexacoordinated. A magnetic moment of $\mu_{eff} = 5.82\ \mu_B$ was found at 298°C. The IR spectrum of the complex in KBr reveals bands of the $\nu_{as}(PO_2)$ vibration at 1140 and 1110 cm^{-1}, that of $\nu_s(PO_2)$ at 1050 and 1022 cm^{-1}, which are typical for the $R_2PO_2^-$ ligands. A weaker $(PO)_2$ combination band exists at 1802 cm^{-1}. The spectrum of the complex in Nujol shows bands at 435, 408, and 335 cm^{-1}, assigned to ν(Mn–O) vibration modes. The solid state electronic spectrum (Nujol) shows absorption maxima at 417, 355, and 350 nm. The complex was found to have a pinkish tint under UV excitation, which is characteristic of fluorescent octahedral Mn^{II} compounds [1, 2].

$\mathbf{[Mn^{II}(C_7H_8O_2P)_2(CH_3NO)]\cdot CH_3NO}$. The mixed ligand complex with formamide, $HC(O)NH_2$ $(=CH_3NO)$, precipitated within several weeks from a solution of $Mn(ClO_4)_2$ (0.127 g) and methyl phenylphosphinic acid (0.156 g) in 25 mL of formamide, which had been placed for evaporation in an open beaker under a hood. The colorless crystals belong to the monoclinic system, space group $C2/m{-}C_{2h}^3$ (No. 12), with the lattice constants a = 22.446(3), b = 8.041(1), c = 11.612(2) Å, β = 92.20(4)°; Z = 4. The structure was refined to R = 0.077 by using anisotropic thermal parameters for all non-hydrogen atoms of the $[Mn(C_7H_8O_2P)_2(CH_3NO)]$ complex and isotropic ones for the uncoordinated formamide molecule. The structure consists of polymeric chains, propagating along the b axis, in which two phosphinato groups form double bridges and the formamide oxygen atom forms a μ-O bridge between two Mn atoms, see **Fig. 30**. Each Mn atom resides at a crystallographic center of symmetry. The $Mn_2O[O_2P(CH_3)(C_6H_5)]_2$ unit is bisected by a crystallographic mirror plane passing through O(3), P(1), P(2), the methyl carbons C(2) and C(6), and the phenyl carbons C(1), C(5), C(7), and C(10). The geometry about the Mn atom is octahedral with four oxygen atoms of four phosphinato groups having Mn–O distances of 2.127(6) and 2.148(6) Å and with two formamide oxygen atoms having Mn–O = 2.304(4) Å. The interchain Mn–Mn distance is half of the b axis, which is 4.020(1) Å. Positional parameters, bond lengths and angles are given in the paper [3].

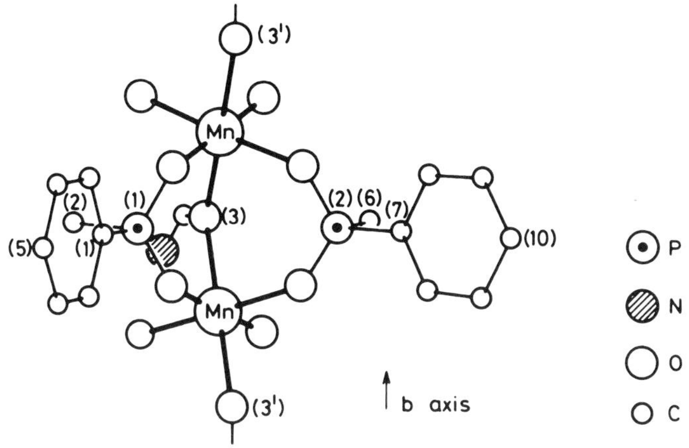

Fig. 30. A section of the polymeric chain of $[Mn^{II}(C_7H_8O_2P)_2(CH_3NO)]$. Only one orientation of N is shown [3].

References:

[1] Mikulski, C. M.; Unruh, J.; Rabin, R.; Iaconianni, F. J.; Pytlewski, L. L.; Karayannis, N. M. (Inorg. Chim. Acta **44** [1980] L77/L79).
[2] Mikulski, C. M.; Unruh, J.; Delacato, D. F.; Iaconianni, F. J.; Pytlewski, L. L.; Karayannis, N. M. (J. Inorg. Nucl. Chem. **43** [1981] 1751/4).
[3] Betz, P.; Bino, A. (Inorg. Chim. Acta **149** [1988] 171/5).

39.7.4 With Aminomethyl(methyl)phosphinic Acid or Related Compounds

$H_2NCH_2(CH_3)P(O)OH$
ligand 1
$(=C_2H_8NO_2P=HL)$

$(C_6H_5NHCH_2)_2P(O)OH$
ligand 2
$(=C_{14}H_{17}N_2O_2P=HL)$

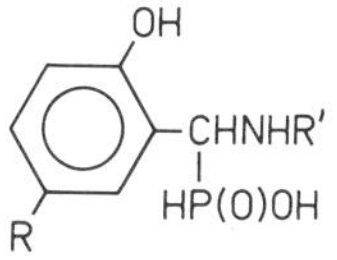

ligands 3 to 8 $(=H_2L)$

ligand	3	4	5
R	H	H	H
R′	C_6H_5	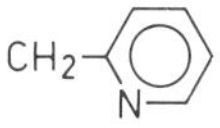	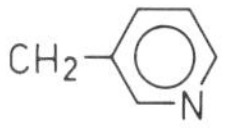
formula	$C_{13}H_{14}NO_3P$	$C_{13}H_{15}N_2O_3P$	$C_{13}H_{15}N_2O_3P$
ligand	6	7	8
R	H	NO_2	NO_2
R′	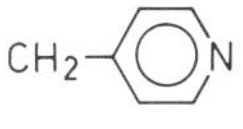	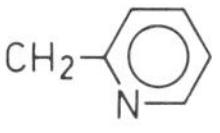	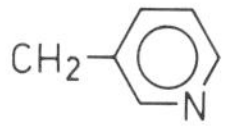
formula	$C_{13}H_{15}N_2O_3P$	$C_{13}H_{14}N_3O_5P$	$C_{13}H_{14}N_3O_5P$

Complexes in Solution. The stability constants of $Mn^{II}L$ complexes with H_2L = ligand 4 or 5, log $K_1 = 5.60$, or with ligand 6, log $K_1 = 5.70$, were determined pH-potentiometrically in aqueous solution under nitrogen at $20 \pm 0.5°C$ and $I = 0.1 M$ $(NaClO_4)$ [1]. The stability constants of complexes with ligands 3, 7, and 8, log $K_1 = 5.05$, 5.75, and 5.70, were determined in aqueous medium containing 15% (v/v) of ethanol [2]. All values are lower than those of corresponding complexes with aminoalkylphosphonic acids; see p. 125 [1, 2].

$[Mn^{II}(C_2H_8NO_2P)_2Cl_2(H_2O)_2]$ was obtained from an aqueous solution containing $MnCl_2$ and aminomethyl(methyl)phosphinic acid in a stoichiometric ratio. X-ray diffraction data showed the crystals to be monoclinic, space group $P2_1/c$–C^5_{2h} (No. 14), with the lattice parameters a = 6.045(1), b = 7.302(1), c = 17.044(3) Å, β = 104.53(1)°; Z = 2. The structure was refined anisotropically to R = 0.044. As shown in **Fig. 31** the complex is centrosymmetric and the Mn atom is coordinated to two oxygen atoms from the phosphinato groups, two chlorine atoms, and two water molecules. The coordination forms a slightly distorted *trans,trans,trans* octahedron. The Mn–Cl, Mn–O, and $Mn–O_{H_2O}$ bond lengths are 2.569(2), 2.141(4), and 2.195(4) Å, respectively. For other interatomic distances and bond angles, see [3]. The bond distances and angles in the coordination polyhedron agree well with those found in $Mn(C_2H_5NO_2)_2Cl_2$ and $Mn(C_2H_5NO_2)_2Cl_2 \cdot 2H_2O$ complexes ($C_2H_5NO_2$ = glycine); see "Manganese" D 4, 1985, pp. 256 and 257, respectively. Ligand 1 exists as a zwitterion. The three hydrogen atoms of the NH_3^+ group are involved in the intermolecular hydrogen bonds. The density, D = 1.73 g/cm³, measured by flotation in a mixture of $C_2H_4Br_2$ and $CHCl_3$, agrees with the calculated value of 1.733 [3].

$[Mn^{II}(C_2H_8NO_2P)_2(H_2O)_2]Br_2 \cdot 2H_2O$. Preparation not given in the paper. Colorless prisms suitable for X-ray analysis are triclinic, space group $P\bar{1}$–C^1_i (No. 2), with a = 10.183(2), b = 9.676(2), c = 11.056(3) Å, α = 112.69(2)°, β = 93.74(1)°, γ = 117.75(2)°; Z = 2. The structure was refined anisotropically to a final R = 0.055. The molecular structure of the cation viewed along c is shown in **Fig. 32**. The manganese atom is octahedrally coordinated by four oxygen atoms from four phosphinato groups and by two water oxygen atoms which are in *cis* positions. The Mn–O distances (in Å) are: Mn–O(1) = 2.137(7), Mn–O(2) = 2.213(8), Mn–O(3) = 2.170(8), Mn–O(4) = 2.129(7), $Mn–O_{H_2O(1)}$ = 2.184(7), $Mn–O_{H_2O(2)}$ = 2.213(8). The octahedron around the

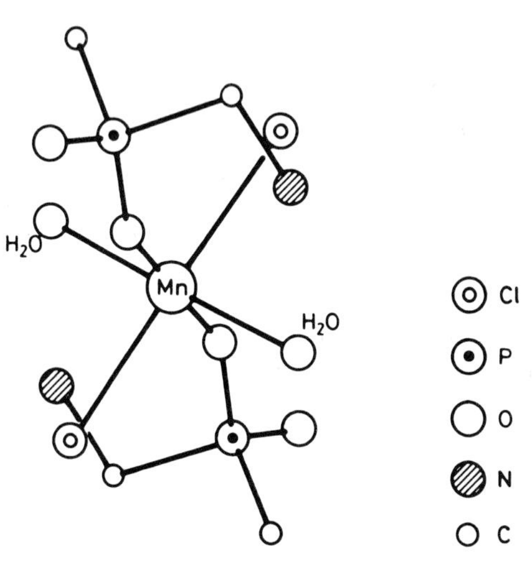

Fig. 31. Molecular structure of $[Mn^{II}(C_2H_8NO_2P)_2Cl_2(H_2O)_2]$, viewed along the b axis. H atoms are omitted [3].

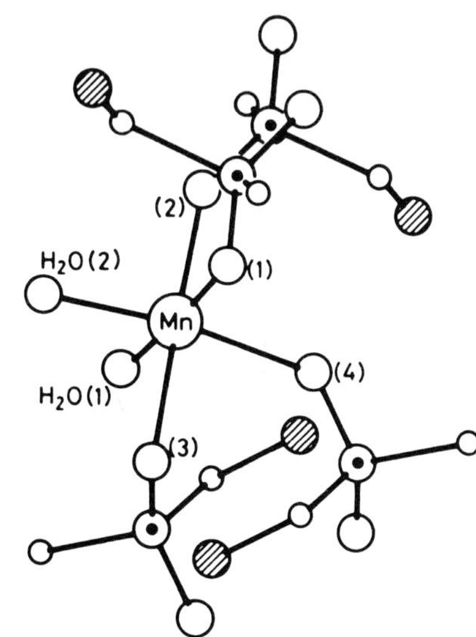

Fig. 32. Configuration about the Mn atom in $[Mn^{II}(C_2H_8NO_2P)_2(H_2O)_2]^{2+}$, viewed along the c axis [4].

Mn atom is moderately distorted. The O–Mn–O angles for pairs of contiguous oxygen atoms range from 83.8(3)° to 101.3(3)°, whereas those involving pairs of opposite oxygen atoms range from 170.2(3)° to 171.9(3)°. The bromine ions do not enter the coordination sphere. The phosphinato oxygen atoms are bonded to different Mn atoms forming Mn–O–P–O–Mn bridges [5]. Similar bridges were found in $Mn(C_2H_5NO_2)_2Br_2 \cdot 2H_2O$; see "Manganese" D 4, 1985, p. 259. All four water molecules form hydrogen bonds; the noncoordinated H_2O molecules participate in hydrogen bonds with bromine atoms and with the oxygen atoms of other water molecules and phosphinato groups. The amino group is also involved in hydrogen-bond formation. The density, D=1.99 g/cm³, was measured by flotation; the calculated value is 1.98 g/cm³ [4].

$Mn^{II}(C_{14}H_{16}N_2O_2P)_2$ was obtained from an aqueous Mn^{II} salt solution containing the stoichometric amount of the potassium salt of ligand 2. The precipitate was washed with water and dried in vacuum. The X-ray powder pattern indicates it to be isomorphous with the corresponding Co^{II} complex. IR bands at 3375 and 3220 cm^{-1} were assigned to ν(NH) modes of noncoordinated and coordinated NH groups, respectively, those at 492 and 470 cm^{-1} to Mn–N stretching modes. A polymeric structure with symmetrically bridging O,O′ phosphinato groups and octahedral coordination around the manganese atom is suggested [5].

References:

[1] Siepak, J. (Polish J. Chem. **59** [1985] 955/69; C.A. **106** [1987] No. 108761).
[2] Siepak, J. (Polish J. Chem. **59** [1985] 651/63; C.A. **106** [1987] No. 176495).
[3] Glowiak, T.; Sawka-Dobrowolska, W. (Acta Cryst. B **33** [1977] 2763/6).
[4] Glowiak, T.; Sawka-Dobrowolska, W. (Acta Cryst. B **33** [1977] 2648/50).
[5] Gillman, H. D.; Eichelberger, J. L. (Inorg. Chim. Acta **24** [1977] 31/4).

39.7.5 With Esters of Phosphinic Acids

ligand 1	$(CH_3)_2P(O)OCH_3$	$(=C_3H_9O_2P)$
ligand 2	$CH_3(C_6H_5)P(O)OCH_3$	$(=C_8H_{11}O_2P)$

$Mn^{II}(C_3H_9O_2P)Br_2$. Formation of the methyl dimethylphosphinate complex as an intermediate on heating an $Mn\{P(CH_3)_3\}Br_2 \cdot O_2$ film (see p. 43) in the presence of dioxygen is indicated by a new IR band at 1040 cm^{-1}. The spectrum of the free ligand 1 shows a prominent band at 1044 cm^{-1} [1].

$[Mn^{II}(C_8H_{11}O_2P)_4(H_2O)](ClO_4)_2$ was prepared by reacting warm solutions of ligand 2 and manganese(II) perchlorate in triethyl orthoformate in a 6:1 mole ratio. Precipitation of a waxy solid required overnight stirring of the reaction mixture at room temperature. The complex was washed with ether and stored over $CaCl_2$. A magnetic moment, $\mu_{eff}=6.14\ \mu_B$, results from measurements of the molar susceptibility, $\chi_m(corr)=15833 \times 10^{-6}$ cm³/mol at 295 K. Coordination of the phosphoryl oxygen is indicated in the IR spectrum by a shift of the ν(PO) band of the ligand (at 1207 cm^{-1}) to a lower wavenumber (1163 cm^{-1}). A $\nu(POCH_3)$ band exists at 1018 cm^{-1}. The presence of ionic ClO_4^- groups is demonstrated by single bands of $\nu_3(ClO_4)$ and $\nu_4(ClO_4)$ at 1095 and 615 cm^{-1}, respectively, and a weak absorption in the $\nu_1(ClO_4)$ region at 920 cm^{-1}. Bands at 404 and 421 cm^{-1} were assigned to ν(Mn–O) vibrations of the ligand and coordinated water molecules, respectively. Pentacoordination is also suggested by the appearance of an absorption maximum in the solid state electronic spectrum (Nujol mull) at 321 nm. Additional bands were observed at 303, 287, 263, and 223 nm. The aqua ligand is rather tenaciously held, neither desiccation for over a month nor overnight heating at 70°C under reduced pressure can remove the coordinated water. The complex is soluble in various organic solvents. It decomposes upon dissolution in water. The molar conductivity of a 0.001M solution in nitromethane, $\Lambda=165\ cm^2\cdot\Omega^{-1}\cdot mol^{-1}$, shows that the complex behaves as a 1:2 electrolyte [2].

References:

[1] Newberry, V. F.; Burkett, H. D.; Worley, S. D.; Hill, W. E. (Inorg. Chem. **23** [1984] 3911/7).

[2] Mikulski, C. M.; Unruh, J.; Pytlewski, L. L.; Karayannis, N. M. (Transition Metal Chem. [Weinheim] **4** [1979] 98/103).

39.8 Complexes with the Dimethyl Ester of Phenylphosphonous Acid

$C_6H_5P(OCH_3)_2$ (= $C_8H_{11}O_2P$)

$[Mn(NO)_2(C_8H_{11}O_2P)_2X]$ (X = Cl, Br). The complexes were prepared under an inert atmosphere by passing nitrogen monoxide through the refluxing benzene solution of $[Mn(C_8H_{11}O_2P)_2(CO)_3X]$. After 4 h the benzene was removed under reduced pressure and the remaining oil was recrystallized from a mixture of ether and hexane. The orange chloro complex was recrystallized from hot ethanol, the red-brown bromo complex from a dichloromethane-cyclohexane mixture. The compounds melt at 136 to 137°C and 143 to 147°C, respectively [1].

Two polymorphic forms of the chloro complex are known; monoclinic crystals were obtained from benzene, triclinic crystals from a mixture of ethanol-dichloromethane [1, 2].

The monoclinic crystals of $[Mn(NO)_2(C_8H_{11}O_2P)_2Cl]$ belong to the space group C2/c–C_{2h}^6 (No. 15), with the lattice constants a = 25.86(1), b = 11.863(5), c = 14.563(5) Å, β = 90.96(5)°; Z = 8. An anisotropic refinement to R = 0.041 for all atoms except H atoms shows that the coordination about the manganese atom is trigonal-bipyramidal with the phosphonoato groups in *trans* positions. The two NO groups are ordered and the atoms of the $Mn(NO)_2Cl$ group are coplanar with the NO groups bent in toward each other, see **Fig. 33** [3].

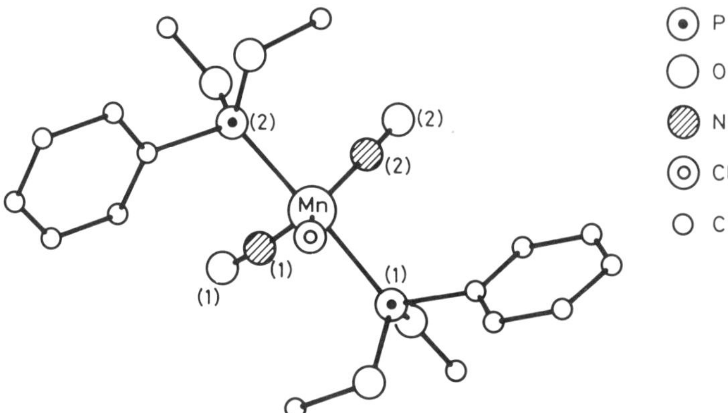

Fig. 33. Molecular structure of $[Mn(NO)_2(C_8H_{11}O_2P)_2Cl]$ in the monoclinic form. Hydrogen atoms are omitted [3].

Bond lengths (in Å) and angles (in °) of interest are:

Mn–Cl	2.351(5)	Mn–N(2)	1.665(10)	Mn–N(1)–O(1)	166(1)
Mn–P(1)	2.294(5)	N(1)–O(1)	1.18(1)	Mn–N(2)–O(2)	163(1)
Mn–P(2)	2.291(5)	N(2)–O(2)	1.18(1)	N(1)–Mn–N(2)	111.5(5)
Mn–N(1)	1.650(10)			O(1)–Mn–O(2)	99.0(5)
				Cl–Mn–N(1)	124.6(5)
				Cl–Mn–N(2)	123.9(5)

The phosphonito ligands bend toward the Cl atom, away from the NO groups: the P(1)–Mn–P(2) angle is 166.9°, those of P(1)–Mn–Cl or P(2)–Mn–Cl are 83.4° and 83.5°. The P–Mn–N angles are between 91.3° and 95.6°. Additional bond lengths or angles as well as fractional atomic coordinates and anisotropic thermal parameters are given in the paper [3].

The triclinic form of $[Mn(NO)_2(C_8H_{11}O_2P)_2Cl]$ belongs to the space group $P\bar{1}-C_i^1$ (No. 2) with a = 15.727(5), b = 15.198(5), c = 9.405(5) Å, α = 90.97(5)°, β = 89.04(5)°, γ = 97.21(5)°; Z = 4 (two molecules per asymmetric unit). Refinement to R = 0.040 reveals that the coordination about the Mn is trigonal-bipyramidal in both molecules of the asymmetric unit with different conformations of the phosphonito ligands. Comparison of **Fig. 34** a and b shows that one molecule ("2") approximates symmetry 2 with the twofold axis coincident with the Mn–Cl bond, while the other ("M") approximates symmetry m with the atoms of the $Mn(NO)_2Cl$ group defining the mirror plane. In both isomers the NO groups are ordered, and the Mn(NO)Cl group is coplanar and the P atoms are bent toward the Cl atom [4]. The geometry of the phosphonito groups is the same as found in the monoclinic form [3].

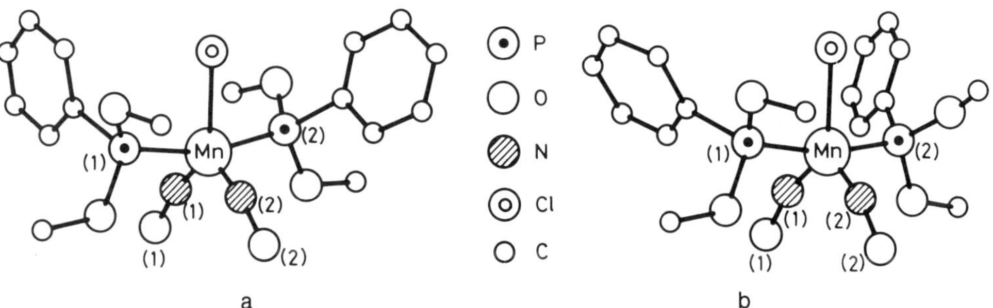

Fig. 34. Molecular structure of $[Mn(NO)_2(C_8H_{11}O_2P)_2Cl]$ in the triclinic form, showing a) the "2" molecule with staggered phenyl rings, *trans* or *anti*; b) the "M" molecule with phenyl rings eclipsed *cis* or *syn* [4].

The atomic coordinates, anisotropic thermal parameters, bond lengths, and angles for both triclinic $[Mn(NO)_2(C_8H_{11}O_2P)_2Cl]$ molecules are given in [4]. Selected data are the following:

distances in Å	molecule "2"	molecule "M"
Mn–Cl	2.360(5)	2.352(5)
Mn–P(1)	2.285(5)	2.298(5)
Mn–P(2)	2.295(5)	2.297(5)
Mn–N(1)	1.657(10)	1.633(10)

distances in Å	molecule "2"	molecule "M"
Mn–N(2)	1.644(10)	1.642(10)
N(1)–O(1)	1.19(1)	1.19(1)
N(2)–O(2)	1.18(1)	1.19(1)

angles in °	molecule "2"	molecule "M"	angles in °	molecule "2"	molecule "M"
P(1)–Mn–P(1)	163.9(5)	167.6(5)	Cl–Mn–N(1)	129.2(5)	123.0(5)
N(1)–Mn–N(2)	113.5(5)	112.1(5)	Cl–Mn–N(2)	117.3(5)	124.9(5)
O(1)–Mn–O(2)	101.9(5)	99.8(5)	Mn–N(1)–O(1)	167(1)	166(1)
P(1)–Mn–Cl	82.3(5)	83.4(5)	Mn–N(2)–O(2)	165(1)	165(1)
P(2)–Mn–Cl	82.4(5)	84.5(5)			

It is assumed that the triclinic isomers do not differ greatly in energy; in solution they both must be present [4]. From X-ray photoelectron spectral data the following binding energies (referred to C1s = 284.0 eV) were calculated: N1s = 400.6 (X = Cl) and 400.2 eV (X = Br), P2p = 131.2 eV (X = Cl or Br). The IR spectra of the complexes in dichloromethane solutions show ν(NO) bands at 1723 and 1674 cm^{-1} (X = Cl), at 1729 and 1678 cm^{-1} (X = Br). The ^1H NMR spectra of the complexes, dissolved in acetone-d_6, show distorted triplets with J = 6.26 and 6.30 Hz, assigned to the methyl resonances of the chloro and bromo complex, respectively. In acetone solution both complexes are nonelectrolytes. $[Mn(NO)_2(C_8H_{11}O_2P)_2Br]$ requires an inert atmosphere for extended storage [1].

$[Mn(NO)_2(C_8H_{11}O_2P)_3]BF_4$ was prepared at room temperature by adding $AgBF_4$ (2.5 mmol) to a solution of $[Mn(NO)_2(C_8H_{11}O_2P)_2Cl]$ (2.5 mmol) in 25 mL acetone containing 3 mmol of $C_6H_5P(OCH_3)_2$. After 2 h the silver chloride was filtered off, and the solvent removed under reduced pressure to give an oil, which was taken up in 5 mL of ethanol. Dropwise addition of ether yielded the complex. Orange prisms were obtained by recrystallization from dichloromethane-ether [1]. The structure determination by X-ray diffraction reveals the monoclinic space group $P2_1/c–C^5_{2h}$ (No. 14) with the lattice constants a = 14.585(5), b = 13.085(5), c = 18.610(10) Å, β = 110.45(5)°; Z = 4. The structure was refined anisotropically for all atoms (except hydrogen) to R = 0.064. The coordination at Mn, see **Fig. 35**, is trigonal-bipyramidal with the two NO groups equatorial. The NO groups are ordered and are bent in toward each other. The two axial phosphonito groups bend toward the equatorial one. The atoms of the equatorial $Mn(NO)_2P$ group are coplanar within 0.1 Å. Atomic coordinates, anisotropic thermal parameters, bond lengths and angles are given in [5]. Values of interest are:

distances in Å		angles in °	
Mn–P(1)	2.312(5)	P(1)–Mn–P(2)	175.5(5)
Mn–P(2)	2.303(5)	N(1)–Mn–N(2)	116.5(5)
Mn–P(3)	2.356(5)	O(1)–Mn–O(2)	107.4(5)
Mn–N(1)	1.649(10)	P(3)–Mn–N(1)	114.0(5)
Mn–N(2)	1.649(10)	P(3)–Mn–N(2)	129.4(5)
N(1)–O(1)	1.18(1)	P(1)–Mn–N(1)	93.5(5)
N(2)–O(2)	1.19(1)	P(1)–Mn–N(2)	89.1(5)
		P(2)–Mn–N(1)	91.0(5)
		P(2)–Mn–N(2)	89.7(5)

The relative conformation of the pairs of axial phosphonito ligands is almost identical. The angles of P(3)–Mn–P(1) and P(3)–Mn–P(2) are 87.7(5)° and 89.8(5)°, respectively; those of Mn–N(1)–O(1) and Mn–N(2)–O(2) are 170(1)° and 168(1)°. The IR spectra of solid $[Mn(NO)_2(C_8H_{11}O_2P)_3]BF_4$ show ν(NO) bands at 1760 and 1712 cm^{-1} [5], in dichloromethane solution at 1772 and 1722 cm^{-1} [1]. The X-ray photoelectron spectra reveal the binding energies N1s = 400.2 and P2p = 131.4 eV, referred to C1s = 284.0 eV. The ^1H NMR spectrum measured in acetone-d_6 solution shows two methyl resonances, a distorted triplet (J = 6.31 Hz) for the two strongly coupled *trans*-axial phosphonito groups, and a doublet (J = 6.83 Hz) for the equatorial one. The electrical conductivity of the complex in dichloromethane is $\Lambda = 139\ cm^2 \cdot \Omega^{-1} \cdot mol^{-1}$. Storage of the complex or of its solutions requires an inert atmosphere [1].

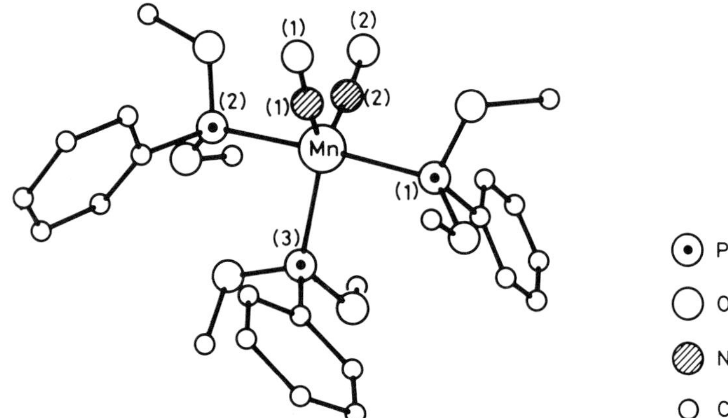

Fig. 35. Structure of the $[Mn^{I}(NO)_2(C_8H_{11}O_2P)_3]^+$ cation in $[Mn(NO)_2(C_8H_{11}O_2P)_3]BF_4$. Hydrogen atoms are omitted [5].

$[Mn(NO)_2(C_8H_{11}O_2P)_2(C_5H_9N)]BF_4$ was prepared by refluxing the chloroform solution of $[Mn(NO)_2(C_8H_{11}O_2P)_3]BF_4$, see p. 118, with an excess of *tert*-butyl isocyanide (= C_5H_9N). Removal of the solvent under reduced pressure gave an oil, which was recrystallized from ethanol-hexane. The yellow plates obtained decompose at 153 to 157°C. The IR spectrum of the complex dissolved in dichloromethane shows a ν(CN) band at 2193 cm^{-1} and bands of ν(NO) at 1762 and 1719 cm^{-1}. The ^1H NMR spectrum of the solution in acetone-d_6 reveals a distorted triplet (J = 6.08 Hz), assigned to phosphonito methyl resonances, and a singlet (J = 8.50 Hz). It was suggested that the equatorial ligand had been displaced and that this complex has the same ligand configuration as the $[Mn(NO)_2(C_8H_{11}O_2P)_2X]$ species, see p. 116. The electrical conductivity of an acetone solution is $\Lambda = 141\ cm^2 \cdot \Omega^{-1} \cdot mol^{-1}$ [1].

References:

[1] Copperthwaite, R. G.; Reimann, R. H.; Singleton, E. (Inorg. Chim. Acta **28** [1978] 107/11).
[2] Laing, M.; Reimann, R. H.; Singleton, E. (Inorg. Nucl. Chem. Letters **10** [1974] 557/60).
[3] Laing, M.; Reimann, R. H.; Singleton, E. (Inorg. Chem. **18** [1979] 324/7).
[4] Laing, M.; Reimann, R. H.; Singleton, E. (Inorg. Chem. **18** [1979] 1648/53).
[5] Laing, M.; Reimann, R. H.; Singleton, E. (Inorg. Chem. **18** [1979] 2666/70).

39.9 Complexes with Phosphonic or Phosphono Carboxylic Acids

General. The importance of interactions between metal salts and organylphosphonic acids for water desalination has long been recognized. Increasing attention has been paid to the study of the complex-forming properties of organophosphorus complexones. Alkylenepolyaminepoly(phosphonic acids) have been shown to have an enhanced coordination capacity compared to their carboxylic acid analogs. The increased stability of the complexes may be due to the greater polarizability and high nucleophilicity of the tetrahedral phosphonic acid group as compared to the planar carboxylic acid group. Increasing the chain length between the two basic nitrogen atoms was found to decrease the stability of the complexes. A review of organophosphorus complexones, specifically modification of their structures, their complexing properties, and their most interesting applications, is given by Kabachnik, M. I.; Medved', T. Ya.; Dyatlova, N. M.; Rudomino, M. V. (Usp. Khim. **43** [1974] 1554/74; Russ. Chem. Rev. **43** [1974] 733/44).

39.9.1 With Alkyl- or Arylphosphonic Acids

ligand 1 $HOCH_2P(O)(OH)_2$ $(=CH_5O_4P)$

ligand 2 $C_6H_5P(O)(OH)_2$ $(=C_6H_7O_3P)$

$Mn^{II}(CH_3O_4P)$. The hydroxymethylphosphonate crystallizes well from an aqueous solution of Mn^{II} salt and ligand 1, if maintained in a sealed tube for several hours at 140°C. As shown by Debye-Scherrer diagrams the compound is isomorphous with corresponding phosphonates of Zn^{II}, Ni^{II}, Co^{II}, Fe^{II}, and Cu^{II}. The IR spectrum reveals bands of ν(OH) at 3300 to 3200, δ(OH) at 1400 and γ(OH) between 800 and 680 cm^{-1}, slightly different from those observed in the spectrum of sodium hydroxymethylphosphonate. Contrary to the sodium salt, splittings of degenerate PO_3 vibration modes were observed with bands of $\nu_d(PO_3)$ at 1150 and between 1100 and 1040 cm^{-1} or $\delta_d(PO_3)$ at 525 and 495 cm^{-1}. Bands at 290 and 270 cm^{-1} were assigned to ν(Mn–O). Electronic absorption maxima at 28100, 27700, and 27170 cm^{-1} show splitting of the $T_{2g}(^4D)$ niveau; 10 Dq≈8100 cm^{-1}. The results are explained by a compact structure of distorted Mn–O octahedra, connected by phosphonate ions with an internal hydrogen bond. The compound is insoluble in water and organic solvents, soluble in acidic medium [1].

$Mn^{II}(C_6H_5O_3P)\cdot H_2O$ precipitated within a few minutes, after addition of 10 mmol Mn^{II} chloride dissolved in 75 mL ethanol to an equimolar solution of phenylphosphonic acid in the same amount of ethanol. The pale pink complex was washed with ethanol and dried [2]. The compound precipitated slowly at room temperature from aqueous solutions of $MnCl_2$ and phenylphosphonic acid, or its mono- or disodium salt. More crystalline samples were obtained by addition of $MnCl_2$ dissolved in H_2O to the aqueous solution of phenylphosphonic acid at 100°C. The platelike crystals show pronounced cleavage. They appear to be orthorhombic or pseudoorthorhombic. The magnetic moments $\mu_{eff}=5.28\ \mu_B$ at 20°C and 4.75 μ_B at −180°C and the Θ value (−42) are indicative of antiferromagnetic interactions [3]; $\mu_{eff}=5.82\ \mu_B$ was found at 298 K [2]. Spectral data indicate octahedral coordination; the phenylphosphonate ion acts as a bridging ligand between the Mn^{2+} ions, giving cross-linked polymers. Absorption bands in the visible spectrum (33330 to ~19400 cm^{-1}) and the IR region (1436 to 959 cm^{-1}) are given in the papers [2, 3]. The water molecule, assumed to be coordinated, is lost on heating at 130°C [3]. The complex is insoluble in all common solvents and dissolves with decomposition in mineral acids [2].

References:

[1] Brun, G.; Jourdan, G. (Rev. Chim. Minerale **12** [1975] 139/46).

[2] Grinonneau, W. C.; Chapman, P. L.; Menke, A. G.; Walmsley, F. (J. Inorg. Nucl. Chem. **33** [1971] 3011/7).

[3] Cunningham, D.; Hennelly, P. J. D.; Deeney, T. (Inorg. Chim. Acta **37** [1979] 95/102).

39.9.2 With Phosphonoalkanoic Acids

ligand 1	$HOOCP(O)(OH)_2$	$(=CH_3O_5P)$
ligand 2	$HOOCCH_2P(O)(OH)_2$	$(=C_2H_5O_5P)$
ligand 3	$HOOCCH_2CH_2P(O)(OH)_2$	$(=C_3H_7O_5P)$

Complexes in Solution. Formation constants of $Mn^{II}L^-$ and $Mn^{II}HL$ complexes with the deprotonated or monoprotonated ligands 1 to 3 ($=H_3L$) in aqueous solution at $I=0.05$ M ($N(C_2H_5)_4ClO_4$) and pH 8 or 4, respectively, were determined by ESR spectroscopy:

ligand	1	2	3
$\log K_1$	5.34 ± 0.05	5.25 ± 0.06	3.15 ± 0.04
$\log K_{MnHL}$	2.57 ± 0.05	2.97 ± 0.06	1.60 ± 0.2

The values for the complexes with phosphonoacetic acid were extrapolated to zero ionic strength in order to determine the thermodynamic constants, $\log K_1^\circ = 6.30 \pm 0.1$ for the complex $Mn(C_2H_2O_5P)^-$ and $\log K_{MnHL}^\circ = 3.50 \pm 0.2$ for $Mn(C_2H_3O_5P)$.

Reference:

Farmer, R. M.; Popov, A. I. (Inorg. Chim. Acta **59** [1982] 87/91).

39.9.3 With Alkanediylbis(phosphonic Acids)

ligand 1	$(HO)_2P(O)CH_2P(O)(OH)_2$	$(=CH_6O_6P_2)$
ligand 2	$(HO)_2P(O)(CH_2)_6P(O)(OH)_2$	$(=C_6H_{16}O_6P_2)$
ligand 3	$(HO)_2P(O)C(CH_3)(OH)P(O)(OH)_2$	$(=C_2H_8O_7P_2)$

Complexes in Solution. The existence of protonated or neutral mono- and binuclear complexes in aqueous solution has been assumed from the results of potentiometric titrations of solutions of the bis(phosphonic acids) in the presence of Mn^{2+} ions (mole ratio of $Mn^{2+}:H_4L=$ 1:1 and 2:1). At a mole ratio of Mn^{2+}:ligand = 1:2 only mononuclear species were observed by [1], but with an excess of ligand there is the possibility of forming bis(complexonates) [3]. Formation constants of $Mn^{II}L^{2-}$, $Mn^{II}HL^-$, and $Mn^{II}H_2L$ complexes formed by reaction of Mn^{2+} with L^{4-} ions or the protonated species, HL^{3-} and H_2L^{2-}, at 25°C are shown below:

ligand	I in mol/L	$\log K_1$	$\log K^{Mn}_{MnHL^-}$	$\log K^{Mn}_{MnH_2L}$	Ref.
1	0.1 (KCl)	12.95	7.20	—	[2]
2	0.1 (KCl)	—	5.82	—	[2]
3	0.1 (KCl)	9.16	5.26	—	[2]
3	0.1 (KNO_3)	6.94	4.42	3.18	[1]

A large number of complexes can be formed with ligand 3 in aqueous solution: Mn_2L, MnL^{2-}, $MnHL^-$, MnH_2L, $MnHL_2^{5-}$, $Mn(HL)_2^{4-}$, $MnH_3L_2^{3-}$, $Mn(H_2L)_2^{2-}$, $Mn(H_3L)_2$. The neutral species have been isolated as hydrates, the anions as potassium salts [3]; see below. Equilibrium constants for the reactions $2Mn^{2+}+L^{4-} \rightleftharpoons Mn_2L$ with H_4L = ligands 1 to 3 have been determined at 25°C, $I=0.1$ M (KCl): $\log K=15.85$, 12.51, and 13.23, respectively. Equilibrium constants for the reactions $2Mn^{2+}+HL^{3-} \rightleftharpoons Mn_2HL^+$ under the same conditions are $\log K=9.60$, 9.62, and 8.06, respectively [2]; see also [4]. An equilibrium constant, $\log K=19.64$, was reported in [2, 4] for the reaction $2Mn^{2+}+L^{5-} \rightleftharpoons Mn_2L^-$, assuming that ligand 3 is fivefold deprotonated, whereas no evidence was gained by [1] from $\bar{n}_H$ calculations to account for the formation of MnL^{3-} species with ligand 3. The hydroxy group is assumed to participate in coordination with retention of its proton. The Mn^{2+} ion is probably located between the two negatively charged phosphonate oxygen atoms in an approximately tetrahedral position, with water acting as the fourth ligand [1]. Dissociation of the proton from the hydroxy group of ligand 3 was also assumed in the case of the Mn^{III} complex species. Formation constants of $Mn^{III}(H_4L)^{2+}$ and $Mn^{III}(H_5L_2)^{2-}$ complexes with 1-hydroxy-1,1-ethanediylbis(phosphonic acid) formed by the reactions $Mn^{3+}+H_5L \rightleftharpoons MnH_4L^{2+}+H^+$ and $Mn^{3+}(H_4L)^{2+}+H_3L^{2-} \rightleftharpoons Mn(H_5L_2)^{2-}+2H^+$ were determined spectrophotometrically: log $K=3.55\pm0.10$ and 15.52 ± 0.5, respectively [5].

Isolated Compounds (with H_4L = ligand 3). Preparation of Mn^{II} complexes was effected by reaction of a solution of 1-hydroxyethylidenebis(phosphonic acid) or its salts with a manganese(II) salt (e.g., soluble $MnCl_2$ or insoluble $MnCO_3$) and also by neutralizing a solution of protonated complexes with various alkalies. The conditions for the preparations and the compositions of several isolated compounds are shown below [3]:

compound	conditions	mole ratio	pH
$Mn_2L \cdot 4\,H_2O$	reaction of $MnCO_3$ with H_4L in hot aqueous solution	2:1	3.0
$MnH_2L \cdot 5\,H_2O$	mixing of saturated H_4L with soluble Mn^{II} salt	1:1	1.0 to 3.0
	or reaction of hot aqueous H_4L with $MnCO_3$	1:1 to 1:2	1.0
$Mn(H_3L)_2 \cdot 9\,H_2O$	as above	1:3	1.0
	or mixing of saturated H_4L and MnH_2L	1:1	1.0
$Mn(H_3L)_2 \cdot 5\,H_2O$	from filtrate of the H_4L and $MnCO_3$ reaction after separation of $MnH_2L \cdot 5\,H_2O$ (see above)	1:2	1.0
$KMnHL \cdot 2\,H_2O$	addition of the calculated quantity of alkali to solution of MnH_2L	1:1	6.0
$K_2MnL \cdot 4\,H_2O$	reaction of aqueous H_4L with $MnCO_3$	1:2	8.5
	mixing of saturated H_4L and soluble Mn^{II} salt	1:1	7.5
$K_2Mn(H_2L)_2 \cdot 4\,H_2O$	addition of the calculated quantity of alkali to solution of $Mn(H_3L)_2$	1:2	3.5
$K_5MnHL_2 \cdot 8\,H_2O$	as above	1:2	7.5

$MnH_2L \cdot H_2O$ was precipitated by dropwise addition of acetone to an aqueous solution of Mn^{II} salt and ligand 3 (1:1 mole ratio). The white precipitate was washed with water containing acetone, then dried at 70 to 80°C. It does not melt up to 270°C and shows a magnetic moment of $\mu_{eff}=5.67\,\mu_B$. A polymeric structure is suggested for the compound [7].

According to the X-ray diffraction results, all the complexes with the exception of $K_2MnL \cdot 4\,H_2O$ and $(NH_4)_2MnL \cdot 6\,H_2O$ (preparation and properties of the ammonium salt not described in the paper) are crystalline [3]. The difference between the degree of hydration of $Mn_2L \cdot 4\,H_2O$ [3] and that of $Mn_2L \cdot 2\,H_2O$ described in [6], or between $MnH_2L \cdot 5\,H_2O$ [3] and $MnH_2L \cdot H_2O$ [7], is probably a result of the different methods used in drying the product.

The IR spectra of the complexes in mineral oil or hexachlorobutadiene are characterized by a distinctive group of absorption bands in the stretching vibration region of the phosphorus–oxygen bond between 1200 and 900 cm^{-1}. The spectra of $Mn(H_3L)_2 \cdot 5\,H_2O$ and $Mn(H_3L)_2 \cdot 9\,H_2O$ are similar. The ν(P=O) bond was found in both cases at 1220 cm^{-1}. The unambiguous assignment of the bands relating to $\nu_{as}(PO_2)$, $\nu_s(PO_2)$, ν_{as}(P–OH), ν_s(P–OH) observed between

1175 and 940 cm^{-1} is difficult on account of vibration interactions. A band of ν(P–OH) was observed at 925 or 930 cm^{-1}, respectively. The spectra are similar to those of $Cu(H_3L)_2 \cdot 12H_2O$ and $Ni(H_3L)_2 \cdot 4H_2O$. A symmetrical arrangement is suggested, where each ligand molecule has one PO_3H_2 and one PO_3H^- group. The spectrum of $MnH_2L \cdot 5H_2O$ is similar to that of $NiH_2L \cdot 6H_2O$ or $ZnH_2L \cdot 4H_2O$. The three intense bands of $\nu_{as}(PO_2)$ at 1160 cm^{-1}, $\nu_s(PO_2)$ at 1080 cm^{-1}, and ν(P–OH) at 930 cm^{-1} point to a symmetrical structure for the phosphonato group [3]. The IR spectrum of $MnH_2L \cdot H_2O$ in KBr reveals bands of ν(P=O) at 1150 cm^{-1}, $\nu_{as}(PO_2)$ at 1090 cm^{-1}, and $\nu_s(PO_2)$ at 990 and 920 cm^{-1} [7]. The spectra of $KMnHL \cdot 2H_2O$ and $NaMnHL \cdot 2H_2O$ differ from the spectrum of $NaCuHL \cdot 6H_2O$ in that the bands for $\nu_{as}(PO_2)$ and $\nu_d(PO_3)$ are well-separated in the manganese complex. Only bands pertaining to monoprotonated and ionized phosphonato groups are observed. The spectra of $Mn_2L \cdot 4H_2O$ or $K_2MnL \cdot 4H_2O$, and $(NH_4)_2MnL \cdot 6H_2O$ show $\nu_d(PO_3)$ and $\nu_s(PO_3)$ bands in the 1200 to 900 cm^{-1} region that result from completely ionized phosphonato groups; e.g., $\nu_d(PO_3)$ at 1010 cm^{-1} and $\nu_s(PO_3)$ at 960 cm^{-1} in the case of $Mn_2L \cdot 4H_2O$ [3].

The solubilities of $MnH_2L \cdot 5H_2O$ and $Mn_2L \cdot 4H_2O$ in water at 20°C are 0.1 and 0.085 mol/L, respectively. The solubilities are higher than those of analogous Ni or Cu complexes [3]. $MnH_2L \cdot H_2O$ was found to be insoluble in water and common organic solvents [7].

Differential thermal analysis demonstrates stepwise dehydration. In most cases two or three dehydration stages were observed, the first one between ~40 and 165°C. Pyrolysis in air and in nitrogen are essentially different: e.g., in air at 220 to 300°C decomposition of $MnH_2L \cdot 5H_2O$ or $Mn_2L \cdot 4H_2O$ and oxidation of manganese is superimposed on the dehydration process. In a nitrogen atmosphere dehydration and decomposition processes do not overlap. In comparison with the corresponding Ni or Cu complexes those of Mn with the same stoichiometric composition are more stable to thermal decomposition [3].

References:

[1] Rizkalla, E. N.; Zaki, M. T. M.; Ismail, M. I. (Talanta **27** [1980] 715/9).

[2] Kabachnik, M. I.; Lastovskii, R. P.; Medved', T. Ya.; Medyntsev, V. V.; Kolpakova, I. D.; Dyatlova, N. M. (Dokl. Akad. Nauk SSSR **177** [1967] 582/5; Dokl. Chem. Proc. Acad. Sci. USSR **172/177** [1967] 1060/3).

[3] Kushikbaeva, B. Kh.; Mitrofanova, N. D.; Martynenko, L. I.; Spitsyn, V. I. (Izv. Akad. Nauk SSSR Ser. Khim. **1987** 2649/54; Bull. Acad. Sci. USSR Div. Chem. Sci. **36** [1987] 2455/60).

[4] Sillén, L. G.; Martell, A. E. (Stability Constants, Chem. Soc. [London] Spec. Publ. No. 25 [1971] 1/865, 274).

[5] Popova, T. V. (Izv. Vysshikh Uchebn. Zavedenii Khim. Khim. Tekhnol. **21** [1978] 778/81; C.A. **89** [1978] No. 139657).

[6] Puri, D. M.; Palta, N. (Indian J. Chem. A **21** [1982] 624/5).

[7] Puri, D. M.; Rao, B. V.; Palta, N.; Dubey, S. N. (J. Indian Chem. Soc. **61** [1984] 899/901).

39.9.4 With Aminoalkylphosphonic Acids

No.	ligand	formula	No.	ligand	formula
1	$H_2NCH_2P(O)(OH)_2$	CH_6NO_3P	5	$H_2NC(CH_3)_2P(O)(OH)_2$	$C_3H_{10}NO_3P$
2	$H_2NCH_2CH_2P(O)(OH)_2$	$C_2H_8NO_3P$	6	$H_2NCH_2CH(OH)P(O)(OH)_2$	$C_2H_8NO_4P$
3	$H_2NCH_2CH_2CH_2P(O)(OH)_2$	$C_3H_{10}NO_3P$	7	$(HOCH_2CH_2)_2NCH_2P(O)(OH)_2$	$C_5H_{14}NO_5P$
4	$H_2NCH(CH_3)P(O)(OH)_2$	$C_2H_8NO_3P$	8	$H_2NCH_2CH_2NHCH_2P(O)(OH)_2$	$C_3H_{11}N_2O_3P$

Complexes in Solution. Stability constants of MnL and MnL_2^{2-} complexes with aminoalkylphosphonic acids in aqueous solution (determined pH-potentiometrically at 25°C and various ionic strengths) are tabulated below, together with stability constants of $MnHL^+$ species formed by the reaction $Mn^{2+} + HL^- \rightleftharpoons MnHL^+$:

ligand	I in mol/L	log K_1	log K_2	$\log^{Mn}_{MnHL^+}$	Ref.
1	0.1 ($NaClO_4$)	3.90	3.56	—	[1]
2	0.1 ($NaClO_4$)	4.25	3.13	2.88	[1]
2	0.2 (KNO_3)	—	—	2.12±0.2	[2]
3	0.1 ($NaClO_4$)	4.34	3.91	2.84	[1]
4	0.2 (KNO_3)	3.5±0.3	—	1.97±0.3	[2]
5	0.1 (KCl)	4.03	3.40	2.94	[3]
6	0.1 (KCl)*)	4.72	—	2.74	[4]
7	0.1 ($KClO_4$)	4.61	—	—	[5]
8	0.1 (KCl)	5.15	—	2.2	[6]

*) At 20°C.

The stability constants of different metal complexes with ligands 1 to 3 were found to be in the order $Mn^{II} < Co^{II} < Ni^{II} < Cu^{II} < Zn^{II}$ and were influenced by ligand basicity and chelate size. Aminophosphonic acids show comparatively low selectivity to the bivalent metal ions used. The stability constants, log K_{MnL} (log K_1), lie between those of glycine (see "Manganese" D 4, 1985, pp. 248/50) and adenosine triphosphate (see "Manganese" D 4, 1985, pp. 31/3) complexes [1]. The low values for the formation constants of $MnHL^+$ species with 1-aminoethylphosphonic acid (ligand 4) are consistent with monodentate phosphonate binding with the proton fixed to the amine group. At high pH the ML and ML_2^{2-} complexes constitute the major species. The magnitudes of their formation constants suggest bidentate ligand coordination for Zn^{II}, Mn^{II}, Ni^{II}, Co^{II}, and Cu^{II}. The $MnHL^+$ complexes with 2-aminoethylphosphonic acid (ligand 2) are more stable than the complexes with ligand 4 in accordance with the higher basicity of the phosphonate moiety in the 2-aminoethyl derivative [2]. The complex $MnHL^+$ with H_2L = ligand 8 probably exists in two isomeric forms: either a bidentate coordination of the anion HL^- via the two nitrogen atoms of the diamine group and a free PO_3H^- group or an O, N coordination via the aminomethylphosphonate part with a free $-CH_2CH_2\overset{+}{N}H_3$ group [6].

$Mn^{II}(CH_5NO_3P)_2 \cdot 2H_2O$ precipitates from an aqueous solution containing equimolar amounts of manganese(II) sulfate and potassium aminomethylphosphonate. The crystals were washed with cold water and ethanol. The IR spectrum reveals shifts of the PO_3 vibrations compared to uncoordinated aminomethylphosphonic acid (in parentheses): $\nu_d(PO_3)$ 1125 (1216, 1172); $\nu_s(PO_3)$ 990 (932). A band at 389 cm^{-1} was assigned to ν(Mn–O). The spectrum was found to be similar to that of $Zn(CH_5NO_3P)_2 \cdot 4H_2O$, which crystallizes in the orthorhombic system (space group $Pca2_1$–C_{2v}^5 (No. 29)) with Zn tetrahedrally coordinated by 4 oxygen atoms of 4 phosphonato groups. A similar type of coordination was therefore assumed for the manganese complex [7].

Na[Mn^{II}($C_2H_7NO_3P$)$_2$Cl(H_2O)] precipitates from a solution of $MnCl_2 \cdot 4H_2O$ (1.45 g) and aminoethylphosphonic acid (1.84 g) in 40 mL H_2O, after adjusting the pH to 6 with 0.1M NaOH and addition of acetone. The solid was washed with 75% acetone. It is soluble in H_2O. The magnetic moment at room temperature is $\mu_{eff} = 5.91\ \mu_B$. The IR spectrum shows that the bands at 1655 and 1560 cm^{-1}, assigned to NH_3^+ deformations of the aminoethylphosphonic acid zwitterion, are much weaker and occur at lower frequencies. This indicates displacement of a proton from the NH_3 group and probable coordination of the nitrogen to the metal. Coordination of Mn to the PO_3 group of the ligand is suggested by observed changes of the $\nu_{as}(PO_3)$ and $\nu_s(PO_3)$ vibration modes in the 1200 to 1000 and 1000 to 900 cm^{-1} regions. Thermal analysis showed weight loss equivalent to the coordinated water between 90 and 170°C with further decomposition above 215°C [8].

References:

[1] Sakurai, H.; Okumara, H.; Takeshima, S. (Yakugaku Zasshi **96** [1976] 242/5; C.A. **84** [1976] No. 141449).
[2] Mohan, M. S.; Abbott, E. H. (J. Coord. Chem. **8** [1978] 175/82).
[3] Dyatlova, N. M.; Medyntsev, V. V.; Balashova, T. M.; Medved', T. Ya.; Kabachnik, N. I. (Zh. Obshch. Khim. **39** [1969] 329/33; J. Gen. Chem. [USSR] **39** [1969] 309/13).
[4] Baranov, G. M.; Perekalin, V. V.; Pomerantseva, O. G.; Speranskii, E. M.; Shilov, S. M. (Koord. Khim. **13** [1987] 741/2; C.A. **107** [1987] No. 103793).
[5] Baseggio, A. A.; Grassi, R. L. (J. Inorg. Nucl. Chem. **43** [1981] 3275/6).
[6] Achilles, W.; Uhlig, E. (Z. Anorg. Allgem. Chem. **390** [1972] 225/33, 227).
[7] Fenot, P.; Darriet, J.; Garrigou-Lagrange, C.; Cassaigne, A. (J. Mol. Struct. **43** [1978] 49/60).
[8] Menke, A. G.; Walmsley, F. (Inorg. Chim. Acta **17** [1976] 193/7).

39.9.5 With Aryl or Pyridinyl Derivatives of Aminoalkylphosphonic Acids

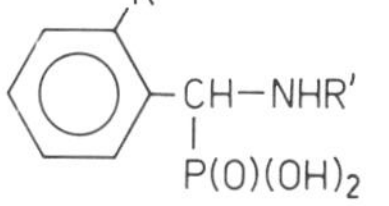

R
CH–NHR'
P(O)(OH)$_2$

ligand 1 (=H_2L)
ligand 2 to 7 (=H_3L)

ligand	R	R′	formula
1	H	$CH_2C_6H_5$	$C_{14}H_{16}NO_3P$
2	OH	$CH_2C_6H_5$	$C_{14}H_{16}NO_4P$
3	OH	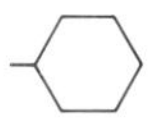	$C_{13}H_{20}NO_4P$
4	OH	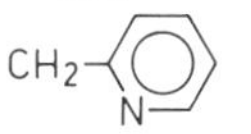 CH_2– N	$C_{13}H_{15}N_2O_4P$
5	OH	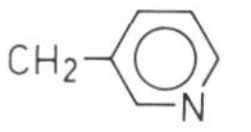 CH_2– N	$C_{13}H_{15}N_2O_4P$
6	OH	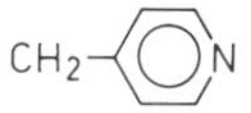 CH_2– N	$C_{13}H_{15}N_2O_4P$
7	OH	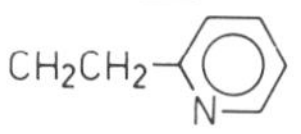 CH_2CH_2– N	$C_{14}H_{17}N_2O_4P$

Complexes in Solution. The stepwise formation of MnHL and MnL^- complexes, according to the reactions $Mn^{2+} + H_3L \rightleftharpoons MnHL + 2H^+$ and $MnHL \rightleftharpoons MnL^- + H^+$, was studied potentiometrically at 20.0 ± 0.5°C and I = 0.1M ($NaClO_4$). The MnHL complex is formed within the pH range of 2.7 to 6.2, the MnL^- complex within the pH range of 7.0 to 10.0 [1]. Calculated values of log K_1 and log K_{MnHL} are:

ligand	1	2	3	4	5	6	7
log K_1	6.30	6.60	6.60	7.70	7.70	7.80	7.60
log K_{MnHL}	3.05*)	3.30	3.30	3.60	3.60	3.60	3.50
Ref.	[2]	[2]	[2]	[1]	[1]	[1]	[2]

*) Value of log K_{MnHL^+}.

The complexes are more stable than those with the corresponding aminoalkylphosphinic acids, see p. 113. Complex stability is increased by the presence of the phenolic group and the pyridine nitrogen atom [2].

References:

[1] Siepak, J. (Polish J. Chem. **59** [1985] 651/63, 660; C.A. **106** [1987] No. 176495).
[2] Siepak, J. (Polish J. Chem. **59** [1985] 955/69, 961; C.A. **106** [1987] No. 108761).

39.9.6 With Amino-1,1-alkanediylbis(phosphonic Acids) or Related Compounds

No.	ligand (= H_4L)	formula
1	$H_2NC(CH_3)[P(O)(OH)_2]_2$	$C_2H_9NO_6P_2$
2	$(CH_3)_2NCH[P(O)(OH)_2]_2$	$C_3H_{11}NO_6P_2$
3	$CH_3(C_6H_5CH_2)NCH[P(O)(OH)_2]_2$	$C_9H_{15}NO_6P_2$
4	$H_2NC(C_6H_5)[P(O)(OH)_2]_2$	$C_7H_{11}NO_6P_2$
5	$(CH_3)_2NCH_2CH_2C(OH)[P(O)(OH)_2]_2$	$C_5H_{15}NO_7P_2$
6	$RCH[P(O)(OH)_2]_2$ R = (piperidino ring) N–	$C_6H_{15}NO_6P_2$
7	$RCH[P(O)(OH)_2]_2$ R = O (morpholino ring) N–	$C_5H_{13}NO_7P_2$

Complexes in Solution. Stability constants of MnL^{2-} complexes (log K_1) and values for the monoprotonated species, $MnHL^-$, formed by the reaction $Mn^{2+} + HL^{3-} \rightleftharpoons MnHL^-$ in aqueous solution (1:1 mole ratio), were determined potentiometrically (glass electrode) at 25°C and ionic strength I = 0.1M (KCl):

ligand	2	3	4	5	6	7
log K_1	7.26	7.03	9.93	8.33	7.05	7.75
log K_{MnHL^-}	6.71	6.38	7.29	8.09	6.58	6.27
Ref.	[1]	[1]	[2]	[3]	[1]	[1]

A formation constant, log K = 15.41, for a 2:1 complex with ligand 4, formed by the reaction $2Mn^{2+} + L^{4-} \rightleftharpoons Mn_2L$, was determined under the same conditions [2]. The nature of the substituents on the nitrogen atom of the ligands have little influence on the stabilities of the complexes, but it influences their solubilities considerably. A characteristic feature is the small difference in stability between the MnL^{2-} and $MnHL^-$ complexes [1].

$Mn^{II}_2(C_2H_5NO_6P_2)\cdot 2H_2O$ and $Mn^{II}_2(C_7H_7NO_6P_2)\cdot H_2O$. The complexes with ligand 1 or 4 were prepared by mixing aqueous solutions of an Mn^{II} salt (0.002 mol) and the corresponding ligand (0.001 mol). In the case of ligand 1 it was followed by the dropwise addition of a 0.004M solution of NaOH. Addition of acetone-methanol to the reaction mixtures improved the yield of the precipitated compounds, which were washed with 50% acetone and dried at 60 to 70°C. Reactions of Mn^{II} salt and ligand in a 1:1 mole ratio also resulted in the formation of Mn_2L complexes. The magnetic moment of the light pink complex, $Mn_2(C_2H_5NO_6P_2)\cdot 2H_2O$, (4.12 μ_B) and the value for the white compound, $Mn_2(C_7H_7NO_6P_2)\cdot H_2O$, (4.04 μ_B) indicate that there is a considerable interaction between the manganese ions. Principal bands (in cm^{-1}) observed in the IR spectra of KBr pellets were assigned as follows (vibration modes of the free ligands are given in parentheses):

complex	ν(OH) + ν(NH_2)	ν(PO)	$\nu_{as}(PO_3)$	$\nu_s(PO_3)$
$Mn_2(C_2H_5NO_6P_2)\cdot 2H_2O$	3320(3400), 3210(3160)	1625(1630), 1110(1200)	1080, 1050, 1020	990, 910
$Mn_2(C_7H_7NO_6P_2)\cdot H_2O$	3000(3400)	1190(1280)	1050, 1030	980

Coordination between the phosphoryl oxygen atoms and manganese is assumed. Participation of the NH_2 group in coordination could not be inferred from the IR spectra, due to the C–N bands being very weak and the masking of the NH bands by broad OH bands. A polymeric structure is suggested for the compounds. They are insoluble in water or common organic solvents and do not melt even up to 270°C [4].

References:

[1] Gross, H.; Medved', T. Ya.; Costisella, B.; Bel'skii, F. I.; Kabachnik, M. I. (Zh. Obshch. Khim. **48** [1978] 1914/6; J. Gen. Chem. [USSR] **48** [1978] 1746/7).

[2] Dyatlova, N. M.; Medyntsev, V. V.; Balashova, T. M.; Medved', T. Ya.; Kabachnik, N. I. (Zh. Obshch. Khim. **39** [1969] 329/33; J. Gen. Chem. [USSR] **39** [1969] 309/13).

[3] Kabachnik, M. I.; Bel'skii, F. I.; Komarova, M. P.; Shcherbakov, B. K.; Matrosov, E. I.; Polikarpov, Yu. M.; Dyatlova, N. M.; Medved', T. Ya. (Izv. Akad. Nauk SSSR Ser. Khim. **1979** 1726/30; Bull. Acad. Sci. USSR Div. Chem. Sci. **1979** 1591/5).

[4] Venkateswara Rao, B.; Palta, N.; Dubey, S. N.; Puri, D. M. (J. Indian Chem. Soc. **61** [1984] 951/5).

39.9.7 With Phosphono or Bisphosphono Amino Carboxylic Acids

No.	ligand	formula
1	$HOOCCH(NH_2)CH_2P(O)(OH)_2$	$C_3H_8NO_5P$ (= H_3L)
2	$HOOCCH_2NHCH_2P(O)(OH)_2$	$C_3H_8NO_5P$ (= H_3L)
3	$(HOOCCH_2)_2NCH_2P(O)(OH)_2$	$C_5H_{10}NO_7P$ (= H_4L)
4	$HOOCCH_2CH_2N(CH_2COOH)CH_2P(O)(OH)_2$	$C_6H_{12}NO_7P$ (= H_4L)
5	$HOOCCH(NH_2)CH_2CH_2P(O)(OH)_2$	$C_4H_{10}NO_5P$ (= H_3L)
6	$HOOCCH_2CH(CH_2NH_2)P(O)(OH)_2$	$C_4H_{10}NO_5P$ (= H_3L)
7	$HOOCCH_2C(NH_2)[P(O)(OH)_2]_2$	$C_3H_9NO_8P_2$ (= H_5L)
8	$HOOCCH_2N[CH_2P(O)(OH)_2]_2$	$C_4H_{11}NO_8P_2$ (= H_5L)

Complexes in Solution. Formation constants of MnL^- complexes in aqueous solution at a mole ratio of $Mn^{2+}:H_3L = 1:1$ have been determined pH-potentiometrically at 25°C: log K_1 = 4.90 [1] for the complex with ligand 1 at I = 0.2 M (KNO_3), log K_1 = 5.47 [2] and 5.53 [3] for the complex with ligand 2 at I = 0.1 M(KNO_3). Values of log K_1 = 4.06 and 7.88 have been determined at 20°C, I = 0.1 M(KCl) for the complexes with ligands 5 and 6, respectively [4].

Stability constants of MnHL complexes formed by different reactions are shown below:

ligand (= H_3L)	1	2	2	5	6
log K_{MnHL}	2.60$^{a)}$	12.30$^{b)}$	6.92$^{c)}$	2.14$^{a)}$	2.98$^{a)}$
I in mol/L	0.2(KNO_3)	0.1(KNO_3)	0.1(KNO_3)	0.1(KCl)	0.1(KCl)
Ref.	[1]	[2]	[3]	[4]	[4]

$^{a)}$ $K_{MnHL} = [MnHL]/[Mn^{2+}][HL^{2-}]$. – $^{b)}$ $K_{MnHL} = [MnHL]/[Mn^{2+}][H^+][L^{3-}]$. – $^{c)}$ $K_{MnHL} = [MnHL]/[H^+]$-$[MnL^-]$.

In the case of the MnHL complex the proton probably resides on the nitrogen donor group. The phosphonato oxygen atoms are involved in the formation of the chelates, but carboxylate coordination is unlikely. For the complex with ligand 2, MnL^-, an approximately planar arrangement is proposed; this minimizes charge repulsions between the *trans*-oriented negatively charged phosphonato and carboxylato groups [3]. Formation of an MnL_2^{4-} complex with ligand 2 was observed at pH 11 to 12: log β_2 = 7.80 [2]. A value of log K = 4.30 was determined at I = 0.1M(KNO_3) and 25°C for the species $MnL(OH)^{2-}$, formed by the reaction $MnL^- + OH^- \rightleftharpoons MnL(OH)^{2-}$. Ligand 2 is assumed to be tridentate [3].

MnL^{2-} complexes are formed with H_4L = ligand 3 or 4 in aqueous solution. The stability constants were determined pH-potentiometrically: log K_1 = 8.01 (mean value) at 30°C and ionic strength I = 0.1M(KCl) for the complex with ligand 3 [5], log K_1 = 7.24 at 25°C and I = 0.1M (KNO_3) for the complex with ligand 4 [6]. For the $MnHL^-$ complex with ligand 3, formed by the reaction $Mn^{2+} + HL^{3-} \rightleftharpoons MnHL^-$, a stability constant of log K = 3.5 was determined by paper electrophoresis at 20°C and I = 0.1 M(KNO_3). The complex is formed in alkaline solution [7].

The formation constant, log K_1 = 7.00 ± 0.5, for the MnL^{3-} complex with ligand 8 (= H_5L) was determined pH-potentiometrically at 25°C and I = 0.1 M(KNO_3) [6]. The stepwise formation of MnH_2L^-, $MnHL^{2-}$, and MnL^{3-} with ligand 8 (pH range 1.4 to 5.3, $Mn^{2+} + H_3L^{2-} \rightleftharpoons MnH_{3-n}L^{n-} + nH^+$) was investigated by potentiometric pH titration at 25°C and I = 0.1 M(KCl): log $K_{MnH_2L^-}$ = 3.87 ± 0.3, log $K_{MnHL^{2-}}$ = 4.75 ± 0.2, log $K_{MnL^{3-}}$ = 8.49 ± 0.2 [9]. Formation of the MnH_2L^- complex with ligand 8 up to the alkaline region was shown by paper electrophoresis [7].

The reaction of Mn^{3+} with ligand 8 has been studied by spectrophotometric and kinetic methods. The formation constant of the complex MnL^{2-}, formed at pH 4.5 by the reaction $Mn^{3+} + H_5L \rightleftharpoons MnL^{2-} + 5H^+$, was determined at 20°C and ionic strength 1.0 M($NaClO_4$): log K = 2.79 ± 0.13 [10].

For the separation of metal ion mixtures by ion exchange with ligand 3 or 8, see [8].

$Mn_5^{II}(C_3H_4NO_8P_2)_2 \cdot 2H_2O$ and **$Mn_5^{II}(C_4H_6NO_8P_2)_2 \cdot 5H_2O$** complexes were prepared by reaction of stoichiometric amounts of Mn^{II} salt solutions with ligand 7 or 8, respectively. Addition of either sodium carbonate or acetone helped to improve the yield of $Mn_5(C_3H_4NO_8P_2)_2 \cdot 2H_2O$, and separation was easier. The precipitate was washed first with hot water, then with 50% acetone, finally with 90% acetone. The isolated solid was dried at 70 to 80°C in an air oven [11]. In the case of the complex with ligand 8, $Mn_5(C_4H_6NO_8P_2)_2 \cdot 5H_2O$, a solution of KOH was added to the reaction mixture. The polynuclear compound is formed only over a narrow pH range (4 to 5). The precipitate was washed with water and dried to constant weight over P_4O_{10} [12].

The observed magnetic moment of $Mn_5(C_3H_4NO_8P_2)_2 \cdot 2H_2O$, μ_{eff} = 3.9 μ_B at 306 K, indicates strong antiferromagnetic exchange interactions, due to either direct metal-metal interactions or super exchange through phosphonic acid bridges. Principal bands observed in the IR spectrum of the complex were assigned as follows (bands of 3-amino-3,3-bis(phosphono)-

propanoic (ligand 7) are given in parentheses): $\nu_{as}(COO)$ 1630(1660); $\nu_s(COO)$ 1400(1300); $\nu(PO)$ 1130(1160); $\nu_{as}(PO_3^{2-})$ 1050; $\nu_s(PO_3^{2-})$ 990; ν(Mn–O) 410; ν(Mn–N) 485 cm^{-1}. The shifts of $\nu_{as}(COO)$ and $\nu_s(COO)$ indicate involvement of the carboxylic group in bond formation. The lowering of $\nu_s(COO)$, mainly due to the $\nu(C{=}O)$ vibration of the COOH group and the difference between $\nu_{as}(COO)$ and $\nu_s(COO)$, clearly suggest the coordination of the C=O moiety of the COOH group to manganese. Coordination by the phosphonato group is shown by the negative shift of $\nu(PO)$, $\nu_s(PO_3^{2-})$, and $\nu_{as}(PO_3^{2-})$ bands. The electronic spectrum shows bands at 8700, 14710, 18180, 21280, 22220, and 25000 cm^{-1}. The diffuse reflectance spectrum of the complex does not provide much information about its oscillator strength [11].

The solubility of $Mn_5(C_4H_6NO_8P_2)_2 \cdot 5H_2O$ at 25°C and I=0.1M (KCl) is $S_{av}=0.00369$ mol/L and the solubility product is $K_{so}=1.10\times10^{-35}$ [12]. As shown by thermal analysis, dehydration of the complex is overlapped by decomposition. Splitting out of water was observed in the range 55 to 245°C; decomposition begins at 295°C [13].

References:

[1] Mohan, M. S.; Abbott, E. H. (J. Coord. Chem. **8** [1978] 175/82).
[2] Motekaitis, R. J.; Martell, A. E. (J. Coord. Chem. **14** [1985] 139/49).
[3] Lundager Madsen, H. E.; Christensen, H. H.; Gottlieb-Petersen, C. (Acta Chem. Scand. A **32** [1978] 79/83).
[4] Baranov, G. M.; Deiko, L. I.; Perekalin, V. V.; Pomerantseva, O. G.; Speranskii, E. M.; Shilov, S. M. (Koord. Khim. **13** [1987] 592/4; C.A. **107** [1987] No. 65591).
[5] Ockerbloom, N.; Martell, A. E. (J. Am. Chem. Soc. **80** [1958] 2351/4).
[6] Westerback, S.; Rajan, K. S.; Martell, A. E. (J. Am. Chem. Soc. **87** [1965] 2567/72).
[7] Majer, J.; Trinh Van Quy; Valášková, I. (Chem. Zvesti **34** [1980] 637/44; C.A. **94** [1981] No. 37256).
[8] Muchová, A.; Pikulíková, Z. (Chem. Zvesti **40** [1986] 37/43; C.A. **105** [1986] No. 17301).
[9] Nikitina, L. V.; Karmazina, L. D.; Dyatlova, N. M. (Zh. Neorgan. Khim. **19** [1974] 3058/63; Russ. J. Inorg. Chem. **19** [1974] 1671/5).
[10] Kurochkina, L. V.; Pechurova, N. I.; Snezhko, N. I.; Spitsyn, V. I. (Zh. Neorgan. Khim. **23** [1978] 2676/80; Russ. J. Inorg. Chem. **23** [1978] 1481/4).

[11] Venkateswara Rao, B.; Palta, N.; Dubey, S. N.; Puri, D. M. (Syn. React. Inorg. Metal-Org. Chem. **15** [1985] 1199/217).
[12] Karmazina, L. D.; Nikitina, L. V.; Rudomino, M. V.; Dyatlova, N. M. (Koord. Khim. **1** [1975] 905/10; Soviet J. Coord. Chem. **1** [1975] 765/9).
[13] Karmazina, L. D.; Dyatlova, N. M.; Nikitina, L. V.; Gurevich, M. Z. (Koord. Khim. **1** [1975] 804/9; Soviet J. Coord. Chem. **1** [1975] 678/82).

39.9.8 With Iminobis(methylphosphonic Acids)

$(HO)_2(O)PCH_2N(R)CH_2P(O)(OH)_2$ (= H_4L)

ligand	1	2	3	4	5	6
R	H	CH_3	C_2H_5	$C(CH_3)_3$	$CH_2C_6H_5$	cyclohexyl (structure)
formula ...	$C_2H_9NO_6P_2$	$C_3H_{11}NO_6P_2$	$C_4H_{13}NO_6P_2$	$C_6H_{17}NO_6P_2$	$C_9H_{15}NO_6P_2$	$C_8H_{19}NO_6P_2$

Complexes in Solution. Formation constants of MnL^{2-} and $MnHL^-$ complexes, formed by the reaction of Mn^{2+} with the anions L^{4-} or HL^{3-} of ligands 1 or 3 to 6 in aqueous solution (1:1 mole ratio), were determined potentiometrically at I=1.0M (KNO_3) and 25°C:

ligand	1	1	3	4	5	6
log K_1	—	6.26[a)]	6.94	6.39	6.54	6.11
log K_{MnHL^-}	2.95	2.71[a,b)]	3.29	4.57	3.23	3.35
Ref.	[1]	[2]	[1]	[1]	[1]	[1]

a) At I=0.1M (KNO_3) – b) Cited in [3].

Stability constants of $MnHL^-$ and MnH_2L complexes with ligand 1, formed by the equilibria $Mn^{2+} + nH^+ + L^{4-} \rightleftharpoons MnH_nL^{2-n}$ (log K=14.42 for n=1, log K=19.28 for n=2), and a constant (log K=−4.4) for the hydrolysis reaction $Mn^{2+} + L^{4-} \overset{H_2O}{\rightleftharpoons} Mn(OH)L^{3-} + H^+$ have been calculated by [2]. Protonation constants of $MnHL^-$ and MnH_2L complexes with ligand 1 or ligand 3 to 6 are reported in [3].

The IR spectrum of the MnL^{2-} complex with ligand 1 in aqueous solution reveals bands of $\nu_{as}(PO_3)$ at 1102 cm^{-1} and $\nu_s(PO_3)$ at 980 cm^{-1} [4]. The reaction with ligand 2 may be useful for the spectrophotometric determination of Mn^{II} in the UV region [5].

References:

[1] Bel'skii, F. I.; Goryunova, I. B.; Petrovskii, P. V.; Medved', T. Ya.; Kabachnik, M. I. (Izv. Akad. Nauk SSSR Ser. Khim. **1982** 103/9; Bull. Acad. Sci. USSR Div. Chem. Sci. **31** [1982] 93/9).

[2] Motekaitis, R. J.; Martell, A. E. (J. Coord. Chem. **14** [1985] 139/49).

[3] Smith, R. M.; Martell, A. E. (Critical Stability Constants, Vol. 6, 2nd Suppl., Plenum, New York – London 1989, pp. 1/643, 292/4).

[4] Zhadanov, B. V.; Polyakova, I. A.; Tsirul'nikova, N. V.; Sushitskaya, T. M.; Temkina, V. Ya. (Koord. Khim. **5** [1979] 1614/9; Soviet J. Coord. Chem. **5** [1979] 1254/9).

[5] Luaces Pazos, M. C.; Bermejo Barrera, A.; Bermejo Martinez, F. (Acta Quim. Compostelana **6** [1982] 1/8; C.A. **99** [1983] No. 98320).

39.9.9 With Ethylenediamine- or 1,2-Phenylenediamine-N,N'-bis(alkylphosphonic Acids)

No.	ligand (= H_4L)	formula
1	$(HO)_2(O)PCH_2NHCH_2CH_2NHCH_2P(O)(OH)_2$	$C_4H_{14}N_2O_6P_2$
2	$(HO)_2(O)PC(CH_3)_2NHCH_2CH_2NHC(CH_3)_2P(O)(OH)_2$	$C_8H_{22}N_2O_6P_2$
3	$(HO)_2(O)PCH_2NHC_6H_4NHCH_2P(O)(OH)_2$	$C_8H_{14}N_2O_6P_2$
4	$(HO)_2(O)PC(CH_3)_2NHC_6H_4NHC(CH_3)_2P(O)(OH)_2$	$C_{12}H_{22}N_2O_6P_2$

Complexes in Solution. MnL^{2-} and MnH_2L complexes are formed in aqueous solution with the bis(alkylphosphonic acids) of ethylenediamine. The stability constants, log K_1 = 7.55 [1] or 7.25 [2] for the MnL^{2-} complex with ligand 1 and log K_1 = 8.00 [1] for the complex with ligand 2, have been determined potentiometrically (glass electrode) at 25°C and I=0.1M (KCl). Values of log K=3.63 and 3.57 have been determined under the same conditions for the MnH_2L complexes, formed by the reaction of Mn^{2+} with H_2L^{2-} ions of ligands 1 and 2, respectively [1]. Polarographic half-wave potentials for the complexes with ligands 1 and 2 at pH 2.5, 9.3, and 12 are given in [3].

$Mn^{II}_2L \cdot nH_2O$. Complexes with n=4 for H_4L=ligand 1 and n=2 for H_4L=ligand 3 were precipitated from Mn^{II} salt solution by slow addition of aqueous solutions of ligand 1 or 3, respectively, in a 2:1 mole ratio, followed by addition of 1 N NaOH solution. The yield was improved by the addition of acetone or methanol to the mixture. The precipitates were washed with 25% acetone-water mixture then dried at 60 to 70°C. The almost colorless compounds are stable in air, nonhygroscopic, and insoluble in water or common organic solvents. They do not melt up to 300°C [4]. $Mn_2L \cdot 2H_2O$ complexes with ligand 2 or 4 were prepared in a similar way, but the addition of sodium hydroxide was omitted. The precipitates were washed with hot water, then with 50% acetone, and finally with 90% acetone. They do not melt up to 270°C [5]. The magnetic moments at 298 K, $\mu_{eff} = 4.54$ and 4.60 μ_B for the complexes with ligands 1 and 3 [4] or $\mu_{eff} = 4.52$ and 4.38 μ_B for the complexes with ligand 2 or 4 [5], suggest appreciable antiferromagnetic exchange interactions, probably arising from super-exchange through PO_3^{2-} groups. The electronic spectrum of the complex with ligand 2 shows absorptions at 8330, 17680, and 31750 cm^{-1} [5]. The compounds and similar complexes of Co^{II}, Ni^{II}, Cu^{II}, Zn^{II}, Cd^{II}, or Pb^{II} are polymeric and probably have octahedral geometry, according to IR and electronic spectral data, magnetic moments, and thermogravimetric studies. The ligands are assumed to be tetradentate for each metal atom; coordination is taking place through phosphoryl oxygen and amino nitrogen atoms [4, 5].

References:

[1] Dyatlova, N. M.; Kabachnik, M. I.; Medved', T. Ya.; Rudomino, M. V.; Belugin, Yu. F. (Dokl. Akad. Nauk SSSR **161** [1965] 607/10; Dokl. Chem. Proc. Acad. Sci. USSR **160/165** [1965] 307/11).

[2] Achilles, W.; Uhlig, E. (Z. Anorg. Allgem. Chem. **390** [1972] 225/33).

[3] Dyatlova, N. M.; Belugin, Yu. F.; Medved', T. Ya.; Rudomino, M. V. (Tr. Vses. Nauchn. Issled. Inst. Khim. Reaktivov Osobo Chist. Khim. Veshchestv No. 29 [1966] 192/201; C.A. **68** [1968] No. 90386).

[4] Palta, N.; Rao, B. V.; Dubey, S. N.; Puri, D. M. (Indian J. Chem. A **23** [1984] 397/400).

[5] Venkateswara Rao, B.; Palta, N.; Dubey, S. N.; Puri, D. M. (Transition Metal Chem. [Weinheim] **10** [1985] 60/2).

39.9.10 With N-Phosphonomethyl Derivatives of Ethylenediamine-N,N'-diacetic Acids

$HOOCCH_2NHCH_2CH_2N(CH_2COOH)CH_2P(O)(OH)_2$
ligand 1 ($= C_7H_{15}N_2O_7P = H_4L$)

$(HO)_2(O)PCH_2(HOOCCH_2)NCH_2CH_2N(CH_2COOH)CH_2P(O)(OH)_2$
ligand 2 ($= C_8H_{18}N_2O_{10}P_2 = H_6L$)

Complexes in Solution. Complex formation of ligand 1 with divalent and trivalent cations was investigated by paper electrophoresis. The shape of electrophoretic mobility curves for Mn^{II}, Co^{II}, Ni^{II}, Zn^{II}, and Cd^{II} indicates the formation of MHL^- complexes in the pH range from 7.5 to 9.0. The anionic mobility decreases at pH > 9.0, probably due to the formation of ML^{2-} complexes [1]. Stability constants of MnL^{4-} and MnH_2L^{2-} complexes formed with the anions of ligand 2, log $K_1 = 13.63$ and log $K_{MnH_2L^{2-}} = 7.00$, have been determined potentiometrically at I = 0.1 M (KCl) and 25°C [2]; see also [3, 4]. According to [5] precipitation takes place above pH 8.5. A stability constant of the protonated complex, log $K_{MnH_2L^{2-}} = 3.9 \pm 0.1$, was determined potentiometrically at I = 1.0 M (KNO_3) and 25°C. The manganese atom is assumed to be coordinated through the nitrogen atoms and the deprotonated carboxylato group [5]. A polarographic half-wave potential of the complex with ligand 2, determined at pH 9.3, $E_{1/2} = -1.58$ V versus NCE, is given in [2].

References:

[1] Butvin, B.; Valášková, I.; Kotouček, M.; Majer, J. (Chem. Zvesti **38** [1984] 341/91; C.A. **102** [1985] No. 46023).
[2] Dyatlova, N. M.; Belugin, Yu. F.; Medved', T. Ya.; Rudomino, M. V. (Tr. Vses. Nauchn. Issled. Inst. Khim. Reaktivov Osobo Chist. Khim. Veshchestv No. 29 [1966] 192/201; C.A. **68** [1968] No. 90386).
[3] Kabachnik, M. I.; Medved', T. Ya.; Dyatlova, N. M.; Rudomino, M. V. (Usp. Khim. **43** [1974] 1554/74; Russ. Chem. Rev. **43** [1974] 733/44, 737).
[4] Kabachnik, M. I.; Dyatlova, N. M.; Medved', T. Ya. (Pure Appl. Chem. **44** [1975] 269/83).
[5] Rajan, K. S.; Murase, I.; Martell, A. E. (J. Am. Chem. Soc. **91** [1969] 4408/12).

39.9.11 With Oxa-, Aza-, or Thiapentanediamine-N,N'-bis(alkylphosphonic Acids)

$(HO)_2(O)PC(CH_3)_2NHCH_2CH_2–Y–CH_2CH_2NHC(CH_3)_2P(O)(OH)_2$ ($=H_4L$)

ligand 1 with Y=O	ligand 2 with Y=NH	ligand 3 with Y=S
($=C_{10}H_{26}N_2O_7P_2$)	($=C_{10}H_{27}N_3O_6P_2$)	($=C_{10}H_{26}N_2O_6P_2S$)

Complexes in Solution. Formation constants of MnL^{2-} and MnH_2L complexes, formed by reaction of Mn^{2+} with L^{4-} or H_2L^{2-} ions in aqueous solution, have been determined pH-potentiometrically at I=0.1M (KCl) and 25°C:

ligand	1	2	3
log K_1	7.45	9.82	8.57
log K_{MnH_2L}	—	3.43	3.11

A comparison with the stability of corresponding complexes of ethylenediaminebis(alkylphosphonic acids), see p. 130, reveals that the presence of the heteroatoms O, N, or S in the phosphorylated amine derivatives leads to an increase in the complex-forming ability of the ligands [1].

$Mn^{II}_2(C_{10}H_{23}N_3O_6P_2)\cdot 3H_2O$. The dark brown complex was prepared by dropwise addition of Mn^{II} salt solution to an alkaline solution of ligand 2 (2:1 mole ratio). The precipitate was allowed to settle overnight, filtered, then washed several times with hot water, aqueous acetone, and 90% acetone. It was kept in vacuum over P_4O_{10} for about a month. The low magnetic moment of 4.9 μ_B reveals the presence of antiferromagnetism. A polymeric structure is suggested from comparison of IR data and electronic spectra of the analogous Ni^{II}, Co^{II}, Cu^{II} complexes. The metal atom is assumed to be hexacoordinated by oxygen atoms of phosphoryl and deprotonated phosphonic acid groups in addition to imino nitrogen atoms and one water molecule [2].

References:

[1] Kabachnik, M. I.; Medved', T. Ya.; Dyatlova, N. M.; Rusina, H. N.; Rudomino, M. V. (Izv. Akad. Nauk SSSR Ser. Khim. **1967** 1501/6; Bull. Acad. Sci. USSR Div. Chem. Sci. **1967** 1450/4).
[2] Venkateswara Rao, B.; Puri, D. M. (Indian J. Chem. A **26** [1987] 1048/50).

39.9.13 With Tris(alkylphosphonic Acids)

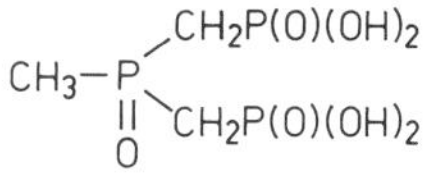

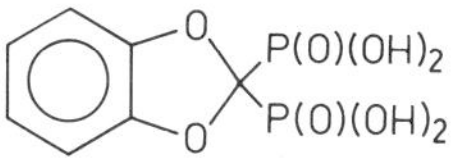

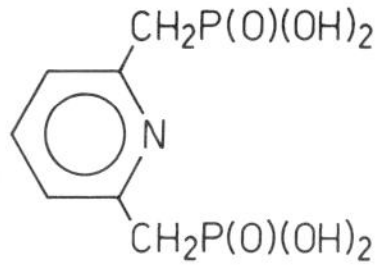

ligand 1 ($=C_3H_{11}O_7P_3$) ligand 2 ($=C_7H_8O_8P_2$) ligand 3 ($=C_7H_{11}NO_6P_2$)

Complexes in Solution. Stability constants of manganese(II) complexes, formed in aqueous solution by reaction of Mn^{2+} with L^{4-} ions or the protonated species HL^{3-} and H_2L^{2-}, have been determined potentiometrically at 25°C:

ligand	I in mol/L	log K_1	log K_{MnHL^-}	log K_{MnH_2L}	log K_2	log K_{Mn_2L}	Ref.
1	0.1 (KCl)	5.06	3.98	—	8.78	8.5	[1]
2	0.5 ($N(CH_3)_4Cl$)	7.25	4.10	—	—	—	[2]
3	0.1 (KNO_3)	6.66	4.02	2.39	—	—	[3]

The bond between ligand 2 and metal ions is predominantly ionic. The substituent of the diphosphonic acid probably has no donor properties. The Mn^{II} complex was found to be more stable than the Co^{II} complex. A comparison with other complexes of divalent metals shows that the stabilities do not conform to the Irving-Williams series [2]. Ligand 3 is apparently tridentate and capable of forming two six-membered chelate rings with the participation of the pyridine ring nitrogen atom and oxygen atoms of the phosphonic acid groups [3].

References:

[1] Kabachnik, M. I.; Dyatlova, N. M.; Medved', T. Ya.; Medyntsev, V. V.; Polikarpov, Yu. M.; Tsareva, Z. I. (Izv. Akad. Nauk SSSR Ser. Khim. **1968** 2058/61; Bull. Acad. Sci. USSR Div. Chem. Sci. **1968** 1955/8).
[2] Gross, H.; Medved', T. Ya.; Nowak, S.; Bel'skii, F. I.; Keitel, I.; Kabachnik, M. I. (Zh. Obshch. Khim. **55** [1985] 734/8; J. Gen. Chem. [USSR] **55** [1985] 654/7).
[3] Kabachnik, M. I.; Polikarpov, Yu. M.; Bodrin, G. V.; Bel'skii, F. I. (Izv. Akad. Nauk SSSR Ser. Khim. **1988** 921/4; Bull. Acad. Sci. USSR Div. Chem. Sci. **37** [1988] 800/3).

39.9.13 With Tris(alkylphosphonic Acids)

$N(CH_2P(O)(OH)_2)_3$

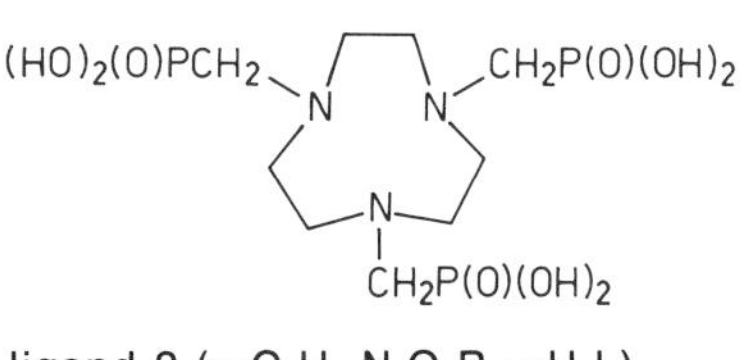

ligand 1 ($=C_3H_{12}NO_9P_3=H_6L$) ligand 2 ($=C_9H_{24}N_3O_9P_3=H_6L$)

Complexes in Solution. The formation of MnL^{4-} complexes or three protonated species $MnH_nL^{(4-n)-}$ with both ligands in aqueous solution according to the reaction $Mn^{2+}+H_nL^{(6-n)-} \rightleftharpoons MnH_nL^{(4-n)-}$ was shown by pH-potentiometric measurements at 25°C [1, 3]. The stability constants are tabulated below together with values determined polarographically at 25°C [2]:

ligand	method	I in mol/L	log K_1	log $K_{MnHL^{3-}}$	log $K_{MnH_2L^{2-}}$	log $K_{MnH_3L^-}$	Ref.
1	pot	0.1 (KNO_3)	10.20 ± 0.2	5.64 ± 0.2	4.42 ± 0.2	3.54 ± 0.3	[1]
1	pol	0.1 ($NaClO_4$)	—	5.7 ± 0.1	4.4 ± 0.1	3.3 ± 0.2	[2]
2	pot	0.1 (KNO_3)	16.6	10.8	7.3	4.4	[3]

The protonated species are formed at pH 6.5 [2]. Protonation constants, log K = 7.5 for the reaction $MnL^{4-} + H^+ \rightleftharpoons MnHL^{3-}$ and log K = 6.1 for $MnHL^{3-} + H^+ \rightleftharpoons MnH_2L^{2-}$ with H_6L = ligand 1, are reported in [4]. The complexation selectivity of ligand 2 was studied by a theoretical conformational analysis. The results of the calculations were compared with the experimental data on the stability of complexes with alkaline earth and transition metal cations [5].

$Mn^{II}_3(C_3H_6NO_9P_3)\cdot 4H_2O$ was obtained by reaction of the completely deprotonated anion of ligand 1 with $MnCO_3$ in a 1:1 mole ratio. As shown by thermogravimetric analysis, the complex is dehydrated in the range 60 to 290°C. Dehydration is overlapped by decomposition [6].

References:

[1] Morozova, S. S.; Nikitina, L. V.; Dyatlova, N. M.; Serebryakova, G. V. (Zh. Neorgan. Khim. **20** [1975] 413/8; Russ. J. Inorg. Chem. **20** [1975] 228/31).
[2] Nozaki, T.; Kabata, T.; Yamashita, H. (Nippon Kagaku Kaishi **1988** 1017/20; C.A. **109** [1988] No. 157412).
[3] Kabachnik, M. I.; Medved', T. Ya.; Polikarpov, Yu. M.; Shcherbakov, B. K.; Bel'skii, F. I.; Matrosov, E. I.; Pasechnik, M. P. (Izv. Akad. Nauk SSSR Ser. Khim. **1984** 835/43; Bull. Acad. Sci. USSR Div. Chem. Sci. **33** [1984] 769/76).
[4] Martell, A. E.; Smith, R. M. (Critical Stability Constants, Vol. 5, 1st Suppl., Plenum, New York 1982, pp. 1/604, 277).
[5] Kabachnik, M. I.; Dashevskii, V. G.; Medved', T. Ya.; Baranov, A. P. (Teor. Eksperim. Khim. **21** [1985] 660/9; Theor. Exptl. Chem. [USSR] **21** [1985] 660/9).
[6] Karmazina, L. D.; Dyatlova, N. M.; Nikitina, L. V.; Gurevich, M. Z. (Koord. Khim. **1** [1975] 804/9; Soviet J. Coord. Chem. **1** [1975] 678/82).

39.9.14 With Diamine-N,N,N',N'-tetrakis(methylphosphonic Acids)

$[(HO)_2(O)PCH_2]_2N{-}R{-}N[CH_2P(O)(OH)_2]_2$ (= H_8L)

ligand	1	2	3	4	5
R	C_2H_4	$CH_2CH(OH)CH_2$	C_6H_{12}		
formula	$C_6H_{20}N_2O_{12}P_4$	$C_7H_{22}N_2O_{13}P_4$	$C_{10}H_{28}N_2O_{12}P_4$	$C_{10}H_{26}N_2O_{12}P_4$	$C_{10}H_{20}N_2O_{12}P_4$

Complexes in Solution. Stability constants of MnL^{6-} complexes with ligands 1 to 4, formed in aqueous solution at an Mn^{2+}:ligand mole ratio = 1:1, have been determined pH-potentiometrically at 25°C:

ligand	1	1	2	3	4
I in mol/L	0.1(KNO_3)	0.1(KCl)	0.1(NaCl)	0.1(KNO_3)	0.1($NaClO_4$)
log K_1	9.40	12.70	11.36	6.69	8.49
Ref.	[1]	[2]	[3]	[4]	[5]

The potentiometric studies indicate that not all the phosphonato groups of ligand 1 are deprotonated until pH's around 9 to 10 are reached [1]. NMR studies show that the hydration number of the MnL^{6-} complex with ligand 1 is between 1 and 2. Plots of water proton relaxation times, T_1 and T_2, at 293 K and 60 MHz are given in the paper. The difference between T_1 and T_2 suggests partial coordination of Mn^{2+} by the ligand below pH 6, multicoordination occurring above pH 7 when $T_1 \approx T_2$. The relatively high value of the rotational correlation time, $T_r = 8.3 \times 10^{-11}$ s, for the MnL^{6-} complex suggests incomplete coordination by the ligand [6]. For the Mn_2L^{4-} complex with ligand 2, obtained at a tenfold Mn^{2+} concentration, log K = 14.09 was calculated [3].

Stability constants of $MnH_nL^{(6-n)-}$ complexes at 25°C, formed in aqueous solution with ligand 1 by the reaction $Mn^{2+} + H_nL^{(8-n)-} \rightleftharpoons MnH_nL^{(6-n)-}$ [2] or with ligand 3 by the stepwise protonation of MnL^{6-} or $MnH_nL^{(6-n)-}$ species [4], are shown below:

ligand	I in mol/L	log $K_{MnHL^{5-}}$	log $K_{MnH_2L^{4-}}$	log $K_{MnH_3L^{3-}}$	log $K_{MnH_4L^{2-}}$	Ref.
1	0.1 (KCl)	9.66	6.99	5.13	3.91	[2]
3	0.1 (KNO_3)	9.66	7.52	6.12	5.43	[4]

Stability constants of $MnH_nL^{(6-n)-}$ complexes with ligand 2, formed by the reaction $Mn^{2+} + nH^+ + L^{8-}$ ions at I = 0.1 M (NaCl), have been calculated by [3]: log $K_\beta = 21.88$ (n = 1), 30.27 (n = 2), 37.44 (n = 3), 43.61 (n = 4), 48.18 (n = 5), 52.11 (n = 6). Increase in the initial concentration of OH^- decreases the degree of protonation of the $MnH_nL^{(6-n)-}$ complexes formed: e.g., MnL^{6-} exists at c_{OH} (initial concentration of OH^-) = 0.006 to 0.01 M, the MnH_6L complex at $c_{OH} = 0$ to 0.003 M [3].

Hydroxo complexes with ligand 2, $Mn(OH)_nL^{(6+n)-}$, exist in strongly alkaline medium at initial OH^- concentrations of 0.007 to 0.01 M. The stability constant of the most significant complex, $Mn(OH)L^{7-}$, is log K = 1.21. The calculated stability constants of $Mn(OH)_2L^{8-}$ (log K = −10.56) and $Mn(OH)_3L^{9-}$ (log K = −21.59) show that the complexes can be detected at the maximum OH^- concentration of ~0.01 M [3].

Homobinuclear $Mn_2(C_4H_5NO_4)_2L^{8-}$ or $Mn_2(C_6H_6NO_6)_2L^{10-}$ complexes with H_8L = ligand 1 are formed by reaction of the MnL^{6-} complex of ethylenediamine-N,N,N',N'-tetrakis(methylphosphonic acid) with the Mn^{II} complexes of iminodiacetic acid, $Mn(C_4H_5NO_4)_2^{2-}$, or nitrilotriacetic acid, $Mn(C_6H_6NO_6)_2^{4-}$; see "Manganese" D 5, 1987, pp. 4 and 15, respectively. Equilibrium constants of $(6 \pm 2) \times 10^3$ L/mol have been determined spectrophotometrically for both reactions at I = 1.5 M and 25°C. It has been shown that the reactions involve the redistribution of the donor atoms of the different anions between two central atoms. In the $Mn_2(C_4H_5NO_4)_2L^{8-}$ complex each central atom binds one nitrogen atom and two oxygen atoms of the PO_3^{2-} groups in ethylenediamine-N,N,N',N'-tetrakis(methylphosphonic acid) and one nitrogen atom and two oxygen atoms in the iminodiacetate ion. The complexes are similar to corresponding compounds of ethylenediamine-N,N,N',N'-tetraacetic acid (H_4edta), $Mn_2(C_4H_5NO_4)_2edta^{4-}$ and $Mn_2(C_6H_6NO_6)_2edta^{6-}$, see "Manganese" D 5, 1987, p. 54, but the tetrakis(methylphosphonato) complexes are more stable, because their formation is accompanied by a large energy gain from the inner-sphere transformations. Equilibrium constants are also given for the corresponding complexes of Co^{II}, Ni^{II}, Cu^{II}, Zn^{II}, and Cd^{II}. A linear correlation was found between the redistribution equilibrium constants of the ethylenediaminetetraacetato and the tetrakis(alkylphosphonato) complexes [8].

The reaction of manganese(III) diphosphate with ligand 1 has been studied by spectrophotometric and kinetic methods. The stability constant, log K = 10.20 ± 0.09, of the MnH_2L^{3-} complex, formed in aqueous solution at pH 5.6, was calculated from the saturation curve at 20°C and ionic strength of 1.0 M ($NaClO_4$) [7].

$Mn^{II}_4(C_6H_{12}N_2O_{12}P_4)\cdot 8H_2O$ and $Mn^{II}_4(C_{10}H_{12}N_2O_{12}P_4)\cdot 2H_2O$. The complexes with ligand 1 or 5, respectively, were precipitated on dropwise addition of 0.002 mol of sodium salt of the ligand to an aqueous solution of 0.008 mol of a manganese salt with constant stirring. Reactions in a 1:1, 2:1, or 3:1 mole ratio always resulted in the formation of the 4:1 metal-ligand complexes [9, 10]. In the case of $Mn_4(C_{10}H_{12}N_2O_6P_4)\cdot 2H_2O$ the yield was improved by addition of acetone or methanol to the mixture [10]. The precipitates were washed with acetone and dried at ~60 to 70°C. The white compounds show very low magnetic moments at room temperature (2.85 and 3.02 μ_B, respectively), due to antiferromagnetic exchange interactions. The IR spectra are discussed together with those of corresponding bivalent metal complexes (absorption bands are given for the copper compound). The complexes are insoluble in water and common organic solvents and do not melt up to 250 or 300°C, respectively. They are suggested to have a polymeric structure [9, 10].

References:

[1] Westerback, S.; Rajan, K. S.; Martell, A. E. (J. Am. Chem. Soc. **87** [1965] 2567/72).
[2] Kabachnik, M. I.; Dyatlova, N. M.; Medved', T. A.; Belugin, Yu. F.; Sidorenko, V. V. (Dokl. Akad. Nauk SSSR **175** [1967] 351/3; Dokl. Chem. Proc. Acad. Sci. USSR **172/177** [1967] 621/3).
[3] Kir'yanov, Yu. A.; Matkovskaya, T. A.; Nikolaeva, L. S.; Mezhonova, E. A.; Evseev, A. M.; Dyatlova, N. M.; Samakaev, R. Kh. (Zh. Neorgan. Khim. **32** [1987] 951/5; Russ. J. Inorg. Chem. **32** [1987] 532/4).
[4] Zaki, M. T. M.; Rizkalla, E. N. (Talanta **27** [1980] 709/13).
[5] Banks, C.; Yerick, R. E. (Anal. Chim. Acta **20** [1959] 301/14).
[6] Oakes, J.; Smith, E. G. (J. Chem. Soc. Dalton Trans. **1983** 601/5).
[7] Kurochkina, L. V.; Pechurova, N. I.; Snezhko, N. I.; Spitsyn, V. I. (Zh. Neorgan. Khim. **23** [1978] 2676/80; Russ. J. Inorg. Chem. **23** [1978] 1481/4).
[8] Polyakova, I. Ya.; Fridman, A. Ya.; Dyatlova, N. M. (Koord. Khim. **13** [1987] 147/51; Soviet J. Coord. Chem. **13** [1987] 81/5).
[9] Palta, N.; Rao, V. B.; Dubey, S. N.; Puri, D. M. (Polyhedron **3** [1984] 527/34).
[10] Palta, N.; Rao, V. B.; Dubey, S. N.; Puri, D. M. (Proc. Indian Acad. Sci. Chem. Sci. **95** [1985] 321/6).

39.9.15 With Other Organotetrakis- or Organopentakis(phosphonic Acids)

$[(HO)_2(O)PCH_2]_2NCH_2CH_2C(OH)[P(O)(OH)_2]_2$
ligand 1 $(=C_5H_{13}NO_{13}P_4=H_8L)$

$[(HO)_2(O)P]_2C(NH_2)CH_2C(NH_2)[P(O)(OH)_2]_2$
ligand 2 $(=C_3H_{14}N_2O_{12}P_4=H_8L)$

(structure: 1,4,7,10-tetraazacyclododecane ring with four $CH_2P(O)(OH)_2$ groups on N: $(HO)_2(O)PCH_2$, $CH_2P(O)(OH)_2$, $(HO)_2(O)PCH_2$, $CH_2P(O)(OH)_2$)
ligand 4 $(=C_{12}H_{32}N_4O_{12}P_4=H_8L)$

$[(HO)_2(O)PCH_2]_2NCH_2CH_2N[CH_2P(O)(OH)_2]CH_2CH_2N[CH_2P(O)(OH)_2]_2$
ligand 3 $(=C_9H_{28}N_3O_{15}P_5=H_{10}L)$

Complexes in Solution. Formation constants of MnL^{6-} complexes with H_8L = ligand 1 or 4 and of $MnH_nL^{(6-n)-}$ protonated species, formed by the reaction of Mn^{2+} with the stepwise deprotonated ligand, have been determined pH-potentiometrically at 25°C and ionic strength I=1.0 M (KNO_3):

ligand	log K_1	log $K_{MnHL^{5-}}$	log $K_{MnH_2L^{4-}}$	log $K_{MnH_3L^{3-}}$	log $K_{MnH_4L^{2-}}$	Ref.
1	11.1	9.9	7.2	5.9	4.8	[1]
4	16.9	12.9	8.8	7.1	4.6	[2]

Stability constants of the MnL^{8-} complex with $H_{10}L$ = ligand 3, log K_1 = 11.15, and those of protonated species, log $K_{MnH_nL^{(8-n)-}}$ = 8.41, 6.31, 5.34, 4.65, and 2.64 for n = 1 to 6, respectively, have been determined potentiometrically (glass electrode) at I = 0.1 M (KCl) and 25°C [5].

The complexes with ligand 1 are more stable than those with 1-hydroxy-3-dimethylamino-1,1-propanediylbis(phosphonic acid), p. 126, or N-ethyliminobis(methylphosphonic acid), p. 129. The manganese atom is assumed to be chelated by the hydroxyalkanediylbis(phosphonic acid) fragment of the ligand, and the resulting complex is stabilized through additional chelate rings with the phosphonic groups and the N atom of the second fragment of the ligand [1]. A theoretical conformational analysis of ligand 3 is given in [3]. For the possible monoplanar conformations, see the paper. The results show that the ligand is not capable of displaying its maximum coordination capacity of 8 when it forms 1:1 complexes [3].

$Mn^{II}_4(C_3H_6N_2O_{12}P_4)\cdot 8H_2O$ and $Mn^{II}_2(C_3H_{10}N_2O_{12}P_4)\cdot 3H_2O$. The complexes were obtained on reaction of an Mn^{II} salt with ligand 2 (4:1 or 2:1 mole ratio, respectively) in aqueous solution. The 4:1 complex requires the addition of an aqueous solution of Na_2CO_3 (0.004 mol) to the reaction mixture. The precipitates were washed with hot water, aqueous acetone, and finally with acetone (90%) and were then dried at 60 to 70°C. A polymeric structure has been inferred for the dirty white complexes on the basis of infrared spectral data, their infusibility up to 280°C, and their insolubility in most of the common organic solvents and water. The deprotonated phosphonate groups showed bands in the 1100 and 990 to 960 cm^{-1} region which could be assigned to $\nu_{as}(PO_3)$ and $\nu_s(PO_3)$, respectively. The electronic spectrum of $Mn_4(C_3H_6N_2O_{12}P_4)\cdot 8H_2O$ with bands at 7690, 15380, 23530(sh), and 33330 cm^{-1} indicates an octahedral geometry at Mn. The temperature dependence of the magnetic susceptibility of $Mn_4(C_3H_6N_2O_{12}P_4)\cdot 8H_2O$ is shown below:

T in K	296	237	189	141	117	93	77
$\chi_m \times 10^6$ in cm^3/mol	55750	66234	78331	95267	108976	129138	149299
μ_{eff} in μ_B	2.87	2.80	2.72	2.59	2.52	2.45	2.39

The negative value of the Weiss constant, $\Theta = -62°$, confirms the presence of antiferromagnetism. The Curie-Weiss law is obeyed [4].

References:

[1] Shcherbakov, B. K.; Bel'skii, F. I.; Komarova, M. P.; Polikarpov, Yu. M.; Medved', T. Ya.; Kabachnik, M. I. (Izv. Akad. Nauk SSSR Ser. Khim. **1982** 560/4; Bull. Acad. Sci. USSR Div. Chem. Sci. **31** [1982] 489/501).

[2] Kabachnik, I. M.; Medved', T. Ya.; Bel'skii, F. I.; Pisareva, S. A. (Izv. Akad. Nauk SSSR Ser. Khim. **1984** 844/9; Bull. Acad. Sci. USSR Div. Chem. Sci. **33** [1984] 777/82).

[3] Kabachnik, M. I.; Dashevskii, V. G.; Medved', T. Ya.; Baranov, A. P. (Teor. Eksperim. Khim. **21** [1985] 660/9; Theor. Exptl. Chem. [USSR] **21** [1985] 630/8).

[4] Venkateswara Rao, B.; Dubey, S. N.; Puri, D. M. (Indian J. Chem. A **25** [1986] 929/33).

[5] Kabachnik, M. I.; Dyatlova, N. M.; Medved', T. A.; Belugin, Yu. F.; Sidorenko, V. V. (Dokl. Akad. Nauk SSSR **175** [1967] 351/3; Dokl. Chem. Proc. Acad. Sci. USSR **172/177** [1967] 621/3).

39.10 Complex with Dimethyldiphosphonic Acid

$$CH_3-\underset{\underset{O}{\|}}{\overset{\overset{OH}{|}}{P}}-O-\underset{\underset{O}{\|}}{\overset{\overset{OH}{|}}{P}}-CH_3 \quad (=C_2H_8O_5P_2)$$

$Mn^{II}_2(C_2H_6O_5P_2)(C_4H_{10}O_3P)_2$. Formation of the mixed ligand complex containing dimethyldiphosphonate and methyl isopropylphosphonate ions ($C_4H_{10}O_3P^-$, see p. 139) was observed on heating a mixture of $MnCl_2$ or $MnI_2 \cdot 4H_2O$ with an excess of diisopropyl methylphosphonate up to 180°C (2°C/min) until precipitation of a crystalline solid occurred. It was washed with ether and dried in vacuum over $CaCl_2$. The compound melts at 330°C. The IR spectrum of the complex in Nujol is characterized by bands attributable to both ligands. The most characteristic absorptions of the dimethyldiphosphonato ligand are those of ν_{as}(POP) and ν_s(POP) vibration modes. In particular the ν_s(POP) vibration which was observed at 655 cm^{-1} has been utilized in order to establish that the complex contains the diphosphonato group. Similar results were obtained for corresponding complexes of Ca, Ni^{II}, and Co^{II}. The complexes exhibit no bands characteristic of uncoordinated PO groups. $Mn_2(C_2H_6O_5P_2)(C_4H_{10}O_3P)_2$ reveals a magnetic moment of $\mu_{eff}=5.75$ μ_B at 301 K. The electronic spectrum of the complex in Nujol shows absorption maxima (shoulders) at 253 and 335 nm. The band at 335 nm was assumed to be suggestive of coordination number five for Mn^{II}. A 0.017 M solution in methanol shows an absorption maximum (sh) at 330 nm with $\varepsilon=19.2$ $L \cdot mol^{-1} \cdot cm^{-1}$.

Reference:

Karayannis, N. M.; Mikulski, C. M.; Strocko, M. J.; Pytlewski, L. L.; Labes, M. M. (Z. Anorg. Allgem. Chem. **384** [1971] 267/79).

39.11 Complexes with Esters of Phosphonic Acids

39.11.1 With Monoesters of Phosphonic or Amino Phosphonic Acids RP(O)(OH)OR′

$CH_3P(O)(OH)(OC_3H_7\text{-}i)$
ligand 1 ($=C_4H_{11}O_3P=HL$)

X
CH−NH
$P(O)(OH)OC_8H_{17}$

ligand 5 with X = H ($=C_{21}H_{30}NO_3P=HL$)
ligand 6 with X = COOH ($=C_{22}H_{30}NO_5P=H_2L$)

$(CH_3)_3C$
HO
$CH_2P(O)(OH)OR'$
$(CH_3)_3C$

ligand 2 with $R'=C_2H_5$ ($=C_{17}H_{29}O_4P=HL$)
ligand 3 with $R'=C_4H_9$ ($=C_{19}H_{33}O_4P=HL$)

OH
$CH-NHCH_2$
X
$P(O)(OH)OC_4H_9$

ligand 7 with X = H ($=C_{18}H_{24}NO_4P=H_2L$)
ligand 8 with X = Cl ($=C_{18}H_{23}ClNO_4P=H_2L$)

CH_3
N
$P(O)(OH)OC_2H_5$

ligand 4 ($=C_8H_{12}NO_3P=HL$)

$C_2H_5O(O)CP(O)(OH)OC_2H_5$
ligand 9 ($=C_5H_{11}O_5P=HL$)

Complexes in Solution. The intermediate formation of a probably polymeric MnL_2 complex with isopropyl hydrogen methylphosphonate (ligand 1) during the reaction of $CH_3P(O)(OC_3H_7\text{-}i)_2$ with MnX_2 salts is assumed by [1, 2]. Attack of hydrogen halide and increasing the temperature leads to the formation of the diphosphonato complex $Mn^{II}_2(C_2H_6O_5P)(C_4H_{10}O_3P)_2$, see p. 138.

Solvent extraction of Mn^{II} from aqueous HCl solutions by ligand 5 dissolved in petroleum ether, CCl_4, or $CHCl_3$ and by ligand 6 dissolved in $CHCl_3$ was investigated by [3, 4]. The composition of the extractable complex depends on the initial metal : ligand ratio and the solvent employed; e.g., the 1:3 complex, $MnL_2 \cdot HL$, is reported to form in petroleum ether during an extraction with a large excess of ligand 5. Further saturation of the organic phase with manganese led to formation of the MnL_2 complex [3]. The carboxylato group of ligand 6 does not appear to coordinate with Mn^{II} [4]. Conditions are described for the separation of Mn^{II} from Co^{II} or Ni^{II} [3] or from Fe^{III} [4], owing to the difference in extraction of these metals.

Colored MnL_2 complexes are formed with ligands 7 and 8 in aqueous solution. At pH 2.5 a characteristic absorption maximum was observed at 510 nm with $\varepsilon=1.2\times10^2$ $L \cdot mol^{-1} \cdot cm^{-1}$. Formation constants, log $\beta_2=7.1$, were determined spectrophotometrically for both complexes, which are extractable with chloroform [5].

$Mn^{II}L_2$ Compounds. The complex with ligand 2, $Mn(C_{17}H_{28}O_4P)_2$, was prepared by mixing the sodium salt of the phosphonate ligand with the stoichiometric amount of $MnCl_2 \cdot 4H_2O$ in aqueous solution at 30°C. The suspension was stirred for 30 min at 30°C, then filtered. The white precipitate was washed with water and dried for 16 h at 80°C under reduced pressure (11 Torr). $Mn(C_{19}H_{32}O_4P)_2$ was prepared similarly by mixing the sodium salt of ligand 3 in ethanol with the aqueous Mn^{II} salt solution. It was washed with ethanol. The preparation of other Mn^{II} alkyl (3,5-dialkyl-4-hydroxyphenyl)methylphosphonato complexes is reported in the paper, but no analyses were given. The compounds are used as light stabilizers for polyamide substrates [6].

In the case of the complex with ligand 4, $Mn(C_8H_{11}NO_3P)_2$, a saturated solution of hydrated manganese(II) chloride in diethyl 6-methyl-2-pyridinylphosphonate was prepared at 50 to 100°C. This solution was subsequently diluted with an equal volume of the phosphonate. The pale pink complex precipitated on raising the temperature to 140°C. It was washed with ether and dried in vacuum over $CaCl_2$. A magnetic moment of $\mu_{eff}=5.78\ \mu_B$ results from susceptibility measurements at 298 K. IR data indicate coordination of the ligand through one of the oxygen atoms of the POO group and the nitrogen atom of the pyridine ring. Bands of $\nu_{as}(PO_2)$ and $\nu_s(PO_2)$ were observed at 1215 and 1086 cm^{-1}, respectively. Bands at 396 and 360 cm^{-1} were assigned to ν(Mn–O) vibration modes, the band at 219 cm^{-1} to ν(Mn–N). The electronic spectrum of the complex in Nujol shows a band at 307 nm, which indicates Mn^{II} to be pentacoordinated. A binuclear structure containing two bridging and two terminal ligand molecules is assumed. The complex is insoluble in common organic solvents [7].

$Mn(C_{21}H_{29}NO_3P)_2$, precipitated from an aqueous Mn^{II} salt solution with the sodium salt of ligand 5, is insoluble in water and organic solvents, but soluble in petroleum ether containing ligand 5 due to $MnL_2 \cdot HL$ formation, see p. 139 [3].

The complex with ligand 9, $Mn(C_5H_{10}O_5P)_2$, was prepared as follows: Hydrated Mn^{II} perchlorate was dissolved in an excess of triethyl orthoformate. The resulting solution was stirred at 50°C for 2 h. After addition of triethyl phosphonoformate (Mn : ligand = 1:7), the mixture was stirred at 50°C until precipitation of the lilac diethyl phosphonoformato complex occurred (80% yield at 50°C, 30% at 20°C). The complex is insoluble in water and common organic solvents and reveals a magnetic moment of $\mu_{eff}=6.14\ \mu_B$ at 300 K. The IR spectrum suggests that half of the ligand molecules coordinate through the carbonyl oxygens, while the other half do not. Hence, a linear doubly-bridged structure is suggested in which one of the ligands is bidentate (bonded by the PO_2 group) and the other is tridentate (coordinating through the carbonyl oxygen and the two PO_2 oxygens). Bands of ν(CO) were observed at 1711 and 1673 cm^{-1}, of $\nu_{as}(PO_2)$ at 1230 and 1211 cm^{-1}, of $\nu_s(PO_2)$ at 1085 and 1062 cm^{-1}, of $\nu(Mn\text{–}O)_{PO_2}$ at 391 and 359 cm^{-1}, and of $\nu(Mn\text{–}O)_{CO}$ at 262 cm^{-1}. The electronic spectrum of the solid compound in Nujol reveals maxima at 200, 217, 257, 276, and 335 nm. The charge-transfer absorption masks the d–d transition bands of the complex [8].

Mn^{II}LX Compounds. The $Mn(C_{17}H_{28}O_4P)Cl$ and $Mn(C_{17}H_{28}O_4P)CH_3COO$ complexes were prepared by reacting ligand 2 with the corresponding manganese salt (1:1 mole ratio) in absolute ethanol. The reaction mixture was refluxed for 30 min and the solution evaporated to dryness (after separation of NaCl in the case of the chloro complex). The residue was purified by extraction with $CHCl_3$ and evaporation of the solvent. The complexes were dried at 80°C in vacuum (11 Torr). The compounds are used as light stabilizers for polyamide substrates [6].

References:

[1] Karayannis, N. M.; Mikulski, C. M.; Pytlewski, L. L.; Labes, M. M. (Z. Anal. Chem. **249** [1970] 380/1).

[2] Karayannis, N. M.; Mikulski, C. M.; Strocko, M. J.; Pytlewski, L. L.; Labes, M. M. (Z. Anorg. Allgem. Chem. **384** [1971] 267/79).

[3] Jagodić, V.; Grdenić, D. (J. Inorg. Nucl. Chem. **26** [1964] 1103/9).

[4] Jagodić, V.; Herak, M. J. (J. Inorg. Nucl. Chem. **33** [1971] 2641/50).

[5] Siepak, J. (Chem. Anal. [Warsaw] **30** [1985] 725/35).

[6] Müller, H.; Moser, P.; Linhart, H. (Swiss 508686 [1971]; C.A. **77** [1972] No. 127443).

[7] Speca, A. N.; Mink, R.; Karayannis, N. M.; Pytlewski, L. L.; Owens, C. (J. Inorg. Nucl. Chem. **35** [1973] 1833/41).
[8] Mikulski, C. M.; Marks, B.; Tuttle, D.; Karayannis, N. M. (Inorg. Chim. Acta **68** [1983] 119/26).

39.11.2 With Diesters of Alkyl- or Acylphosphonic Acids $RP(O)(OR')_2$

ligand	R	R′	formula
1	CH_3	CH_3	$C_3H_9O_3P$
2	CH_3	i-C_3H_7	$C_7H_{17}O_3P$
3	CH_3	i-C_5H_{11}	$C_{11}H_{25}O_3P$
4	C_2H_5	C_2H_5	$C_6H_{15}O_3P$
5	C_4H_9	C_4H_9	$C_{12}H_{27}O_3P$
6	$CH_3(CH_2)_3CH(C_2H_5)CH_2$	$CH_3(CH_2)_3CH(C_2H_5)CH_2$	$C_{24}H_{51}O_3P$
7	CH_3CO	C_2H_5	$C_6H_{13}O_4P$
8	C_6H_5CO	C_2H_5	$C_{11}H_{15}O_4P$

Complexes in Solution. The extraction of Mn^{II} salts from aqueous solution with neat ligand 3 or its 0.8 M solution in CCl_4 was studied by [1]. Manganese(II) nitrate is extracted much more completely than the chloride, which was attributed to the difference in the solvation numbers. The formation of $Mn(NO_3)_2 \cdot 3C_{11}H_{25}O_3P$ and $MnCl_2 \cdot 2C_{11}H_{25}O_3P$ complexes in the organic phase was indicated by dilution experiments [1].

$[Mn^{II}(C_3H_9O_3P)_6](ClO_4)_2$. For preparation the hydrated Mn^{II} perchlorate was heated for ca. 1 h in a slight excess of triethyl orthoformate at 40 to 50°C, then ligand 1 was added (mole ratio Mn : ligand = 1 : 6). Treatment of the oily reaction product with anhydrous ether yields a crystalline white precipitate which was stored in vacuum over $CaCl_2$ [2] or P_4O_{10} [3]. The solid compound melts at 134.5 to 136°C; it is stable in air when dry. A magnetic moment of μ_{eff} = 5.87 μ_B results from susceptibility measurements (Faraday method) at 297 K [2]; μ_{eff} = 5.9 μ_B was determined in solution by the method of Evans, using benzene as an inert reference [3]. The negative shift of the PO stretching vibration by ~42 cm^{-1} [2, 3] clearly indicates that coordination to Mn occurs through the phosphoryl oxygen atom of the ligand. A band of $\nu_3(ClO_4)$ was observed at 1085 cm^{-1} [2], and one of ν(Mn–O) at 310 cm^{-1} [10]. The electronic spectrum of a 0.39 M solution of the complex in the neat ligand shows bands at 496, 442, 416, 399, 370, 361, and 298 nm. The molar extinction coefficients are on the order of 0.005 to 0.02 $L \cdot mol^{-1} \cdot cm^{-1}$, whereas in the region <280 nm the ε value becomes >5.00 $L \cdot mol^{-1} \cdot cm^{-1}$. A symmetry lower than O_h was suggested. The molar conductivity of a 10^{-3} M solution in nitromethane at 25°C, Λ = 185.8 $cm^2 \cdot \Omega^{-1} \cdot mol^{-1}$, justifies the formulation of the complex as a 1:2 electrolyte [2]. The methoxy proton spin-spin relaxation rate, $1/T_2 = 1.1 \times 10^4\ s^{-1}$, and the phosphorus spin-lattice relaxation rate, $1/T_1 = 8.7 \times 10^3\ s^{-1}$, were determined for 0.14 to 7.2×10^{-2} M neat ligand solutions in the temperature range 66 to 110°C. The small activation energies, $-E_a$ = 3.9 and 4.0 kcal, respectively, eliminate the chemical exchange time as a likely correlation time; the proton spin-spin relaxation time is dominated by a contact interaction [3].

$[Mn^{II}(C_7H_{17}O_3P)_4ClO_4]ClO_4$ was prepared from Mn^{II} perchlorate and ligand 2 by the procedure described above. The mixture was heated with continuous stirring for 2 to 3 h at 50°C and was allowed to cool slowly at room temperature with stirring. The crystalline precipitate was washed with hexane or cyclohexane and dried over $Mg(ClO_4)_2$ in vacuum. The complex was purified by slow evaporation (~9 weeks) in a closed two-arm glass tube at room temperature, using a temperature gradient of 8°C. The complex melts at 71.2 to 72°C, is moderately soluble in $CHCl_3$ and CCl_4, very soluble in ethanol, acetone, nitromethane, or dioxane, and is decomposed by water [4]. Magnetic measurements at 297 K by the Faraday method reveal a magnetic moment of $\mu_{eff} = 5.80\ \mu_B$. The IR spectrum of the complex in Nujol shows characteristic bands of ν(PO) at 1189 cm^{-1}, of ionic perchlorate groups at 1085 and 622 cm^{-1}, of coordinated $OClO_3$ groups at 1122 and 614 cm^{-1} (having shoulders at 922 and 635 cm^{-1}), and shows a ν(Mn–O) vibration mode at 370 cm^{-1}. Bands attributable to $\nu(Mn-O_{ClO_4})$ could not be identified. They are probably masked by ligand absorptions. Nevertheless, the presence of coordinated perchlorate ions was assumed in the solid complex [5], in contrast to the earlier publication [4] where a lowering of the tetrahedral symmetry of the ClO_4^- ion in the crystal lattice was discussed. The electronic spectrum of a 0.033 M solution in the neat ligand reveals an absorption maximum at 318 nm with $\varepsilon = 4.12\ L \cdot mol^{-1} \cdot cm^{-1}$. The molar conductivity of a 0.003 M solution in nitromethane at 25°C, $\Lambda = 150.9\ cm^2 \cdot \Omega^{-1} \cdot mol^{-1}$, is indicative of a 1:2 electrolyte; i.e., of the existence of $[Mn(C_7H_{17}O_3P_4)_4]^{2+}$ ions in solution [4].

$[Mn^{II}(C_7H_{17}O_3P)_2X_2]$ (X = NO_3, Cl, I). For preparation of the complexes with X = NO_3 and I the manganese(II) salt (1 mmol) was treated with 30 mL of triethyl orthoformate at 50°C for 1 h with stirring. Then 3 mmol of diisopropyl methylphosphonate (ligand 2) was added. In the case of $[Mn(C_7H_{17}O_3P)_2(NO_3)_2]$ the reaction mixture was stirred at 50°C for 2 h. The compound was gradually precipitated in the form of an off-white powder; recrystallization from triethyl orthoformate produced white, nonfluorescent, flaky crystals, which were washed with dichloromethane. $[Mn(C_7H_{17}O_3P)_2I_2]$ precipitated immediately as a bright yellow fluorescent compound upon addition of the ligand to the solution of MnI_2 in triethyl orthoformate. This complex is very deliquescent in the atmosphere and tends to collapse to a viscous liquid even under N_2 or upon storage in vacuum at ambient temperature. It was stored at 0°C under N_2 after being rapidly washed with triethyl orthoformate. $[Mn(C_7H_{17}O_3P)_2Cl_2]$ was prepared by dissolving 1 mmol $MnCl_2$ and 2 mmol of the ligand in THF. The solution was stirred for 1 to 2 h, then allowed to stand at ambient temperature for several days. Pale green, fluorescent crystals were obtained in small yield from the green solution. This compound is also very deliquescent; therefore it was washed with ligroin and stored in vacuum as above. The complex was obtained in good yield by mixing ligand 2 and $MnCl_2$ in a 2:1 mole ratio in CS_2. The mixture was introduced into an H-shaped recrystallization cell. Large crystals were deposited on the warm arm of the cell within 1 d by using a temperature gradient of 8°C. The chloro and iodo complexes cannot be obtained pure if the temperatures specified in the preparation method are exceeded during the syntheses [6]. If $MnCl_2$ reacts with ligand 2 at 150 to 220°C, the complex $Mn_2^{II}(C_2H_6O_5P)(C_4H_{10}O_3P)_2$ is formed [7], see also p. 138.

Chemisorption of ligand 2 onto the surface of $MnCl_2$ or other metal halides was observed by [8, 9] and complex formation was assumed due to the large negative shifts of the PO stretching vibration modes. Pertinent IR data (in cm^{-1}), magnetic moments at 300 K, absorption maxima in the electronic spectra of the solid compounds in Nujol, and molar conductivities of 10^{-3} M solutions in nitromethane at 25°C are tabulated below for the complexes with ligand 2 (L = $C_7H_{17}O_3P$) [6]:

complex	color	μ_{eff} in μ_B	ν(PO)	Δν(PO)	ν(Mn–O)	λ_{max} in nm	Λ*)
$[MnL_2(NO_3)_2]$	off-white	5.89	1183	(−58)	304	—	36
$[MnL_2Cl_2]$	pale green	6.01	1170	(−71)	404	447, 469	3
$[MnL_2I_2]$	yellow	5.82	1165	(−76)	397	486, 502, 525	5

*) In $cm^2 \cdot \Omega^{-1} \cdot mol^{-1}$.

Since the spectrum of the nitrato complex is characterized by a separation of the $(\nu_1 + \nu_4)$ mode ($41\,cm^{-1}$), the presence of exclusively bidentate chelating nitrato ligands was assumed (with Mn^{II} being hexacoordinated). The chloro and iodo complexes are assumed to be pseudotetrahedral [6].

$Mn^{II}L_n(ClO_4)_2 \cdot xH_2O$. Complexes with L = ligands 4 to 8 (for the compositions see the table below) were prepared by the general procedure: 1 mmol of hydrated $Mn(ClO_4)_2$ was dissolved in ca. 10 mL triethyl orthoformate. After stirring for 1 h at 50 to 60°C the ligand was added (mole ratio Mn : ligand = 1:6, but 1:2 in the case of ligand 8). The resulting mixture was stirred for 2 to 3 h at 50 to 60°C. The complex with ligand 5 precipitated in the form of a fine powder upon cooling with continuous stirring [10]. In the case of ligand 7 a crystalline complex precipitated on addition of a large excess of anhydrous diethyl ether or ligroin [12]. With ligand 4 or 8 viscous oily products were obtained and were solidified to tacky solids by treatment with anhydrous diethyl ether [10, 13]. The complex with ligand 6 was precipitated in small yield in the form of a waxy solid by reducing the volume of the reaction mixture to about one half (by heating at 60 to 70°C) and allowing the residue to stand in an evacuated desiccator for several days [10]. The complexes were washed with diethyl ether or ligroin and stored in vacuum over $CaCl_2$ [10, 11] or $Mg(ClO_4)_2$ in the case of ligands 7 and 8 [12, 13]. The compositions and physical properties of the complexes are tabulated below: IR vibration modes and band shifts in comparison to the free ligand (in parentheses) are given in cm^{-1}.

L	complex	color	μ_{eff} in μ_B	ν(PO)	ν(Mn–O)	λ_{max} [c]	Λ [d]	Ref.
4	$[Mn(C_6H_{15}O_3P)_3(H_2O)_3]$-$(ClO_4)_2$	pink	5.93 [a]	1196 (−51)	330	247, 290	173	[10]
5	$[Mn(C_{12}H_{27}O_3P)_4(H_2O)_2]$-$(ClO_4)_2$	white	5.98 [a]	1194 (−50)	330	260, 298	172	[10, 11]
6	$[Mn(C_{24}H_{51}O_3P)(H_2O)_5]$-$(ClO_4)_2$	off-white	5.85 [a]	1181 (−61)	330	255, 284	169	[10]
7	$[Mn(C_6H_{13}O_4P)_3]$-$(ClO_4)_2 \cdot H_2O$	pink	6.18 [b]	1219 (−34)	316	<300, 342, 445	163	[12]
8	$[Mn(C_{11}H_{15}O_4P)_2(H_2O)_2ClO_4]$-$ClO_4$	yellow	5.76 [b]	1222 (−36)		267, 352, 381	62	[13]

[a] At 298 K. – [b] At 296 K. – [c] Absorption maximum (λ_{max} in nm) of the solid complexes in Nujol. – [d] Molar conductivity of a 10^{-3}M nitromethane solution in $cm^2 \cdot \Omega^{-1} \cdot mol^{-1}$.

In the case of the complexes with ligands 4 to 6 or 8 the water molecules are assumed to be coordinated. The compounds cannot be dehydrated in the presence of $CaCl_2$, $Mg(ClO_4)_2$, or P_4O_{10}. Ligand 7 is assumed to be bidentate, chelating through the carbonyl and phosphoryl oxygen atoms [12], whereas in $[Mn(C_{11}H_{15}O_4P)_2(H_2O)_2ClO_4]ClO_4$ one molecule of ligand 8 acts bidentately while the second molecule is monodentate, coordinating only through the phosphoryl oxygen atom. The presence of both ionic and monodentately coordinated perchlorate groups is indicated by corresponding IR bands and is also substantiated by the 1:1 electrolyte nature of the complex in nitromethane [13]. All complexes are sensitive to moisture. They dissolve in organic solvents, such as nitromethane, nitrobenzene, alcohols, ketones, or acetonitrile [10 to 13].

References:

[1] Rozen, A. M.; Skotnikov, A. S.; Martynova, M. K. (Zh. Neorgan. Khim. **29** [1984] 1798/802; Russ. J. Inorg. Chem. **29** [1984] 1031/3).

[2] Karayannis, N. M.; Owens, C.; Pytlewski, L. L.; Labes, M. M. (J. Inorg. Nucl. Chem. **31** [1969] 2767/76).
[3] Frankel, L. S. (Inorg. Chem. **8** [1969] 1784/7).
[4] Karayannis, N. M.; Owens, C.; Pytlewski, L. L.; Labes, M. M. (J. Inorg. Nucl. Chem. **31** [1969] 2059/71).
[5] Karayannis, N. M.; Mikulski, C. M.; Strocko, M. J.; Pytlewski, L. L.; Labes, M. M. (Inorg. Chim. Acta **8** [1974] 91/6).
[6] Karayannis, N. M.; Pytlewski, L. L.; Owens, C. (J. Inorg. Nucl. Chem. **42** [1980] 675/82).
[7] Karayannis, N. M.; Mikulski, C. M.; Strocko, M. J.; Pytlewski, L. L.; Labes, M. M. (Z. Anorg. Allgem. Chem. **384** [1971] 267/79).
[8] Guilbault, G. G. (Anal. Chim. Acta **39** [1967] 260/4).
[9] Guilbault, G. G.; Scheide, E. P. (J. Inorg. Nucl. Chem. **32** [1970] 2959/62).
[10] Karayannis, N. M.; Mikulski, C. M.; Gelfand, L. S.; Pytlewski, L. L. (J. Inorg. Nucl. Chem. **40** [1978] 1513/8).

[11] Karayannis, N. M.; Mikulski, C. M.; Strocko, M. J.; Pytlewski, L. L.; Labes, M. M. (Inorg. Chim. Acta **4** [1970] 557/61).
[12] Mikulski, C. M.; Henry, W.; Pytlewski, L. L.; Karayannis, N. M. (J. Inorg. Nucl. Chem. **40** [1978] 769/72).
[13] Mikulski, C. M.; Henry, W.; Pytlewski, L. L.; Karayannis, N. M. (Transition Metal Chem. [Weinheim] **2** [1977] 135/40).

39.12 Complexes with Esters of Phosphorous Acid

ligand $P(OR)_3$	1	2	3	4	5
R	CH_3	C_2H_5	$C_{10}H_{21}$	C_6H_5	C_6H_4Cl-4
formula	$C_3H_9O_3P$	$C_6H_{15}O_3P$	$C_{30}H_{63}O_3P$	$C_{18}H_{15}O_3P$	$C_{18}H_{12}Cl_3O_3P$

$P(OC_{10}H_{21})_2(OC_6H_5)$ ligand 6 (= $C_{26}H_{47}O_3P$)

[Mn(NO)$_3$L]. For the preparation of complexes with L = ligand 2, 3, 4, or 6 a solution of $[Mn(CO)_5I]$ in cyclohexane was treated first with a stream of nitrogen monoxide at 60°C. After separation of MnI_2 the solution of $[Mn(NO)_3(CO)]$ was reacted with the $P(OR)_3$ ligand [1]. The complex with ligand 4 was also formed on bubbling nitrogen monoxide through a solution of $[Mn(CO)_4(C_{18}H_{15}O_3P)I]$ in benzene [2]. The solvent was removed by evaporation in high vacuum and the green oil purified by extraction with absolute ethanol [1, 2]. The IR spectra of the complexes in cyclohexane show two ν(NO) bands, corresponding to a tetrahedral Mn environment with C_{3v} symmetry having the NO groups approximately in a plane. The bands (in cm^{-1}) observed in cyclohexane and the calculated NO force constants (in mdyn/Å) are [1]:

complex	ν(NO) A$_1$	ν(NO) E	f_{NO}
$Mn(NO)_3(C_6H_{15}O_3P)$	1792	1695	13.14
$Mn(NO)_3(C_{30}H_{63}O_3P)$	1792	1694.5	13.13
$Mn(NO)_3(C_{18}H_{15}O_3P)$	1802	1705.5	13.31
$Mn(NO)_3(C_{18}H_{12}Cl_3O_3P)$	1805	1709.5	13.35
$Mn(NO)_3(C_{26}H_{47}O_3P)$	1792.5	1700, 1694.5	13.13

Preparation, handling, and storage of the complexes require the absence of oxygen and moisture [2].

[Mn(NO)$_2$L$_2$X] (X = Cl, Br, I). The complex with ligand 1, [Mn(NO)$_2$(C$_3$H$_9$O$_3$P)$_2$Cl], was prepared under an inert atmosphere by passing a stream of nitrogen monoxide through the refluxing benzene solution of [Mn(CO)$_3$(C$_3$H$_9$O$_3$P)$_2$Cl]. After 4 h, the solvent was removed under reduced pressure, and the remaining oil was crystallized from ether-hexane to obtain red prisms which melt at 91 to 93°C [3]. [Mn(NO)$_2$L$_2$I] complexes with L = ligand 3, 5, or 6 have been identified only by IR data of their cyclohexane solutions [1]. They were obtained by the method given in [2] for preparation of [Mn(NO)$_2$(C$_{18}$H$_{15}$O$_3$P)$_2$X] with ligand 4 and X = Cl, Br, I or of the iodo complex with ligand 2, [Mn(NO)$_2$(C$_6$H$_{15}$O$_3$P)$_2$I]: The [Mn(CO)$_3$L$_2$X] complex solution, formed by the reaction of [Mn(CO)$_5$X] with the appropriate amount of the corresponding ligand in benzene, was heated to 80°C, and treated with a stream of nitric oxide for 3 to 5 h. The hot mixture was filtered, reduced in volume by evaporation, and the complexes precipitated by adding ethanol or pentane. They were recrystallized from benzene-ethanol or benzene-pentane mixtures and dried in high vacuum. Preparation, handling, and storage of the complexes require the exclusion of oxygen and moisture [1].

The crystals of [Mn(NO)$_2$(C$_{18}$H$_{15}$O$_3$P)$_2$X] are yellow-green (X = Cl), ocher (X = Br), or red-brown (X = I). The iodo complex melts at 129°C with decomposition. [Mn(NO)$_2$(C$_6$H$_{15}$O$_3$P)$_2$I] shows a susceptibility of $\chi = 210 \times 10^{-6}$ cm^3/mol at 298 K. The dipole moments of the [Mn(NO)$_2$-(C$_{18}$H$_{15}$O$_3$P)$_2$X] complexes in 20% benzene solution are $\mu = 2.91 \pm 0.4$ D (X = Br) and 3.05 ± 0.05 D (X = I) [1]. [Mn(NO)$_2$(C$_3$H$_9$O$_3$P)$_2$Cl] is a nonelectrolyte in acetone [3]. [Mn(NO)$_2$-(C$_{18}$H$_{15}$O$_3$P)$_2$I] is monomeric in benzene and a nonelectrolyte in acetone [2]. A trigonal-bipyramidal geometry at Mn (with apical phosphite ligand molecules and the NO groups in *cis* positions) is the interpretation of the IR ν(NO) bands observed in carbon tetrachloride. These are tabulated below together with the calculated NO force constants for the [Mn(NO)$_2$L$_2$X] complexes [1]:

ligand	2	3	4	5	6	4	4
X	I	I	I	I	I	Cl	Br
ν(NO) in cm^{-1}	1737, 1689	1737, 1688	1754, 1703.5	1756, 1705.5	1756, 1706	1750.5, 1699	1752.5, 1702
f_{NO} in mdyn/Å	12.91	12.90	13.15	13.18	13.19	13.09	13.13

The IR spectrum of [Mn(NO)$_2$(C$_3$H$_9$O$_3$P)$_2$Cl] in dichloromethane solution shows bands at 1737 and 1662 cm^{-1}, assigned to the ν(NO) vibration. The frequencies correlate with the π acceptor property of the trimethyl phosphite ligand. The ^{1}H NMR methyl resonances, measured in acetone-d$_6$, appear as a distorted triplet, J = 6.22 Hz, for the two strongly-coupled *trans*-axial ligand molecules in the trigonal-bipyramidal environment at Mn. The X-ray photoelectron spectrogram shows binding energies of N1s = 401.1 and P2p = 132.4 referred to C1s = 284.0 eV for [Mn(NO)$_2$(C$_3$H$_9$O$_3$P)$_2$Cl] [3].

[Mn(NO)$_2$(C$_3$H$_9$O$_3$P)$_3$]BF$_4$ was prepared at room temperature by adding AgBF$_4$ (2.5 mmol) to an acetone solution containing 2.5 mmol of [Mn(NO)$_2$(C$_3$H$_9$O$_3$P)$_2$Cl], see above, and 3 mmol of ligand 1. After 2 h the mixture was filtered, evaporated under reduced pressure, and the remaining oil was taken up in ethanol. Dropwise addition of ether yielded golden needles that decompose on heating at 132 to 134°C. The IR spectrum of the complex in dichloromethane shows bands at 1772 and 1722 cm^{-1}, assigned to ν(NO) vibrations correlating with the π acceptor property of ligand 1. The ^{1}H NMR methyl resonances, measured in acetone-d$_6$, appear as a distorted triplet and a doublet of J = 6.10 and 6.18 Hz, respectively; these correspond to two *trans*-axial ligands and one equatorial ligand in a trigonal bipyramid. The X-ray PE measurements reveal binding energies of N1s = 400.2 and P2p = 132.2 eV referred to C1s = 284.0 eV. The electrical conductivity of the complex in acetone solution is $\Lambda = 139$ cm$^2 \cdot \Omega^{-1} \cdot$ mol^{-1}. The complex requires storage under nitrogen [3].

$[Mn(C_3H_9O_3P)_2(C_6H_{16}P_2)_2]PF_6$ was prepared by dropwise addition of NH_4PF_6 dissolved in tetrahydrofuran to a solution containing the cyclopentadienyl compound *syn*-η^3-C_5H_5Mn-$[(CH_3)_2PCH_2CH_2P(CH_3)_2]_2$ and a large excess of ligand 1 in THF maintained at −78°C. The mixture was warmed to room temperature and stirred for 12 h. The precipitate was washed with pentane, extracted with methanol, and recrystallized from methanol-pentane at −30°C. The *cis* orientation of ligand 1 is inferred from the $^{13}C\{^1H\}$ and ^1H NMR spectra of the complex in dichloromethane-d_2. Several chemical shift values as well as IR data (bending and stretching vibrations of CH, CP, POC, and PF_6) are given in the paper [4].

$[Mn(C_3H_9O_3P)_2(C_{48}H_{36}N_4)]PF_6$ ($C_{48}H_{36}N_4^{2-}$ = anion of 5,10,15,20-tetrakis(4-methylphenyl)-porphyrin, see "Manganese" D 4, 1985, p. 94). The Mn^{III} complex was prepared by adding 0.3 mL of ligand 1 to a refluxing solution of 100 mg $Mn(C_{48}H_{36}N_4)Cl$ in 100 mL of dry toluene. After 1 h the solvent was removed in vacuum and the residue dissolved in dichloromethane; NH_4PF_6 (212 mg) in 3 mL of acetone was added to precipitate the compound. The IR spectrum of the dark green platelets (in KBr) shows bands assigned to ν(COP) at 1045 cm^{-1}, to $\nu(PF_6)$ at 850 and 565 cm^{-1}. In the 384 to 670 nm region of the electronic spectrum in dichloromethane solution eight bands with extinction values between 3.22 and 4.75 $L \cdot mol^{-1} \cdot cm^{-1}$ were observed [5].

References:

[1] Beck, W.; Lottes, K. (Chem. Ber. **98** [1965] 2657/73).
[2] Hieber, W.; Tengler, H. (Z. Anorg. Allgem. Chem. **318** [1962] 136/54).
[3] Copperthwaite, R. G.; Reimann, R. H.; Singleton, E. (Inorg. Chim. Acta **28** [1978] 107/11).
[4] Bleeke, J. R.; Kotyk, J. J. (Organometallics **4** [1985] 194/7).
[5] Buchler, J. W.; Dreher, C.; Lay, K.-L. (Chem. Ber. **117** [1984] 2261/74).

39.13 Complexes with Esters of Phosphoric or Diphosphoric Acid

Remarks. Manganese complexes with certain mono-, di-, or triphosphates of biological importance were already treated together with the N-coordinating, heterocyclic ligands in volume "Manganese" D 4, 1985: Complexes with nucleotides, dinucleotides, polynucleotides, and nucleic acids are described on pp. 12/5, 26/49, and 62; complexes with phosphoryl peptides on p. 346; ternary complexes of manganese with nucleotides and amino acids on pp. 269, 271, 280, and 286. Information on ternary and higher-order complexes of Mn with proteins, particularly enzymes, biological phosphates, and other compounds (mainly substrates and products of enzyme reactions) is included in the section on "Complexes with Proteins", pp. 348/60.

39.13.1 Complexes with Monoesters of Phosphoric Acid

39.13.1.1 With Alkyl or Aryl Dihydrogen Phosphates $(RO)P(O)(OH)_2$ (= H_2L)

ligand	1	2	3	4	5
R	CH_3	C_4H_9	C_8H_{17}	C_6H_5	$C_6H_4NO_2$-4
formula	CH_5O_4P	$C_4H_{11}O_4P$	$C_8H_{19}O_4P$	$C_6H_7O_4P$	$C_6H_6NO_6P$

Complexes in Solution. Mn^{2+} ions form MnL complexes with deprotonated ligands 1 to 5 in aqueous solution. The stability constant, log K_1 = 2.19 at 20°C and 2.55 at 65°C, I = 0.1 mol/L, was determined for the complex with methyl phosphate, $Mn(CH_3O_4P)$, by IR spectroscopy [1]. Potentiometric pH titrations at 25°C, I = 0.1 mol/L ($NaNO_3$), revealed log K_1 = 2.34 ± 0.01 for $Mn(C_4H_9O_4P)$, 2.12 ± 0.03 for $Mn(C_6H_5O_4P)$, and 1.87 ± 0.04 for $Mn(C_6H_4NO_6P)$ [2]. Octyl phosphate (ligand 3) forms micelles (aggregates) above a critical concentration (0.13 mol/L for the sodium monooctyl phosphate used). At concentrations < 0.13 mol/L low values of log K_1 were found for the complexes with the monomeric ligand: 0.663 at pH 5 and 20°C by polarography [3] and 0.505 at 30°C by ^{31}P NMR measurements [4]. The studies show that above the critical micellar concentration very stable 1:2 complexes are formed [3, 4]. For these complexes the residence time of the phosphate O atoms in the first coordination sphere of the Mn^{2+} ions ($\tau_h = 8 \times 10^{-8}$ s) is considerably shortened compared to the complex with the monomeric ligand ($\tau_h = 5 \times 10^{-7}$ s) [4]. The increase of complexing properties of the micellar ligand is interpretable from the electrostatic attraction of the Mn^{2+} ions by the highly charged micelle. The coordination number changes from 1 to 2, owing to the close packing of the complexing polar heads on the surface of the micelles [4].

An entropy change for the formation of the 1:1 complex with ligand 1, $\Delta S = 22$ cal·mol^{-1}·K^{-1}, was calculated from log K_1 values at 20 and 65°C [1]. By use of thermodynamic data taken from the literature an entropy change of $\Delta S = 9$ cal·mol^{-1}·K^{-1} was calculated; this is consistent with the displacement of one water molecule from $Mn(H_2O)_6^{2+}$ on coordination with the phosphate group [5]. Correlations between the complex stability and the ligand basicity (plots of log K_1 vs. log K for $K = [HL^-]/[L^{2-}][H^+]$ are straight lines for ligands 1, 2, 4, 5, and ribose monophosphate) are also in accord with monodentate coordination through one oxygen atom of the phosphate group, or possibly also with the formation of a six-membered "chelate" ring involving a coordinated water molecule and a hydrogen bond to a second oxygen atom of the phosphate group [2].

The IR spectrum of the $Mn(CH_3O_4P)$ complex shows a broad intense band at 1100 cm^{-1} and a weak band at 1150 cm^{-1} which were both assigned to $\nu_{as}(PO_3)$ vibrations. The intensity of the band at 1100 cm^{-1} was found to be decreased, that at 1150 cm^{-1} increased, with respect to the free ligand [1].

References:

[1] Brintzinger, H. (Helv. Chim. Acta **48** [1965] 47/54).
[2] Massoud, S. S.; Sigel, H. (Inorg. Chem. **27** [1988] 1447/53).
[3] Neveu, C. R.; Leicknam, J. P. (Electrochim. Acta **31** [1986] 705/15).
[4] Chevalier, Y.; Chachaty, C. (J. Am. Chem. Soc. **107** [1985] 1102/7).
[5] Aruga, R. (J. Inorg. Nucl. Chem. **43** [1981] 3414/5).

39.13.1.2 With Ribose, Glucose, or Fructose Phosphates

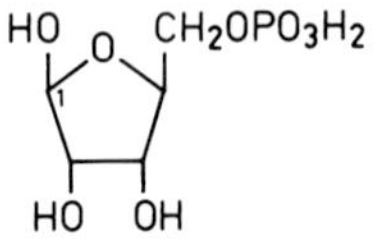

ligand 1
(= D-$C_5H_{11}O_8P$ = H_2L)

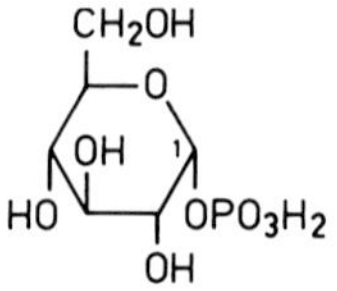

ligand 2
(= α-D-$C_6H_{13}O_9P$ = H_2L)

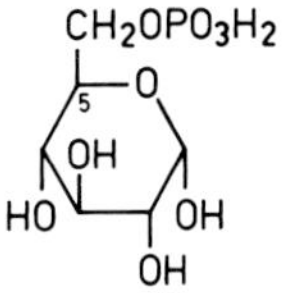

ligand 3
(= α-D-$C_6H_{13}O_9P$ = H_2L)

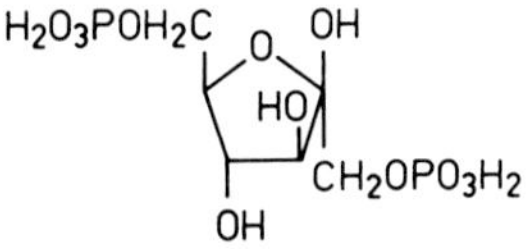

ligand 4
(= D-$C_6H_{14}O_{12}P_2$ = H_4L)

Complexes in Solution. Stability constants of 1:1 and 1:2 complexes of Mn^{II} with the ligands 1 to 4 in aqueous solution were determined by various methods:

ligand	1	2	2	3	3	4
t in °C	25	33	25	25	33	20
I in mol/L	0.1($NaNO_3$)	0.05[a)]	0.1($KClO_4$)[b)]	0.1 NaCl	0.05[a)]	0.08 KCl[c)]
method	pot[d)]	ESR	ion ex[e)]	pot[d)]	ESR	^{1}H NMR[f)], ESR
log K_1	2.20	2.28	2.19	1.86	2.32	4.82
log K_2	—	—	—	2.92	—	~2.22
Ref.	[1]	[2]	[3]	[4]	[2]	[5]

a) Tris buffer pH 7.48 (Tris = $(HOCH_2)_3CNH_2$). – b) Veronal buffer pH 7.2. – c) 0.02 M Tris-HCl buffer pH 7.2 – d) By potentiometric pH measurement with a glass electrode – e) By ion exchange. – f) By measurement of the water proton relaxation rates with the spin echo method.

References:

[1] Massoud, S. S.; Sigel, H. (Inorg. Chem. **27** [1988] 1447/53).
[2] Cohn, M.; Townsend, J. (Nature **173** [1954] 1090/1).
[3] Doody, B. E.; Tucci, E. R.; Scruggs, R.; Li, N. C. (J. Inorg. Nucl. Chem. **28** [1966] 833/44).
[4] Asso, M.; Benlian, D. (Compt. Rend. C **278** [1974] 1373/5).
[5] Fell, D. A.; Peacocke, A. R.; Dwek, R. A. (Eur. J. Biochem. **29** [1972] 128/33).

39.13.1.3 With Phytic Acid

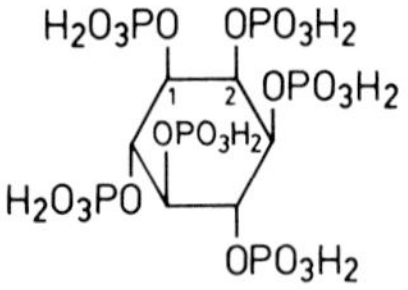

(= $C_6H_{18}O_{24}P_6$)

Potentiometric pH titrations of sodium phytate with standard solutions containing Mn^{2+} ions indicate the formation of the $Mn_5(C_6H_8O_{24}P_6)$ complex in aqueous solution. A series for the relative stability of various analogous metal complexes, including the Mn^{II} complex, was deduced from the different drops in the pH of the solutions on complex formation: $Cu^{II} > Zn^{II} > Ni^{II} > Co^{II} > Mn^{II} > Ca^{II}$ at pH 7.4.

Reference:

Vohra, P.; Gray, G. A.; Kratzer, F. H. (Proc. Soc. Exptl. Biol. Med. **120** [1965] 447/9).

39.13.1.4 With Phosphoglyceric or Phosphoenolpyruvic Acid and Related Compounds

CH_2OH–$CH(OP(O)(OH)_2)$–COOH

ligand 1
(= DL- or D-$C_3H_7O_7P$)

CHR=$C(OP(O)(OH)_2)$–COOH

ligand 2
with R = H (= $C_3H_5O_6P$)

ligand 3
with R = F (= (*E*)- or (*Z*)-$C_3H_4FO_6P$)

CH_3–CH=$C(OP(O)(OH)_2)$–COOH

ligand 4
(= (*E*)- or (*Z*)-$C_4H_7O_6P$)

Solution studies by various methods indicate the formation of 1:1 complexes of Mn^{II} with the ligands (H_3L) [1 to 6]. Stability constants determined for the complexes with ligands 1 and 2 in aqueous solution are:

ligand	1(DL form)	1(D form)	2	2
t in °C	23	25	20	25
I in mol/L	0.056(Na_2SO_4)	0.1(($C_3H_7)_4NI$)	0.08 KCl [a]	0.1(($C_3H_7)_4NI$)
method	ion ex [b]	pot [c]	^{1}H NMR [d], ESR	pot [c]
log K_1	3.36	3.09	2.48	2.75
Ref.	[1]	[2]	[3]	[2]

[a] 0.02 M Tris-HCl buffer pH 7.2 (Tris = $(HOCH_2)_3CNH_2$). – [b] Ion exchange. – [c] Potentiometric pH titration by use of a glass electrode. – [d] Measuring of the water proton relaxation rates with the spin echo method.

An average stability constant of log K_1 = 3.5 at 22°C was determined spectrophotometrically for the Mn^{II} complexes with ligand 1 or 2 in their equilibrium mixture in aqueous solution (0.05 M Tris-HCl buffer, pH 7.2) [4]. Studies of the ^{1}H, ^{19}F, and ^{31}P nuclear magnetic relaxation rates of ligands 2, 3, and 4 for the binary Mn-ligand complexes indicate the formation of inner-sphere complexes in all cases; the phosphate, but not the carboxylate group, is involved in the coordination; Mn–P distances of r = 2.8 to 3.0 Å were calculated. Structural studies of the ternary enzyme-Mn-ligand complexes with the enzyme phosphoenolpyruvate carboxykinase were also performed [5, 6].

A solid Mn^{II} compound containing manganese and phosphoglycerate in a 1:1 ratio was prepared by adding an equimolar amount of $MnCl_2$ to a 0.02 M aqueous solution of Na(DL-$C_3H_5O_7P$) and allowing the mixture to stand for about 1 h at room temperature. Then, absolute ethanol was added until a turbidity formed (13 vol% was required), and the solution was cooled at 4°C for 12 h. The precipitate was centrifuged off and dried in a vacuum over P_4O_{10}. The IR spectra of the Na, Zn, and Ni compounds are recorded and discussed [7].

References:

[1] Malmström, B. G. (Arch. Biochem. Biophys. **49** [1954] 335/42).
[2] Wold, E.; Ballou, C. E. (J. Biol. Chem. **227** [1957] 301/12).
[3] Fell, D. A.; Peacocke, A. R.; Dwek, R. A. (Eur. J. Biochem. **29** [1972] 128/32).
[4] Malmström, B. G. (Arch. Biochem. Biophys. **58** [1955] 381/97).
[5] Hwang, S., H.; Nowak, T. (Arch. Biochem. Biophys. **269** [1989] 646/63).
[6] Duffy, T. H.; Nowak, T. (Biochemistry **24** [1985] 1152/60).
[7] Rosenberg, A.; Malmström, B. G. (Acta Chem. Scand. **9** [1955] 1546/8).

39.13.1.5 With Phosphoethanolamine

CH_2NH_2
|
$CH_2{-}OP(O)(OH)_2$ $(= C_2H_8NO_4P = H_2L)$

The formation of 1:1 and 1:2 complexes of manganese(II) with the ligand was revealed by potentiometric pH titrations at 25°C [1 to 4]. The stability constants, log $K_1 = 3.85$ at $I = 0.1$ mol/L ($NaClO_4$) [1] and 2.55 at $I = 0.15$ mol/L(KCl) [2], were determined for $Mn(C_2H_6NO_4P)$ [1, 2]. No reliable data could be obtained for this complex by [4]; log $K = 1.72$ at $I = 0.15$ mol/L(KCl) [2] and 1.89 at $I = 0.2$ mol/L(KNO_3) [3, 4] for $K = [Mn(C_2H_7NO_4P)^+]/[Mn^{2+}]\cdot[C_2H_7NO_4P^-]$ were obtained for the stability constants of the protonated complex [2 to 4] and log $K_2 = 2.88$ at $I = 0.1$ mol/L($NaClO_4$) for $Mn(C_2H_6NO_4P)_2^{2-}$ [1].

References:

[1] Sakurai, H.; Okumura, H.; Takeshima, S. (Yakugaku Zasshi **96** [1976] 242/5; C.A. **84** [1976] No. 141449).
[2] Österberg, R. (Acta Chem. Scand. **16** [1962] 2434/51).
[3] Mohan, M. S.; Abbott, E. H. (J. Coord. Chem. **8** [1978] 175/82).
[4] Mohan, M. S.; Abbott, E. H. (Inorg. Chem. **17** [1978] 2203/7).

39.13.1.6 With Phosphoserine, Phosphothreonine, or Phospho-2-methylserine

ligand 1	ligand 2	ligand 3
	CH_3 |	
$CH_2{-}OP(O)(OH)_2$ | $CHNH_2$ | COOH	$CH{-}OP(O)(OH)_2$ | $CHNH_2$ | COOH	$CH_2{-}OP(O)(OH)_2$ | $H_3C{-}C{-}NH_2$ | COOH
ligand 1 (= DL-$C_3H_8NO_6P$)	ligand 2 (= DL-$C_4H_{10}NO_6P$)	ligand 3 (= DL-$C_4H_{10}NO_6P$)

Complexes in Solution. The formation of MnL^- and MnHL complexes with ligands 1 to 3 (H_3L) was revealed by pH potentiometric titrations at 25°C [1 to 4]. Stability constants and the protonation constant of the MnL^- complexes ($K^{MnL}_{MnHL} = [MnHL]/[MnL^-]\cdot[H^+]$) are as follows:

ligand	I in mol/L	log K_1	log K^{Mn}_{MnHL}	log K^{MnL}_{MnHL}	Ref.
1	0.15(KCl)	3.92	1.91	7.68	[1, 2]
	0.2(KNO_3)	3.80	2.33	8.25	[3, 4]
2	0.2(KNO_3)	3.81	2.20	8.06	[4]
3	0.2(KNO_3)	3.65	1.92	8.34	[4]

The neutral ligands are amphoteric compounds with one OH of the phosphate group deprotonated and a proton located on the amino group. Stepwise deprotonations at increasing pH are related to the carboxylato group, second OH of the phosphate group, and the amino group. The small stability constants of the MnHL complexes suggest coordination of the ligands only through the phosphate groups, whereas for the MnL complexes chelate formation involving the phosphate, amino, and carboxylate groups is assumed [4].

References:

[1] Österberg, R. (Nature **179** [1957] 476/7).
[2] Österberg, R. (Arkiv Kemi **13** [1959] 393/408).
[3] Mohan, M. S.; Abbott, E. H. (J. Coord. Chem. **8** [1978] 175/82).
[4] Mohan, M. S.; Abbott, E. H. (Inorg. Chem. **17** [1978] 2203/7).

39.13.1.7 With 2-Pyridinylmethyl Dihydrogen Phosphate

N $CH_2OP(O)(OH)_2$ $(=C_6H_8NO_4P=H_2L)$

Complex in Solution. The stability constant, log $K_1 = 2.44 \pm 0.05$ at 25°C and $I = 0.1$ mol/L (KNO_3), was determined by potentiometric pH titrations (glass electrode). The small value suggests coordination of the ligand through the phosphate group and only negligible participation of the pyridyl nitrogen atom.

Reference:

Murakami, Y.; Takagi, M. (J. Phys. Chem. **72** [1968] 116/21).

39.13.1.8 With Pyridoxal or Pyridoxamine Phosphate or a Related Compound

CHO
HO
$CH_2OP(O)(OH)_2$
H_3C
N

ligand 1
$(=C_8H_{10}NO_6P)$

CH_2NH_2
HO
$CH_2OP(O)(OH)_2$
H_3C
N

ligand 2
$(=C_8H_{13}N_2O_5P)$

$CH_2N(CH_2COOH)CH_2$
HO
$CH_2OP(O)(OH)_2$
H_3C
N
2

ligand 3
$(=C_{22}H_{32}N_4O_{14}P_2=H_8L)$

Complexes in Solution. Potentiometric pH titrations of aqueous solutions containing Mn^{2+} ions and one of the ligands reveal the formation of 1:1 complexes in all cases [1, 3]. Stability constants of the complexes with ligand 1 or 2 were determined at 25°C in 0.1 M $(C_2H_5)_4NBr$ [1]. ^1H, ^{13}C, and ^{31}P NMR measurements on the Mn^{II}-pyridoxal phosphate system in aqueous solution confirm the formation of a 1:1 complex and indicate a structure in which the Mn^{2+} ion is coordinated by the phosphate group ($r_{Mn-P}=3.3$ Å) and the aldehyde oxygen atom [2]. The stability constant of the complex with ligand 3, $Mn(C_{22}H_{24}N_4O_{14}P_2)^{6-}$, is log $K_1=15.10\pm0.08$ at 25°C and I = 0.1 mol/L(NaCl). Consecutive protonation constants of this complex, according to the reactions $MnH_{p-1}L^{(7-p)-}+H^+\rightleftharpoons MnH_pL^{(6-p)-}$, were obtained under the same conditions: log K = 9.35 ± 0.02, 8.55 ± 0.04, 6.41 ± 0.01, and 5.76 ± 0.01 for p = 1 to 4. The distribution of the various species in the Mn^{II}-ligand 3 system as a function of pH is shown in **Fig. 36**. The predominant species at pH 7.4, $Mn(C_{22}H_{26}N_4O_{14}P_2)^{4-}$, is assumed to have two protons on the pyridine nitrogen atoms [3]. The spin-only magnetic moment of this complex in D_2O (contain-

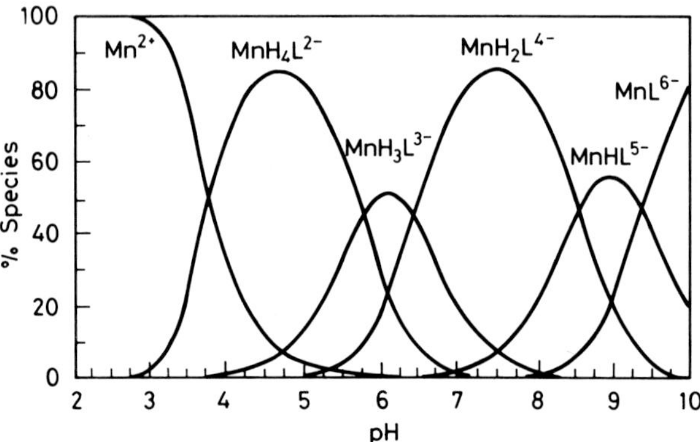

Fig. 36. Distribution of $MnH_pL^{(6-p)-}$ species with ligand 3 (= H_8L) as a function of pH determined in 10^{-3} M solution of $Mn(C_{22}H_{26}N_4O_{14}P_2)^{4-}$ at 25°C and I = 0.10 M(NaCl) [3].

ing 2% acetone) is 5.93 μ_B at pD 6.5, determined by NMR measurements (modified Evans method) [3]. The $Mn(C_{22}H_{26}N_4O_{14}P_2)^{4-}$ complex may be useful as an NMR-imaging contrast agent [3, 4].

$Na_2Ca[Mn^{II}(C_{22}H_{26}N_4O_{14}P_2)]\cdot 21\,H_2O$. Ligand 3 (10 mmol) was dissolved in 10 mL of degassed water containing NaOH (30 mmol). $MnCl_2\cdot 4H_2O$ (10 mmol) was added with stirring to the colorless solution; the pH dropped to ~3 as the solution turned yellow. NaOH (30 mmol) in 10 mL of water was added; the pH increased to 7.5. Solutions containing the $Mn(C_{22}H_{26}N_4O_{14}P_2)^{4-}$ anion prepared in this manner were used for equilibrium and magnetic solution studies. The extremely high water solubility of $Na_4[Mn(C_{22}H_{26}N_4O_{14}P_2)]$ makes the complex difficult to crystallize from aqueous solution. To induce crystallization, the stoichiometric amount of $CaCl_2\cdot 2H_2O$ was added, and the clear orange solution was allowed to stand at room temperature. Orange-yellow crystals of composition $Na_2Ca[Mn(C_{22}H_{26}N_4O_{14}P_2)]\cdot 21\,H_2O$ were isolated by filtration.

An X-ray crystal structure determination revealed the monoclinic system, space group $C2/c-C^6_{2h}$ (No. 15) with a = 24.490(3), b = 14.440(2), c = 30.703(4) Å, β = 112.747(12)°; Z = 8. Final positional and isotropic thermal parameters for all atoms are given in the paper. The structure was solved up to R = 0.0683. The molecular structure of the $[Mn(C_{22}H_{26}N_4O_{14}P_2)]^{4-}$ anion is shown in **Fig. 37**. The Mn^{2+} ion is coordinated by the aliphatic N atoms, two carboxylate O atoms, and the phenolic O atoms in a distorted octahedron. The calcium and sodium coordination spheres include three carboxylate O atoms (O(2), O(3), and O(10)), 11 water

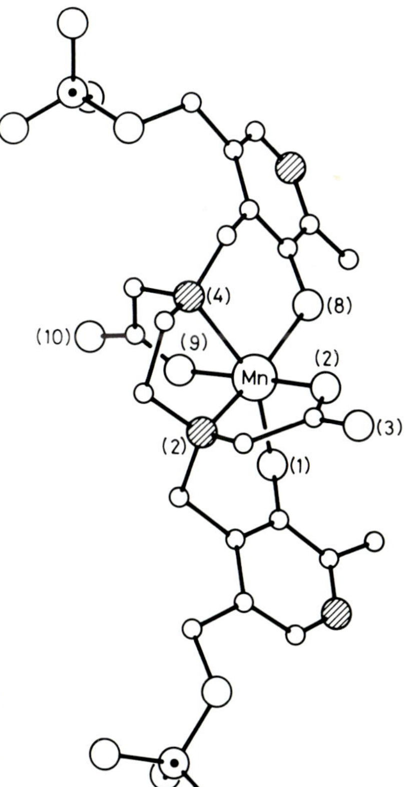

Fig. 37. Structure of the $[Mn(C_{22}H_{26}N_4O_{14}P_2)]^{4-}$ ion in $Na_2Ca[Mn(C_{22}H_{26}N_4O_{14}P_2)]\cdot 21H_2O$. Hydrogen atoms have been omitted for clarity [3].

molecules, and 4 O atoms from adjacent molecular units. The coordination geometry around the Ca^{2+} ion can be approximated as tetragonal antiprismatic. One Na^+ ion is coordinated in a distorted octahedral fashion, while the other Na^+ ion is pentacoordinate in a trigonal bipyramid. The complex anion, the cations, and the water molecules form an infinite polymeric network via bridging carboxylate groups and various hydrogen bonds. Selected bond distances (in Å) are:

Mn–N(2)	2.329(2)	Mn–O(1)	2.095(2)	Mn–O(8)	2.084(3)
Mn–N(4)	2.348(3)	Mn–O(2)	2.160(2)	Mn–O(9)	2.160(2)

The O atoms of the carboxylate groups are *trans* to each other with a bond angle of O(2)–Mn–O(9) = 173.46°. The phenolic O atoms are *trans* to the aliphatic N atoms with O(1)–Mn–N(4) = 173.46° and O(8)–Mn–N(2) = 152.79°. Other bond angles are from 74.44° for O(9)–Mn–N(4) to 115.27° for O(1)–Mn–O(8) [3].

The compound is air-stable. Upon standing in dry air it loses water of hydration [3].

References:

[1] Briggs, T. N. (Case Western Reserve University 1976; Diss. Abstr. B **37** [1977] 6146/7).
[2] Viswanathan, T. S.; Swift, T. J. (Can. J. Chem. **57** [1979] 1050/5).
[3] Rocklage, S. M.; Cacheris, W. P.; Quay, S. C.; Hahn, F. E.; Raymond, K. N. (Inorg. Chem. **28** [1989] 477/85).
[4] Rocklage, S. M.; Quay, S. C. (Eur. Appl. 290047 [1988]; C.A. **110** [1989] No. 154753).

39.13.1.9 With Thiamin Phosphate

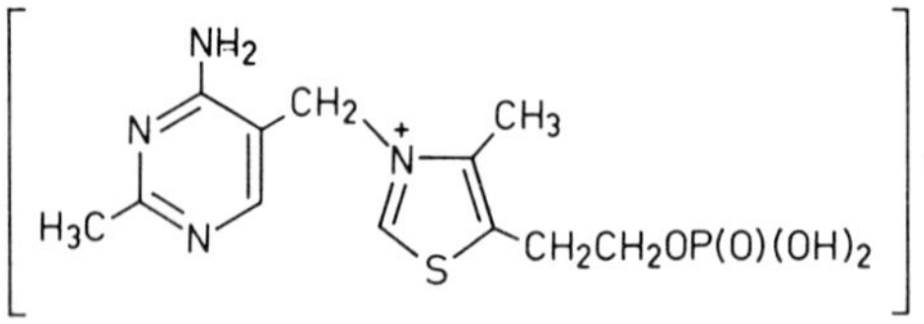

$[\,\ldots\,]Cl^-$ $(= C_{12}H_{18}ClN_4O_4PS = H_2L)$

Complexes in Solution. The formation of 1:1 complexes from Mn^{2+} ions and the ligand in aqueous solution was revealed by potentiometric titrations. Stability constants, log K_1 and log K^{Mn}_{MnHL} for $K^{Mn}_{MnHL} = [Mn(C_{12}H_{17}ClN_4O_4PS)^+]/[Mn^{2+}][C_{12}H_{17}ClN_4O_4PS^-]$, were determined at various temperatures (t) and I = 0.1 mol/L(KNO_3) [1, 2]:

t in °C	5	35	45
log K_1	2.75	2.99	2.85
log K^{Mn}_{MnHL}	1.85	2.16	2.00
Ref.	[1]	[2]	[1]

Thermodynamic parameters were calculated from the log K values at 5 and 45°C. Relatively small ΔH values and large positive ΔS values indicate entropy-driven complex formation reactions [1]. Coordination involving the amino N atom, the thiazole N atom, and one O atom of the phosphate group is assumed [2].

References:

[1] Taqui Khan, B.; Nageswara Rao, P.; Taqui Khan, M. M. (Indian J. Chem. A **20** [1981] 857/9).
[2] Taqui Khan, M. M.; Amara Babu, M. (J. Inorg. Nucl. Chem. **40** [1978] 2110/3).

39.13.2 Complexes with Diesters of Phosphoric Acid

39.13.2.1 With Dialkyl or Diphenyl Hydrogen Phosphates $(RO)_2P(O)OH$ (=HL)

ligand	1	2	3
R	CH_3	C_2H_5	C_3H_7
formula	$C_2H_7O_4P$	$C_4H_{11}O_4P$	$C_6H_{15}O_4P$

ligand	4	5	6
R	C_4H_9	$C_4H_9CH(C_2H_5)CH_2$	C_6H_5
formula	$C_8H_{19}O_4P$	$C_{16}H_{35}O_4P$	$C_{12}H_{11}O_4P$

Complexes in Solution. The $\mathbf{MnL_2}$ complexes with ligands 1 to 6 are polymeric in noncoordinating organic solvents. Molecular weight determinations indicate degrees of aggregation varying considerably with solvent and complex concentration. In coordinating solvents such as pyridine or dimethylformamide (L′), monomeric mixed ligand complexes of the type $\mathbf{MnL_2L'_2}$ are formed [19]. On the basis of polarographic studies, the formation of the $Mn(C_8H_{18}O_4P)_2$ complex in aqueous solutions containing Mn^{2+} ions and ligand 4 was assumed to occur. The stability constant $\log \beta_2=1.43$ was determined at 20°C [1]. Various extraction studies of Mn^{II} from aqueous solutions with ligand 5 in organic solvents and investigations of the extracted species indicate that the monomeric protonated $\mathbf{MnL_2(HL)_2}$ species, $Mn(C_{16}H_{34}O_4P)_2$-$(C_{16}H_{35}O_4P)_2$, is formed when ligand 5 is in excess in the organic phase [3 to 5, 7, 8, 11, 14]. The formation of the $MnL_2(HL)_2$ species was confirmed by the method of continuous variation [4]. At high Mn^{II} concentrations in the organic phase (>15 g/L [7]; Mn^{II}-saturated organic solution [3 to 5]) polymeric species $[Mn(C_{16}H_{34}O_4P)_2]_n$ are formed [4, 7, 8, 12] with the degree of polymerization n depending on both the metal concentration and temperature [12]. Determinations of molecular weights in benzene yielded values for n of, e.g., 2 or 4 [7] and 5 [4, 8]. Magnetic, ESR, and electronic spectral studies of the isolated $[MnL_2]_n$ and $MnL_2(HL)_2$ species (see pp. 156 and 157) indicate that the tetrahedral molecular structure of the solid compounds is retained for the complexes in solution.

The ligand exchange of the $Mn(C_{16}H_{34}O_4P)_2$ complex in a solution of excess ligand 5 in trichloroethylene was studied by measuring of the proton relaxation rate of the uncomplexed ligand in the range −90 to +80°C. The pseudo-first order rate constant $k=5.60\times10^6\ s^{-1}$ at 298 K for the exchange of a ligand atom at a particular Mn site was determined. The enthalpy and entropy for the exchange reaction are: $\Delta H_{298}=8.5$ kcal/mol and $\Delta S=0.9\pm3.3$ cal·mol^{-1}·K^{-1} [9].

Extraction studies of Mn^{II} from aqueous solutions with ligand 5 in various solvents were performed at various pH using $MnCl_2$ in aqueous HCl [4, 10], $MnSO_4$ [7, 12 to 15] in aqueous H_2SO_4 [7, 13] or Na_2SO_4 [12], or $MnNO_3$ in aqueous HNO_3+LiNO_3 [11] as the aqueous phases and 0.1 M to 50% ligand 5 in kerosene [4, 13 to 15], benzene [7, 11], toluene [10], a high-boiling hydrocarbon oil [12], hexane, octane, o-xylene, chloroform, or carbon tetrachloride as the organic phases [11]. The formation of an Mn^{III} complex in the organic phase by air oxidation was assumed by [13]. From the experimental results, the equation for the extraction reaction, $Mn^{2+}(a)+2(HL)_2(o)\rightleftharpoons Mn(HL_2)_2(o)+2H^+(a)$ was derived, where (a) denotes the aqueous and (o) the organic phase [4, 7, 10, 11]. Enthalpy changes for the extraction process were determined from measurements of the temperature-dependence of the distribution coefficients [4, 7, 11, 14]. Extraction constants K were calculated for various solvents and ionic strengths. Extrapolation of the values to zero ionic strengths yielded K=0.1 [11]. The effect of tributyl phosphate, bis(1-methylheptyl) methylphosphonate, and trioctylphosphane oxide on the extraction process was investigated. The synergetic effect of the additive decreased in the order trioctylphosphane oxide>bis(1-methylheptyl) methylphosphonate>tributyl phosphate [14].

The order of extraction Mn > Co is not affected by the system temperature or by the presence of tributyl phosphate [15]. The effect of the reducing agents hydrazine and hydroxylamine on the separation of Mn and Fe by extraction from aqueous solution and re-extraction from the organic phase was studied [13].

The formation of an **Mn^{III} complex**, presumably $Mn(C_{16}H_{34}O_4P)_3$, was revealed by spectrophotometric studies of the organic phase in the extraction of trace amounts of Mn^{III} from 16 M HNO_3 [16], 9 M $HClO_4$, 12 M CH_3COOH, or 22 M HCOOH [17] with 0.75 M [16] or 1.5 M ligand 5 [17] in heptane at 25°C [16, 17]. The distribution ratio for Mn^{III} was found to be greater than for Mn^{II} by a factor of ~300 [16]. The distribution ratio varied from 300 in $HClO_4$ to $>10^3$ in HCOOH and CH_3COOH [17].

Isolated Manganese(II) Compounds. The **MnL_2** complexes with dimethyl and diethyl phosphate, $Mn(C_2H_6O_4P)_2$ and $Mn(C_4H_{10}O_4P)_2$, were prepared by heating anhydrous $MnCl_2$ dissolved in the appropriate trialkyl phosphate at 150 to 200°C until evolution of alkyl halide had ceased [19, 22]. The excess solvent was removed by distillation at diminished pressure. The crystalline product was collected, washed with carbon tetrachloride followed by hexane, and dried at reduced pressure [19] (procedure adopted from [18]). Or, $Mn(C_4H_{10}O_4P)_2$ was precipitated from the solution by addition of absolute ethyl alcohol. The light brown complex melts at 184 to 185°C [22]. Also the formation of the complexes with dipropyl, dibutyl, and diphenyl phosphate by this procedure is reported in [19]. $Mn(C_2H_6O_4P)_2$ was also formed by using $MnBr_2$ instead of $MnCl_2$ in a similar procedure [20] and during the polarographic reduction of $Mn(ClO_4)_2$ in trimethyl phosphate containing 0.1 M $(C_2H_5)_4NClO_4$ [21]. The complex with dibutyl phosphate, $Mn(C_8H_{18}O_4P)_2$, was prepared by adding, in slight excess, $MnCO_3$ to the aqueous solution of ligand 4. The mixture was filtered and evaporated on a water bath until the waxy compound deposited; the deposit was then recrystallized [2]. The complex with diphenyl phosphate was obtained by crystallizing $Mn(C_{12}H_{10}O_4P)_2(dmf)_2$, see p. 157, from acetonitrile [19] (procedure adopted from [23]).

Two methods are reported for the preparation of the complex with bis(2-ethylhexyl) phosphate (ligand 5): 1) A solution of 0.1 M ligand 5 in hexane was contacted with aqueous $MnCl_2$ (20 g/L) and then 0.5 M aqueous NaOH added dropwise. The mixture was centrifuged and the phases separated. The organic phase was heated in vacuum at 50°C to evaporate the solvent [4, 5, 8]. 2) A solution (200 mL) of $MnSO_4$ (15 g/L) in aqueous 3.5 N H_2SO_4 was added to a solution (200 mL) of 0.2 M ligand 5 in benzene. The mixture was agitated until equilibrium was reached and then centrifuged. The resulting organic phase was then repeatedly (7 to 8 times) agitated with equal volumes of fresh $MnSO_4$ solution of the same Mn content and acidity. Benzene was allowed to evaporate from the saturated organic phase . To remove any adhering ligand, the solid mass was washed repeatedly (10 to 12 times) with alcohol. It was then air-dried and dehydrated by use of a high-vacuum pump [7] (procedure adopted from [6]).

Determinations of molecular weights indicate polymeric compounds [4, 7, 8, 12, 19, 25] with the degree of polymerization probably differing from that of the complexes in solution (see p. 155) [12]. A tetrahedral geometry around the Mn atom was inferred from magnetic, ESR, and electronic spectral studies [8, 22].

Measurements of the magnetic susceptibility at room temperature yielded $\mu_{eff} \approx 5.9\ \mu_B$ for all the MnL_2 compounds indicating high-spin Mn^{II} complexes [3 to 5, 8, 22, 25]. ESR X-band spectra of the complexes with ligands 1, 2, and 6, doped into the corresponding Cd salts, show resonances from zero field to above 400 mT. Hyperfine coupling constants A are in the range 0.0084 to 0.0092 cm^{-1} [19]. $Mn(C_2H_6O_4P)_2$ in toluene exhibits a six-line ESR spectrum at 295 K and additional hyperfine structure at 77 K. The parameters g = 2.000 and A = 93 G have been evaluated [24]. A one-line ESR spectrum at room temperature was observed for $Mn(C_{16}H_{34}O_4P)_2$ in the solid state [8] and in hexane [3 to 5]. The parameters $g_{\parallel} = 2.052$ and

$g_{\perp}=1.956$ have been evaluated [8]. The IR spectrum of $Mn(C_4H_{10}O_4P)_2$ in Nujol shows the ν(P=O) band at 1215 cm^{-1} (free ligand at 1271 cm^{-1}). The $ν(O–C_2H_5)$ band has also shifted to lower frequencies, whereas ν(P–O–C) remains almost unaffected [22]. For $Mn(C_8H_{18}O_4P)_2$ (in Nujol) the ν(P=O) band was observed at 1220 cm^{-1} (for $Na(C_8H_{18}O_4P)$ at 1242 cm^{-1}). Shifted bands in the 909 to 730 cm^{-1} region were assigned to the $P–O^-$ group [2]. IR bands of $Mn(C_{16}H_{34}O_4P)_2$ (no state given) were assigned as follows: 1190 to $ν_{as}(PO_2)$, 1102 to $ν_s(PO_2)$ (free ligand at 1230 cm^{-1}), 585, 480 to δ(O–P–O), and 385 cm^{-1} to ν(Mn–O). The splitting of the $ν(PO_2)$ and δ(O–P–O) bands is ascribed to the formation of the polymeric species [8]. Bonding through the phosphoryl O atom [2, 8, 22] and the O atom of the deprotonated hydroxy group was concluded [2]. The electronic spectrum of $Mn(C_4H_{10}O_4P)_2$ in chloroform shows absorption bands at 24000 and 28000 cm^{-1} with the extinction coefficients ε = 7.6 and 8.7 $L \cdot mol^{-1} \cdot cm^{-1}$ [22]. $Mn(C_{16}H_{34}O_4P)_2$ in benzene exhibits the broad and weak band assigned to the transition $^6A_1 \rightarrow {}^4T_1(G)$ at 20000 cm^{-1} and a strong absorption due to the charge-transfer transition at above 25000 cm^{-1}, a similar spectrum as is observed for the tetrahedral $MnCl_4^{2-}$ ion [8].

$Mn(C_4H_{10}O_4P)_2$ is a stable nonhygroscopic compound. It was found to be insoluble in common organic solvents [22].

The protonated $\mathbf{Mn(C_{16}H_{34}O_4P)_2(C_{16}H_{35}O_4P)_2}$ complex was obtained by extraction of Mn^{II} from aqueous solutions of $MnCl_2$ or $MnSO_4$ in HCl or H_2SO_4 with 0.1 or 0.2 M ligand 5 in hexane or benzene and evaporating the solvent at 40°C and 1 Torr [3 to 5, 7]. Excess ligand was removed by several washings with alcohol [7] (procedure adopted from [6]). The ESR spectrum of the complex at room temperature shows a six-line spectrum in the solid state or in hexane solution. The parameters g = 1.996 and A = 77.0 G have been evaluated [3 to 5]. The electronic spectrum shows broad and weak bands at 19400, 20000, and 23500 cm^{-1} with the extinction coefficients ε = 4.4, 4.9, and 2.6 $L \cdot mol^{-1} \cdot cm^{-1}$ assigned to the transitions from $^6A_1(S)$ to $^4T_1(G)$, $^4T_2(G)$, and $^4E_1(G)$ and/or $^4A_1(G)$, and a strong absorption above 25000 cm^{-1} due to the charge-transfer transition. The ligand field parameters 10 Dq = 21000 cm^{-1} and B = 620 cm^{-1} were determined [4]. On the basis of the electronic spectra a tetrahedral structure was proposed for the complex, the manganese atom being chelated by two HL_2^- groups [3 to 5].

Mixed ligand complexes $\mathbf{MnL_2py_2}$ with HL = ligand 1, 2, or 6 were prepared by adding a small amount of pyridine to a hot ethanol solution of the MnL_2 complexes (see p. 156) and then allowing the solution to cool. $\mathbf{Mn(C_{12}H_{10}O_4P)_2(dmf)_2}$ was prepared by heating anhydrous $MnCl_2$ and methyl diphenyl phosphate in dimethylformamide at 140 to 150°C for ~2 h. The solution was cooled and the volume doubled with xylene. The precipitate formed on standing and was washed with small quantities of ethanol and ether [19].

ESR X-band spectra at room temperature of powdered samples of the complexes doped into the corresponding cadmium complexes show six-line spectra with a strong group of lines centered at about g = 2, indicating small anisotropies in g and A tensors. Hyperfine coupling constants are in the range 0.0084 to 0.0093 cm^{-1}. Values for the axial crystal field splitting parameter, D, lie between 0.012 and 0.0255 cm^{-1}. Interpretation of the ESR spectra in terms of three g values implies that the symmetry of the ligand atoms around the Mn^{2+} ion is a distorted octahedron with the oxygen atoms from the phosphate groups lying approximately at the corners of a rectangle in the XY plane, and the two pyridine or DMF molecules along the z axis [19].

A complex $Mn_2(C_4H_{10}O_4P)_2(C_4H_{10}O_7P_2)$ containing diethyl hydrogen phosphate and diethyl diphosphate anions is described on p. 168.

References:

[1] Neveu, C. R.; Leicknam, J. P. (Electrochim. Acta **31** [1986] 705/15).
[2] Smith, T. D. (J. Inorg. Nucl. Chem. **9** [1959] 150/4).

[3] Sato, T.; Nakamura, T.; Kawamura, M.; Ueda, M. (J. Inorg. Nucl. Chem. **39** [1977] 1573/6).
[4] Sato, T.; Kawamura, M.; Nakamura, T.; Ueda, M. (J. Appl. Chem. Biotechnol. **28** [1978] 85/94).
[5] Sato, T.; Nakamura, T.; Kawamura, M. (Spec. Vol. Can. Inst. Mining Metall. **21** [1979] 159/63).
[6] Islam, F.; Biswas, R. K. (J. Inorg. Nucl. Chem. **40** [1978] 559/66).
[7] Islam, F.; Biswas, R. K. (J. Inorg. Nucl. Chem. **43** [1981] 1929/33).
[8] Sato, T.; Ueda, M. (Proc. Intern. Solvent Extract. Conf., Lyon, Fr., 1974, pp. 871/91).
[9] Lafferty, F. L. (Diss. Abstr. Intern. B **36** [1976] 5577; C.A. **85** [1976] No. 37610).
[10] Kimura, K. (Bull. Chem. Soc. Japan **34** [1961] 63/8).

[11] Smelov, V. S.; Drobysh, V. V.; Smyk, Z. A. (Radiokhimiya **14** [1972] 255/9; Soviet Radiochem. **14** [1972] 225/9).
[12] Barnes, J. E.; Setchfield, J. H.; Williams, G. O. R. (J. Inorg. Nucl. Chem. **38** [1976] 1065/7).
[13] Gorbanev, A. I.; Tsvetkova, Z. N.; Sobol', N. L.; Fomin, G. S. (Zh. Neorgan. Khim. **18** [1978] 2205/8; Russ. J. Inorg. Chem. **18** [1978] 1167/9).
[14] Shen, Jinglan; Jiang, Dehua; Sun, Sixiu (Gaodeng Xuexiao Huaxue Xuebao **5** [1984] 7/14; C.A. **100** [1984] No. 127567).
[15] Sun, Sixiu; Gao, Zili; Jiang, Dehua; Shen, Jinglan (Shandong Daxue Xuebao Ziran Kexueban **1985** 75/9; C.A. **103** [1985] No. 184629).
[16] Wood, R. A.; Roscoe, R. V. (J. Inorg. Nucl. Chem. **22** [1970] 1351/5).
[17] Wood, R. A.; Wakakuwa, S. T.; Beck, A.; Rutherford, T. H. (J. Inorg. Nucl. Chem. **33** [1971] 4291/9).
[18] Cookson, D. J.; Wakefield, R. F.; Smith, T. D.; Pilbrow, J. R. (Australian J. Chem. **28** [1975] 1217/29).
[19] Cookson, D. J.; Wakefield, R. F.; Smith, T. D.; Boas, J. F.; Pilbrow, J. R.; Hicks, P. R.; Lamble, D. H. (J. Chem. Soc. Faraday Trans. II **75** [1979] 500/8).
[20] Gutmann, V.; Fenkart, F. (Monatsh. Chem. **99** [1968] 1452/3).

[21] Gutmann, V.; Schmid, R. (Monatsh. Chem. **100** [1969] 1564/73).
[22] Paul, R. C.; Kapila, V. P.; Battu, R. S.; Sharma, S. K. (Indian J. Chem. **12** [1974] 827/9).
[23] Cookson, D. J.; Wakefield, R. F.; Smith, T. D.; Pilbrow, J. R. (Australian J. Chem. **28** [1975] 2515/9).
[24] Garif'yanov, N. S.; Kamenev, S. E.; Kozyrev, B. M.; Ovchinnikov, I. V. (Dokl. Akad. Nauk SSSR **177** [1967] 880/2; C.A. **68** [1968] No. 82425).
[25] Wakefield, R. F.; Smith, T. D. (unpublished results from Cookson, D. J.; Wakefield, R. F.; Smith, T. D.; Boas, J. F.; Pilbrow, J. R.; Hicks, P. R.; Lamble, D. H., J. Chem. Soc. Faraday Trans. II **75** [1979] 500/8).

39.13.2.2 With Phospholipids

Phospholipids are the major constituents of many biological membranes. Dispersed in aqueous solution they form ordered structures, such as micelles or liposomes (vesicles). The vesicles consist of concentrically arranged lipid bilayers around an aqueous interior, the bilayers being separated by layers of water. The polar head groups of the phospholipid molecules are oriented toward the aqueous solution side of the bilayers and are able to bind cations on their surfaces; this interaction greatly affects the stability and permeability of the bilayers. Therefore, aqueous lipid dispersions were investigated as models for biological membrane structures.

Aqueous dispersions of the phospholipids used for equilibrium studies with Mn^{2+} ions, consisting mostly of uni- or multilamellar vesicles, were prepared by shaking the dried lipid with an electrolyte solution [1, 5, 11] and/or exposing the crude dispersions to ultrasonic irradiation [1, 3, 4, 7 to 11], which causes a reduction of the particle size [2]. Investigations were performed in buffered solutions, mostly at pH 7.4, I = 0.1M NaCl, by measurement of the X-band electron spin resonance [4, 7, 8, 11]; proton relaxation rates, mostly of the solvent water [3, 11, 12]; ^{31}P nuclear magnetic resonance [5]; electrophoretic mobility [1, 5, 9, 10]; and the surface potential [6]. From the results, association constants were derived. These, however, because of differing definitions, are not presented here. Differences in the equilibrium equations, for example, are due to the kind of free ligand moieties introduced (concentrations of free phospholipid molecules, polar head groups, or binding sites). A normalization factor was introduced by some authors to consider the fact that only ~2/3 of the phosphate head groups are exposed to the external solution [7, 8], as reported to have been shown by NMR investigations [7]. Definitions of association constants are also dependent on the methods used. Thus, the number of molecules on the surface of the membrane per unit area was introduced into the equilibrium equation for evaluating results from measuring the electrophoretic mobility of the particles, e.g. [9].

Association constants for the interaction of Mn^{2+} ions with dispersed particles of the listed phospholipids are reported in the references cited below:

```
CH2-OC(O)R
|
CH-OC(O)R'
|
CH2-OP(O)OR"
     |
     O-
```

No.	R	R′	R″	formula	Ref.
1[a)]	$C_{17}H_{35}$	$C_{17}H_{33}$	$CH_2CH_2\overset{\oplus}{N}(CH_3)_3$	$C_{44}H_{86}NO_8P$	[1]
2	$C_{15}H_{31}$	$C_{15}H_{31}$	$CH_2CH_2\overset{\oplus}{N}(CH_3)_3$	$C_{40}H_{80}NO_8P$	[3,12]
3	$C_{17}H_{33}$	$C_{17}H_{33}$	$CH_2CH_2\overset{\oplus}{N}(CH_3)_3$	$C_{44}H_{84}NO_8P$	[5]
4	mixtures of various fatty acid alkyls		$CH_2CH_2\overset{\oplus}{N}(CH_3)_3$	b)	[5,11,12]
5	mixtures of various fatty acid alkyls		$CH_2CH_2(\overset{\oplus}{N}H_3)COOH$	c)	[3,4, 6 to 9]
6	mixtures of various fatty acid alkyls		$CH_2CH_2(\overset{\oplus}{N}H_3)COOH$	d)	[7]

ligands 1 to 6

```
CH2-OC(O)R
|
CH-OC(O)R'
|
CH2-OP(O)OCH2-CH-CH2OH
     |        |
     OH       OH
```

ligand 7 with R = R′ = $C_{15}H_{31}$ (= $C_{38}H_{75}O_{10}P$) [7]
ligand 8 with R = R′ = mixtures of various fatty acid alkyls (= egg phosphatidylglycerol) [7, 10]

a) Mixtures of lignd 1 with 1-stearoyl-2-oleoyl-*syn*-glycerol 3-phosphate. – b) Egg phosphatidylcholine. – c) Bovine phosphatidylserine. – d) Hydrogenated phosphatidylserine.

Association constants (K_a) of Mn binding to vesicles of ligand 5 were determined in the presence of various monovalent salts [4] and as a function of Na^+ ion concentration. The results, showing a strong decrease of K_a with $[Na^+]$, are discussed in terms of the relevant theory of ionic binding to charged surfaces [4, 8]. A strong increase of the association constant above pH 7.3 reflects chelate formation involving the serine amino group in complex formation of Mn^{2+} ions with ligand 5 [8]. Enthalpy changes for Mn^{2+} binding to ligands 5, 6, and 8 (from equilibrium studies in the 15 to 65°C range) indicate endothermic reactions (ΔH values between 2.5 and 4.2 kcal/mol). Possible reasons for the endothermic course of the reactions are discussed [7].

References:

[1] Barton, D. G. (J. Biol. Chem. **243** [1968] 3884/90).

[2] Finer, E. G.; Flook, A. G.; Hauser, H. (Biochim. Biophys. Acta **260** [1972] 49/58).
[3] Nolden, D. W.; Ackermann, T. (Biophys. Chem. **3** [1975] 183/91).
[4] Puskin, J. S. (J. Membrane Biol. **35** [1977] 39/55).
[5] McLaughlin, A.; Grathwohl, C.; McLaughlin, S. (Biochim. Biophys. Acta **513** [1978] 338/57).
[6] Ohki, S.; Sauve, R. (Biochim. Biophys. Acta **511** [1978] 377/87).
[7] Puskin, J. S.; Martin, T. (Biochim. Biophys. Acta **552** [1979] 53/65).
[8] Puskin, J. S.; Coene, M. T. (J. Membrane Biol. **52** [1980] 69/74).
[9] McLaughlin, S.; Mulrine, N.; Gresalfi, T.; Vaio, G.; McLaughlin, A. (J. Gen. Physiol. **77** [1981] 445/73).
[10] Lau, A.; McLaughlin, A.; McLaughlin, S. (Biochim. Biophys. Acta **645** [1981] 279/92).

[11] Sabatini, G.; Tiezzi, E.; Valensin, G. (Nuovo Cimento Soc. Ital. Fis. D **2** [1983] 55/63; C.A. **98** [1983] No. 139329).
[12] Radda, G. K. (Biochem. J. **122** [1971] 385/96).

39.13.3 Complexes with Triesters of Phosphoric Acid $P(O)(OR)_3$

39.13.3.1 With Trimethyl, Triethyl, or Tripropyl Phosphate

ligand	1	2	3
R	CH_3	C_2H_5	C_3H_7
formula	$C_3H_9O_4P$	$C_6H_{15}O_4P$	$C_9H_{21}O_4P$

Complexes in Solution. The formation of $[Mn(C_3H_9O_4P)_6]^{2+}$ ions in trimethyl phosphate solutions of $MnBr_2$ was shown by spectrophotometric, potentiometric, and conductometric studies [1]. From ESR X- and S-band spectra the parameters $g = 2.0012 \pm 0.0002$, $A = 97.3 \pm 0.5$ G, and $D = 120$ G were obtained. The line width maximum was observed at −35°C [2]. Potentiometric and conductometric measurements in the presence of N_3^- ions reveal the formation of $[Mn(C_3H_9O_4P)_{6-n}(N_3)_n]^{(2-n)+}$ complexes [3]. Solutions of $MnCl_2$ in trimethyl phosphate probably contain the octahedral complexes $[Mn(C_3H_9O_4P)_2Cl_4]^{2-}$ and $MnCl_6^{4-}$ [4].

Triethyl phosphate forms 1:2 solvates with $MnCl_2$. Dilute solutions of manganese(II) chloride (0.0001 mol/L) in triethyl phosphate have a specific conductance of $7.3 \times 10^{-4}\ \Omega^{-1} \cdot cm^{-1}$, indicating that $MnCl_2$ is dissociated in the neat ligand. This may be due to the formation of $Mn(C_6H_{15}O_4P)_2Cl_2$, $Mn(C_6H_{15}O_4P)_3Cl^+$, $Mn(C_6H_{15}O_4P)Cl_3^-$, $Mn(C_6H_{15}O_4P)_4^{2+}$, and $MnCl_4^{2-}$ ions. The UV and visible spectra of the dilute solution show the presence of $MnCl_4^{2-}$ ions in the solvent. At higher concentrations of solutes (0.1M), adducts having the composition $Mn(C_6H_{15}O_4P)_2Cl_2$ separate out [5]. The ESR spectra of $Mn(NO_3)_2$ in triethyl phosphate measured at 293 K and in the glassy state at 103 K are similar to the corresponding spectra in tributyl phosphate. Measurements in triethyl phosphate in the presence of $LiNO_3$ at various temperatures (257 to 307 K) indicate the formation of at least two complexes, depending on the amount of excess NO_3^-. The formation of mono- and dinitrato compounds is assumed. A solution of $Mn(ClO_4)_2$ in triethyl phosphate with $LiClO_4$ added up to saturation probably contains complexes in which manganese is bonded to two perchlorate groups in addition to solvent molecules. The structure is octahedral with relatively high crystal-field symmetry. The ESR parameters $g = 2.001 \pm 0.0005$, $A = 97.6 \pm 0.5$ G, and $D = 110$ G result from study of X-band and S-band spectra at 293, 233, and 103 K. The line width maximum was observed at −10°C [2]. The nature and stability of the bond in $MnX_2 \cdot xL \cdot yH_2O$ complexes with $X = NO_3$ or Cl and L = trimethyl or tripropyl phosphate is discussed in [6] and an estimation for the extractability of the manganese salts given. The extraction of Mn^{II} chlorides or nitrates with tripropyl phosphate was investigated by [7]. IR data indicate that the Mn ligand bond is weaker than for tributyl phosphate complexes.

$[Mn^{II}(C_3H_9O_4P)_n](ClO_4)_2$ (n = 6 or 5) and **$[Mn^{II}(C_3H_9O_4P)_5(H_2O)](ClO_4)_2$**. The complex with n = 6 was prepared by dehydration of $Mn(ClO_4)_2 \cdot 6H_2O$ with a large excess of triethyl orthoformate at 0°C, followed by the addition of a slight excess of trimethyl phosphate. The complex was precipitated as an oil upon addition of diethyl ether. A solid product was obtained by repeated washings with fresh ether. The compound was dried in vacuum over P_4O_{10} [8]. A similar procedure, dehydration of the Mn^{II} salt with a slight excess of triethyl orthoformate at 40 to 50°C, followed by addition of the ligand (mole ratio of the ligand to salt was slightly in excess of 6:1), was used by [9]. An initially formed pink complex, $[Mn(C_3H_9O_4P)_5(H_2O)](ClO_4)_2$ (not further characterized), was reported to change into $[Mn(C_3H_9O_4P)_5](ClO_4)_2$ if kept in vacuum over $Mg(ClO_4)_2$. The white complex melts at 134 to 134.5°C. It is very hygroscopic. Reversible formation of $[Mn(C_3H_9O_4P)_5(H_2O)](ClO_4)_2$ when exposed to the atmosphere or on addition of water was assumed [9].

The IR spectra of the complexes (in Nujol) show the ν(PO) band shifted from 1280 cm^{-1} (free ligand) to 1260 cm^{-1} [8] or 1235 cm^{-1} [9]. This indicates coordination of the ligand to Mn through the phosphoryl oxygen atom. A shoulder at 1070 cm^{-1} [9] and a band at 619 [10] in the spectrum of $[Mn(C_3H_9O_4P)_5](ClO_4)_2$ are assigned to $\nu(ClO_4)$ vibrations of ionic perchlorate. The electronic spectra of this complex show absorption maxima at 358 and 495 nm (Nujol) or at 309, 363, 412sh, and 490 nm in neat ligand solution (1.3×10^{-2} mol/L), assigned to a square-pyramidal environment at Mn with a symmetry considerably lower than C_{4v} [9]; an octahedral geometry at Mn was assumed for $[Mn(C_3H_9O_4P)_6](ClO_4)_2$. The magnetic moment, $\mu_{eff} = 5.87\ \mu_B$ at 297 K, was determined by the method of Evans using benzene as an inert reference in a solution of the ligand [8]. The electric conductivity of $[Mn(C_3H_9O_4P)_5](ClO_4)_2$ in 10^{-3} M nitromethane solution at 25°C, $\Lambda = 167\ cm^2 \cdot \Omega^{-1} \cdot mol^{-1}$, is that of a 1:2 electrolyte. The complex is soluble in many organic solvents. Dry $[Mn(C_3H_9O_4P)_5](ClO_4)_2$ deliquesces slowly when exposed to the atmosphere to form reversibly the pale pink monohydrate [9].

$Mn^{II}(C_6H_{15}O_4P)_2Cl_2$ was prepared in CCl_4 by refluxing stoichiometric amounts of triethyl phosphate with $MnCl_2$ (dehydrated previously by refluxing with anhydrous $SOCl_2$). The precipitate was washed with petroleum ether and dried in vacuum [5].

References:

[1] Gutmann, V.; Fenkart, K. (Monatsh. Chem. **98** [1967] 286/93).
[2] Burlamacchi, L. (Gazz. Chim. Ital. **106** [1976] 347/58, 352).
[3] Gutmann, V.; Lux, W. (J. Inorg. Nucl. Chem. **29** [1967] 2391/9).
[4] Gutmann, V.; Lux, W. (Monatsh. Chem. **98** [1967] 276/85).
[5] Paul, R. C.; Sharma, O. D.; Bhatia, J. C. (Indian J. Chem. **13** [1975] 692/4).
[6] Afanas'ev, Yu. A.; Akhrimenko, Z. M. (Zh. Fiz. Khim. **48** [1974] 789; Russ. J. Phys. Chem. **48** [1974] 459).
[7] Afanas'ev, Yu. A.; Akhrimenko, Z. M. (Khim. Khim. Tekhnol. Tezisy 2nd Kraev. Nauchn. Tekhn. Konf. Molodykh Uch. Aspir. Spets. Khim. Kubani, Krasnodar 1973, Vol. 2, pp. 196/9; C.A. **85** [1976] No. 153328).
[8] Frankel, L. S. (Inorg. Chem. **8** [1969] 1784/7).
[9] Karayannis, N. M.; Bradshaw, E. E.; Pytlewski, L. L.; Labes, M. M. (J. Inorg. Nucl. Chem. **32** [1970] 1079/88).
[10] Karayannis, N. M.; Mikulski, C. M.; Strocko, M. J.; Pytlewski, L. L.; Labes, M. M. (Inorg. Chim. Acta **8** [1974] 91/6).

39.13.3.2 With Tributyl or Triisobutyl Phosphate $P(O)(OC_4H_9)_3$ ($=C_{12}H_{27}O_4P$)

Complexes in Solution. The use of tributyl phosphate to extract Mn^{II} salts from aqueous solution is well known. A trisolvate is formed with $Mn(NO_3)_2$ [1], a disolvate with $MnCl_2$ [1 to 3]. The nitrate is extracted much more completely than the chloride [1]. Chemical composition and IR data for the hydrato solvates of $Mn(NO_3)_2$ are reported in [4], of $MnCl_2$ in [5]. The IR spectra exhibit deformational vibration bands of H_2O; i.e., the extracted compounds have H_2O along with tributyl phosphate coordinated to manganese. Thermochemical and other physical data for the hydrato solvates in the organic phase are reported in [6, 7]. The spin echo method has been used to measure the spin-spin relaxation of the phosphorus (^{31}P) nuclei in the tributyl phosphate solution of Mn^{II} nitrate. The lifetime of the solvent molecules in the solvate sheath of the complex has been determined. A ligand exchange rate constant of $k=4.5\times10^6\ s^{-1}$ was found at 293 K. In the presence of H_2O the ligand exchange rate is decreased [8]. A magnetic moment of 6.03 μ_B results from susceptibility measurement in tributyl phosphate solution at 77 and 300 K. The ESR spectrum of 0.5M $Mn(NO_3)_2\cdot xH_2O$ in tributyl phosphate, measured at −196°C, can be related to Mn^{II} complexes with strong covalent bonds [9]. The ESR data show that $Mn(NO_3)_2$ in tributyl phosphate forms at least two complexes depending on the concentration of excess NO_3^- ions. The formation of mono- and dinitrato complexes was assumed [10, 11].

The conductivity of a solution of $MnCl_2\cdot2C_{12}H_{27}O_4P$, see p. 163, in tributyl phosphate indicates its ionic character. A solution of $MnCl_2$ (5.0×10^{-5}M) in tributyl phosphate has a specific conductivity of $0.85\times10^{-4}\ \Omega^{-1}\cdot cm^{-1}$. Formation of solvated species with tributyl phosphate (=L) and of $MnCl_4^{2-}$ ions according to $2MnL_2Cl_2+2L\rightleftharpoons[MnL_3Cl]^+[MnLCl_3]^-\rightleftharpoons[MnL_4]^{2+}+MnCl_4^{2-}$ was assumed. The conductivity was found to increase on the addition of LiCl. Conductometric titrations between $(C_2H_5)_4NCl$ and $MnCl_2$ in tributyl phosphate show sharp peaks at 1:1 and 2:1 molar ratios, indicating the occurrence of the following reactions: $MnL_2Cl_2\xrightarrow[-L]{+Cl^-}[MnCl_3L]^-\xrightarrow[-L]{+Cl^-}[MnCl_4]^{2-}$ [12]. The presence of $MnCl_4^{2-}$ is corroborated by ESR data [10] and by the electronic spectrum of $MnCl_2$ solution in tributyl phosphate. It shows a maximum at 420 and a shoulder at 350 nm and is identical with that of $[(C_4H_9)_4N]_2MnCl_4$ [12]. According to [3] Mn^{2+} ions are extracted by tributyl phosphate from concentrated HCl solutions as $[H(H_2O)_4L]_2MnCl_4$. The distribution of trace quantities of Mn^{2+} (2.5×10^{-6} to 10^{-4} mol/L) between tributyl phosphate and aqueous solutions of LiCl has been studied at 20°C using ^{54}Mn γ tracer ($t_{1/2}=291$ d). An increase of the distribution ratio from D=0.001 at 0.1M LiCl to D=100 at 10M LiCl was observed and the existence of $[Li(H_2O)_{4-x}L_x]_2MnCl_4$ complexes with x=0 to 2 suggested [13].

ESR data show that Mn^{II} perchlorate's behavior in tributyl phosphate solution is similar to that in triethyl phosphate solution, see p. 160. In the presence of $LiClO_4$ up to saturation, formation of a complex with two perchlorate groups and two solvent molecules bonded to manganese was suggested. The ESR parameters are $g=1.9982\pm0.0005$ and $A\approx97$ G [10].

The extraction of Mn^{II} from aqueous thiocyanate solutions by tributyl phosphate diluted with kerosene was investigated by [14]. Partition coefficients of manganese have been determined for various concentrations of Mn^{2+}, H^+, and SCN^- ions in the aqueous phase and for various tributyl phosphate concentrations in the organic phase. An instability constant of K=10 mol/L for the extracted $[Mn(C_{12}H_{27}O_4P)_3(NCS)_2]$ complex and an extraction constant of 1.2 have been estimated [14]. For the extraction of Mn^{2+} from thiocyanate solutions with tributyl phosphate, see also [15, 16].

Formation of a mixed ligand complex, $Mn(C_{12}H_{27}O_4P)_2(C_7H_5O_3)_2$, was observed when Mn^{2+} ions were extracted from aqueous solution into tributyl phosphate in the presence of salicylic acid ($=C_7H_6O_3$) [17]. The synergistic effect of tributyl phosphate was studied in the extraction of Mn^{2+} using various diketones (HL) in organic solvents. Adduct-formation constants for the reaction $MnL_2 + nC_{12}H_{27}O_4P \rightleftharpoons Mn(C_{12}H_{27}O_4P)_nL_2$ with n = 1 or 2 have been determined by distribution measurements at 25°C: log β_2 = 10.20 or 6.90 for the $Mn(C_{12}H_{27}O_4P)_2(C_8H_4F_3O_2S)_2$ complex with 4,4,4-trifluoro-1(2-thienyl)-1,3-butanedione in cyclohexane or benzene, respectively [18], log β_1 = 4.9 and log β_2 = 9.4 for the $Mn(C_{12}H_{27}O_4P)_n(C_5HF_6O_2)_2$ complexes with 1,1,1,5,5,5-hexafluoro-2,4-pentanedione in CCl_4 [21], log β_1 = 7.90 and log β_2 = 10.22 for the $Mn(C_{12}H_{27}O_4P)_n(C_{17}H_{13}N_2O_2)_2$ complexes with 4-benzoyl-5-methyl-2-phenyl-2,4-dihydro-3 H-pyrazol-3-one in benzene [19].

$[Mn^{II}(C_{12}H_{27}O_4P)_4(H_2O)ClO_4]ClO_4$ and **$[Mn^{II}(C_{12}H_{27}O_4P)_4(H_2O)](ClO_4)_2$** complexes with tributyl or triisobutyl phosphate, respectively, were prepared similarly to the complex with trimethyl phosphate (p. 161): $Mn(ClO_4)_2 \cdot 6H_2O$ was treated with triethyl orthoformate at 40 to 50°C, followed by addition of excess ligand. The resulting mixture was heated and stirred at 50°C for 3 to 4 h, then allowed to cool slowly. The complexes were obtained in crystalline form by stirring the viscous reaction mixture at room temperature for 1 to 6 d. The white compounds, being sensitive to atmospheric moisture, were thoroughly washed with ether in a drybox and subsequently stored over $CaCl_2$ in an evacuated desiccator; they are not completely dehydrated by drying agents ($CaCl_2$, $Mg(ClO_4)_2$, or P_4O_{10}). The complex with tributyl phosphate collapses to a viscous liquid in the desiccator and could not be recrystallized from a number of solvents. A magnetic moment of μ_{eff} = 5.96 μ_B was calculated for both complexes from susceptibility measurements at 300 K. Bands observed in the IR spectra (in cm^{-1}) were assigned as follows, shifts of ν(PO) in comparison to the free ligands are given in parentheses:

complex	ν(PO)	ν(OH)	δ_{HOH}	$\nu_3(ClO_4)_{ionic}$	$\nu_2(OClO_3)$
$[Mn(C_{12}H_{27}O_4P)_4(H_2O)ClO_4]ClO_4$	1228 (−47)	3478	1622	1103	913
$[Mn(C_{12}H_{27}O_4P)_4(H_2O)](ClO_4)_2$	1230 (−36)	3405	1634	1105	—

Additional vibration modes of the ligand in the 920 to 882 cm^{-1} range are given in the paper. The electronic spectra of the complexes in Nujol show maxima at 302 and 260 nm or 300 and 259 cm^{-1}, respectively. From the physical data the cationic complex with tributyl phosphate is assumed to be hexacoordinated, probably involving a monodentate perchlorate group, whereas the complex with the bulky triisobutyl phosphate groups seems to involve pentacoordinated complex cations. The molar electrical conductivity of the tributyl phosphate complex in 10^{-3} M nitromethane solution at 25°C, $\Lambda = 177\ cm^2 \cdot \Omega^{-1} \cdot mol^{-1}$, indicates that in solution the perchlorato group is displaced by a solvent molecule. The electrical conductivity of $[Mn(C_{12}H_{27}O_4P)_4(H_2O)](ClO_4)_2$, $\Lambda = 166\ cm^2 \cdot \Omega^{-1} \cdot mol^{-1}$, shows it to be a 1:2 electrolyte [20].

$Mn^{II}(C_{12}H_{27}O_4P)_2Cl_2$ was prepared by refluxing stoichiometric amounts of anhydrous $MnCl_2$ and tributyl phosphate in carbon tetrachloride, similarly to the complex with triethyl phosphate, $Mn(C_6H_{15}O_4P)_2Cl_2$, see p. 161 [12].

References:

[1] Rozen, A. M.; Skotnikov, A. S.; Martynova, M. K. (Zh. Neorgan. Khim. **29** [1984] 1789/802; Russ. J. Inorg. Chem. **29** [1984] 1031/3).

[2] Jeżowska-Trzebiatowska, B.; Kopach, S. (Radiokhimiya **8** [1966] 145/53; Soviet Radiochem. **8** [1966] 137/43).

[3] Kopacz, S. (Pr. Nauk Isnt. Chem. Nieorg. Met. Pierwiastkow Rzadkich Politech. Wroclaw No. 16 [1973] 278/325; C.A. **81** [1974] No. 54856).

[4] Afanas'ev, Yu. A.; Panyushkin, V. T.; Akhrimenko, Z. M. (Zh. Neorgan. Khim. **19** [1974] 281/3; Russ. J. Inorg. Chem. **19** [1974] 154/5).

[5] Afanas'ev, Yu. A.; Panyushkin, V. T.; Akhrimenko, Z. M. (Zh. Neorgan. Khim. **18** [1973] 1992/4; Russ. J. Inorg. Chem. **18** [1973] 1056/7).
[6] Afanas'ev, Yu. A.; Akhrimenko, Z. M.; Aslanyan, L. A. (Zh. Fiz. Khim. **48** [1974] 788/9; Russ. J. Phys. Chem. **48** [1974] 458/9).
[7] Afanas'ev, Yu. A.; Akhrimenko, Z. M. (Zh. Fiz. Khim. **48** [1974] 789; Russ. J. Phys. Chem. **48** [1974] 459).
[8] Vashman, A. A.; Pronin, I. S.; Vereshchagina, T. Ya. (Zh. Neorgan. Khim. **15** [1970] 3124/7; Russ. J. Inorg. Chem. **15** [1970] 3124/7).
[9] Vashman, A. A.; Vereshchagina, T. Ya.; Pronin, I. S. (Zh. Strukt. Khim. **11** [1970] 433/6; J. Struct. Chem. [USSR] **11** [1970] 397/400).
[10] Burlamacchi, L. (Gazz. Chim. Ital. **106** [1976] 347/58).

[11] Burlamacchi, L. (J. Chem. Phys. **55** [1971] 1205/12).
[12] Paul, R. C.; Singh, M.; Sharma, O. D.; Bhatia, J. C. (Indian J. Chem. A **14** [1976] 379/83).
[13] Morris, D. F.; Short, E. L.; Slater, D. N. (Electrochim. Acta **8** [1963] 289/300).
[14] Gorbanev, A. I.; Tsvetkova, Z. N.; Fomin, G. S.; Sobol', N. L. (Zh. Neorgan. Khim. **18** [1973] 3043/5; Russ. J. Inorg. Chem. **18** [1973] 1618/20).
[15] Brigevich, R. F.; Kuznetsov, R. A. (Radiokhimiya **10** [1968] 243/6; Soviet Radiochem. **10** [1968] 229/32).
[16] Różycki, C.; Lachowicz, E. (Chem. Anal. [Warsaw] **15** [1970] 255/69).
[17] Agett, J.; Evans, D. J.; Hancock, R. (J. Inorg. Nucl. Chem. **30** [1968] 2529/38, 2534).
[18] Nakamura, S.; Imura, H.; Suzuki, N. (J. Radioanal. Nucl. Chem. **82** [1984] 33/44, 43).
[19] Li, Le Min; Hsu, Kwang Hsien (Gaodeng Xuexiao Huaxue Xuebao **2** [1981] 265/74; C.A. **95** [1981] No. 193123).
[20] Karayannis, N. M.; Mikulski, C. M.; Strocko, M. J.; Pytlewski, L. L.; Labes, M. M. (Inorg. Chim. Acta **4** [1970] 557/61).

[21] Sekine, T.; Murai, R.; Takahashi, K.; Iwahori, S. (Bull. Chem. Soc. Japan **50** [1977] 3415/6).

39.13.3.3 With Poly(triallyl Phosphate) $[P(O)(OCH_2CH{=}CH_2)_3]_n$ (= $(C_9H_{15}O_4P)_n$)

The polymeric $\mathbf{Mn^{II}L_4(ClO_4)_2}$ and $\mathbf{Mn^{II}L_2X_2}$ complexes (L = one unit of the polymer ligand; X = NO_3, Cl, Br, I) were prepared by reaction of the polymer ligand with the corresponding manganese(II) salt in H_2O(?), ethanol, or acetone. In the case of ethanol or acetone the appropriate amount of trimethyl or triethyl orthoformate was added for dehydration. After stirring the mixture, the precipitates were washed with acetone and/or ether and dried in vacuum over P_4O_{10}. The perchlorate and the complexes with X = NO_3 or Cl are white, those with X = Br or I are yellow. The compounds isolated from ethanol or acetone are hygroscopic. IR data indicate that the polymer ligand coordinates to Mn via the phosphoryl oxygen atom; the ν(PO) vibration mode of the ligand (at 1270 cm^{-1}) is shifted to lower energy in the complexes (1250 to 1230 cm^{-1}). The existence of ν_4 as a split band in the 640 to 620 cm^{-1} region shows the ClO_4^- ion to be coordinated to the metal. The spectra also indicate that the NO_3^- ions are most likely bidentate. A band at 235 cm^{-1} in the spectrum of the bromo complex was assigned to ν(Mn–Br).

Reference:

de Bolster, M. W. G.; de Jong, G.; Smit, C. N.; Snoeks, J. M. (Z. Anorg. Allgem. Chem. **463** [1980] 179/84).

39.13.3.4 With Tritolyl Phosphate

$(CH_3-C_6H_4-O)_3PO$ $(=C_{21}H_{21}O_4P)$

$[Mn^{II}(C_{21}H_{21}O_4P)_4](ClO_4)_2$ and $[Mn^{II}(C_{21}H_{21}O_4P)_2Cl_2]$. When precipitating the complexes from solution it is important to use precise ligand:salt molar ratios (4:1 or 2:1, respectively): A solution of 2 mmol ligand in 50 mL methanol was combined with the stoichiometric amount of the Mn^{II} salt in 50 mL methanol or methanol-acetone (4:1 v/v). The solution was stirred at 50°C until the volume was reduced to ~15 mL, then 50 mL triethyl orthoformate was added. Heating at 50°C was continued until near dryness. The solid residue was separated and stored in vacuum over $CaCl_2$. Washing, while still wet, with anhydrous diethyl ether was found to lead to partial decomposition; washing with ligroin or hexane causes the formation of an oil. Therefore, the residual solvent was removed by filtering with suction. $[Mn(C_{21}H_{21}O_4P)_2Cl_2]$ may also be prepared by heating the ligand (~5 g) in a beaker up to ~80°C; $MnCl_2$ is added and dissolves slowly in the melt. The solution was stirred and the temperature raised at a rate of 2°C/min. The complex precipitated between 150 and 200°C. The mixture was allowed to cool to ~100°C and the solid complex separated by filtration through a heated funnel. The complex was pressed between sheets of filter paper and stored in an oven at 90°C. The procedure must be repeated until the IR spectrum no longer exhibits the band at 1310 cm^{-1} (due to uncoordinated phosphate ligand). The white perchlorate melts between 70 and 85°C; the white chloro complex does not melt below 250°C. Magnetic moments of $\mu_{eff}=5.89$ μ_B for the perchlorate and 5.93 μ_B for the chloro complex result from susceptibility measurements (Faraday method) at 300 K. Pertinent IR bands (in cm^{-1}) and bands observed in the electronic spectra of the complexes in Nujol (λ_{max} in nm) are shown below:

complex	ν(PO)	ν(Mn–O)	ν(Mn–Cl)	λ_{max}
$[Mn(C_{21}H_{21}O_4P)_4](ClO_4)_2$	1296, 1276 sh	404, 339	—	482, 440 sh, 400 sh, 381 sh, 321
$[Mn(C_{21}H_{21}O_4P)_2Cl_2]$	1300 sh, 1296	390, 334	277 sh	488, 450 sh, 411, 379 sh, 308

Additional ligand vibration bands are reported in the paper. The uncoordinated ligand reveals ν(PO) bands at 1310 and 1300 cm^{-1}. The molar conductivity of 10^{-3} M solutions in nitromethane at 25°C is 151 $cm^2\cdot\Omega^{-1}\cdot mol^{-1}$ for the perchlorate and 24 $cm^2\cdot\Omega^{-1}\cdot mol^{-1}$ for the chloro complex. The compounds appear to be binuclear, containing both terminal and bridging ligands (L) and pentacoordinated manganese. These binuclear complexes, $[L_3MnL_2MnL_3](ClO_4)_4$ and $[Cl_2LMnL_2MnLCl_2]$, do not seem to undergo any significant structural changes in nitromethane solution. The lower IR frequency and the electronic spectral evidence suggest that tris(4-methylphenyl) phosphate is almost as strong a ligand as triorganylphosphane oxides and significantly stronger than trimethyl phosphate. The increased donor strength of tris-(4-methylphenyl) phosphate is probably due to a combination of inductive and steric effects. The complexes are soluble in alcohols, acetone, and nitromethane.

Reference:

Mikulski, C. M.; Pytlewski, L. L.; Karayannis, N. M. (Inorg. Chim. Acta **32** [1979] 263/72).

39.13.4 Complexes with Esters of Diphosphoric Acid

39.13.4.1 With Alkenyl or Alkadienyl Trihydrogen Diphosphate (= H_3L)

$CH_2{=}C(CH_3){-}CH_2{-}CH_2{-}O{-}P(O)(OH){-}O{-}P(O)(OH){-}OH$ ligand 1 (= $C_5H_{12}O_7P_2$)

$(CH_3)_2C{=}CH{-}CH_2{-}CH_2{-}C(CH_3){=}C(R')R$

ligand 2 with R = $CH_2{-}O{-}P(O)(OH){-}O{-}P(O)(OH){-}OH$, R′ = H

(= (*E*)-$C_{10}H_{20}O_7P_2$ = geranyl diphosphate)

ligand 3 with R = H, R′ = $CH_2{-}O{-}P(O)(OH){-}O{-}P(O)(OH){-}OH$

(= (*Z*)-$C_{10}H_{20}O_7P_2$ = neryl diphosphate)

Complexes in Solution. Spectrophotometric [1] and kinetic studies [2 to 4] reveal the formation of MnL^- complexes with ligands 1 to 3 and Mn_2L^+ complexes with ligands 2 and 3 in aqueous solutions containing Mn^{2+} ions and one of the ligands. The formation of the Mn_2-((*E*)-$C_{10}H_{17}O_7P_2)^+$ complex (ligand 2) was confirmed by kinetic measurements using Job's method [3]. The stability constants of the MnL^- complexes are: $\log K_1 = 3.92$ for $Mn(C_5H_9O_7P_2)^-$ and $\log K_1 = 4.52$ for Mn((*E*)-$C_{10}H_{17}O_7P_2)^-$ at 25°C, I = 0.05 mol/L (Tes buffer, pH 7.0; Tes = N-(tris(hydroxymethyl)methyl)-2-aminoethanesulfonic acid), determined spectrophotometrically (at 360 nm) by the competition method using 8-quinolinol as an auxiliary ligand [1, 2]. The stability constants of Mn_2L^+ complexes formed from MnL^- and Mn^{2+} ions are: $\log K \approx 2.60$ [3] or ≈ 2.82 [4], both at 40°C, I = 0.1 mol/L (Tes buffer, pH 7) for Mn_2-((*E*)-$C_{10}H_{17}O_7P_2)^+$ [3, 4] and $\log K \approx 0.046$ at 40°C, I = 0.1 mol/L (Tes buffer, pH 6.75) for Mn_2-((*Z*)-$C_{10}H_{17}O_7P_2)^+$ [4]. A far lower value, $\log K \approx 0.15$ at 55°C, I = 4.8 mol/L (NaCl) had been found for the (*E*)-isomer in an earlier work [2]. All log K values were derived from kinetic measurements of the Mn^{2+}-catalyzed decomposition reactions of geranyl and neryl diphosphate [2 to 4].

The products of these C–O bond fission reactions are terpene alcohols and hydrocarbons, with a larger portion of cyclic monoterpenes for neryl diphosphate; the Mn_2L^+ complexes were found to be the most reactive species [3, 4]. Rate constants are: $k = 1.5 \times 10^{-5}\ s^{-1}$ at 40°C, I = 0.1 mol/L (Tes, pH 7) for Mn_2((*E*)-$C_{10}H_{17}O_7P_2)^+$ [3, 4] and $k = 2.1 \times 10^{-4}\ s^{-1}$ at 40°C, I = 0.5 mol/L (Tes, pH 6.75) for Mn_2((*Z*)-$C_{10}H_{17}O_7P_2)^+$. While in the case of geranyl diphosphate the 1:1 complex does not participate in the reaction, the Mn((*Z*)-$C_{10}H_{17}O_7P_2)^-$ complex decomposes with $k = 1.2 \times 10^{-5}\ s^{-1}$ under the above conditions [4]. From the temperature dependence of the reaction rate the activation energy for the decomposition of Mn_2((*E*)-$C_{10}H_{17}O_7P_2)^+$ was E = 31 kcal/mol [3]. The kinetic data indicate that, after C–O bond cleavage of the complex, the neutral $Mn_2P_2O_7$ and the 3,7-dimethyl-2,6-octadiene-1-carbocation are formed and the latter reacts further to give the observed products [3]. Reasons for the increased elimination and cyclization reactions of the carbocations in the presence of metal ions are discussed [4]. The reactions were studied as models for the metal-ion-dependent enzymatic biosynthesis of cyclic monoterpenes [3].

References:

[1] King, H. L.; Rilling, H. C. (Biochemistry **16** [1977] 3815/9).
[2] Brems, D. N.; Rilling, H. C. (J. Am. Chem. Soc. **99** [1977] 8351/2).
[3] Vial, M. V.; Rojas, C.; Portilla, G.; Chayet, L.; Pérez, L. M.; Cori, O. (Tetrahedron **37** [1981] 2351/7).
[4] Chayet, L.; Rojas, M. C.; Cori, O.; Bunton, C. A.; McKenzie, D. C. (Bioorg. Chem. **12** [1984] 329/38).

39.13.4.2 With Thiamin Diphosphate

$(=C_{12}H_{18}N_4O_7P_2S=H_2L)$

Complexes in Solution. The formation of 1:1 complexes from Mn^{2+} ions and the ligand in aqueous solution was revealed by potentiometric titration [1 to 3] and ESR studies [4]. Stability constants, log K_1 and log K^{Mn}_{MnHL}, for $K^{Mn}_{MnHL}=[Mn(C_{12}H_{17}N_4O_7P_2S)^+]/[Mn^{2+}]\cdot[(C_{12}H_{17}N_4O_7P_2S)^-]$, were determined at various temperatures:

t in °C	5	25	25	35	45
I in mol/L	0.1(KNO_3)	0.1($(CH_3)_4NBr$)	0.1Tris-HCl	0.1(KNO_3)	0.1(KNO_3)
log K_1	3.82	4.30±1.44	3.70	4.30	4.03
log K^{Mn}_{MnHL}	2.51	3.56±2.54	—	2.84	2.65
Ref.	[1]	[3]	[4]	[2]	[1]

The stability constants of the Mn^{II} complexes and other divalent metal complexes (Mg^{II}, Co^{II}, Ni^{II}, Cu^{II}, and Zn^{II}) do not follow the usual Irving-Williams sequence. Instead, the order of stability is similar to that generally found with oxygen donor ligands, being especially marked for phosphates [3]. Enthalpy (ΔH) and entropy (ΔS) changes for the formation reactions have been evaluated from stability constants at 5 and 45°C. Relatively small ΔH values (ca. −1 kcal/mol) and large positive ΔS values (ca. 10 $cal\cdot mol^{-1}\cdot K^{-1}$) indicate that the formation of the complexes is mainly favored by a large positive entropy contribution [1].

^1H and ^{13}C NMR investigations show that the Mn^{2+} ion is chelated by the diphosphate group [5] through three oxygen atoms [4]. Longitudinal proton relaxation times, T_1 (measurements in D_2O, pH 6.6 at 27 and 60°C, by the progressive saturation and Carr-Purcel methods), were used to evaluate internuclear distances between the Mn^{2+} ion and the various protons in the complex. From these calculated distances, a folded structure, shown schematically in **Fig. 38**, p. 168, was derived. In the structure, the pyrimidine ring lies perpendicular to the plane of the paper, and the thiazolium ring, with the CH_3 group pointing backwards, lies almost within this plane, so that the 2′-proton is situated above the C(4)–C(5) bond of the pyrimidine ring. The diphosphate group is folded over and behind the thiazolium ring. Thus, the Mn^{2+} ion is situated above the 6-proton, and this proton, the Mn^{2+} ion, and the 4′-CH_3 group are in the same plane [4]. The folding of the diphosphate group onto the thiazolium ring was also derived from other ^1H NMR studies (broadening and shifting data) [5, 6].

Thiamin diphosphate is of biochemical importance because of its functions as the coenzyme in numerous enzyme systems [3, 4]. Metal ion bonding properties and conformation of the ligand (which are significant in enzyme reactions) have been studied using the paramagnetic manganese complexes [3 to 6].

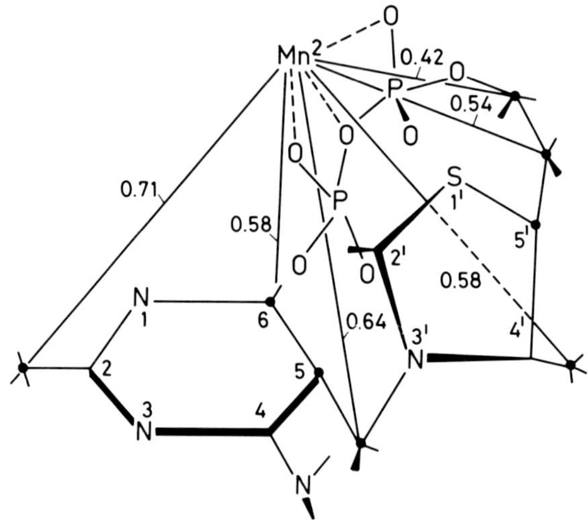

Fig. 38. Schematic drawing of the manganese(II) complex with thiamin diphosphate $Mn(C_{12}H_{16}N_4O_7P_2S)$. The Mn^{2+}–proton nuclear distances are given in nm [4].

References:

[1] Taqui Khan, B.; Nageswara Rao, P.; Taqui Khan, M. M. (Indian J. Chem. A **20** [1981] 857/9).
[2] Taqui Khan, M. M.; Amara Babu, M. (J. Inorg. Nucl. Chem. **40** [1978] 2110/3).
[3] Katz, H. B.; Kustin, K. (Biochim. Biophys. Acta **313** [1973] 235/48).
[4] Grande, H. J.; Houghton, R. L.; Veeger, C. (Eur. J. Biochem. **37** [1973] 563/9).
[5] Jordan, F.; Mariam, Y. H. (Microchem. J. **22** [1977] 182/200).
[6] Gallo, A. A.; Sable, H. Z. (J. Biol. Chem. **250** [1975] 4986/91).

39.13.4.3 With Diethyl Dihydrogen Diphosphate

```
           OH    OH
           |     |
C2H5O—P—O—P—OC2H5    (=C4H12O7P2)
           ||    ||
           O     O
```

$Mn^{II}_2(C_4H_{10}O_7P_2)(C_4H_{10}O_4P)_2$. The mixed ligand complex, also containing anions of diethyl hydrogen phosphate, see p. 155, was prepared by dissolving the hydrated manganese(II) halide or suspending the anhydrous salt in a large excess of triethyl phosphate. The temperature of the mixture was increased at a rate of 2 to 3°C/min until the white complex deposited. Prior to precipitation the evolution of a mixture of ethyl chloride, hydrogen chloride, and ethylene was observed. The complex was washed with anhydrous diethyl ether and dried in an evacuated desiccator over $CaCl_2$. $Mn_2(C_4H_{10}O_7P_2)(C_4H_{10}O_4P)_2$ does not melt or decompose up to 300°C. The magnetic moment determined by susceptibility measurement (Faraday method) at 300 K is $\mu_{eff}=5.62\ \mu_B$. The IR spectrum of the mixed ligand complex exhibits bands characteristic of both the diethyl dihydrogen diphosphate and diethyl hydrogen phosphate ligands. Thus, in addition to the absorptions of $\nu_{as}(PO_2)$ at 1241 cm^{-1} and $\nu_s(PO_2)$ at 1137 cm^{-1}, bands of $\nu_{as}(POP)$ at 956 cm^{-1} and $\nu_s(POP)$ at 724 cm^{-1} were observed. The spectrum shows no absorptions associated with POH or uncoordinated P=O groups. Two ligand bands observed at 550 and 375 cm^{-1} mask the ν(Mn–O) modes. The absorption maximum observed in the electronic spectrum of the complex (in Nujol) at 337 nm is suggestive of a pentacoordinated configuration at Mn^{II}. A linear polynuclear backbone structure of the type shown below is assumed.

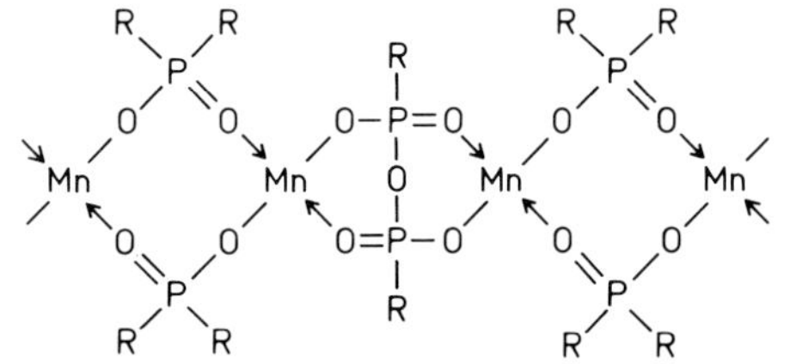

R = OC_2H_5

The complex probably attains coordination number five by formation of oxygen bridges between linear polymeric units. The complex is insoluble in many organic solvents but dissolves in water or methanol.

Reference:

Mikulski, C. M.; Pytlewski, L. L.; Karayannis, N. M. (Z. Anorg. Allgem. Chem. **403** [1974] 200/10).

39.14 Complex with Triphenylphosphane Imide $(C_6H_5)_3PNH$ $(=C_{18}H_{16}NP)$

$[Mn^{III}(C_{18}H_{15}NP)(C_{48}H_{36}N_4)] \cdot H_2O$. The complex also containing a 5,10,15,20-tetrakis(4-methylphenyl)porphyrin anion, $C_{48}H_{36}N_4^{2-}$, was prepared by reaction of the nitrido manganese(V) complex, $[Mn(N)(C_{48}H_{36}N_4)]$, with triphenylphosphane: 656 mg of $(C_6H_5)_3P$ were added to the solution of 369 mg of $[Mn(N)(C_{48}H_{36}N_4)]$ in 300 mL of dry toluene. The mixture was refluxed under a stream of nitrogen for 4 h then the solvent was evaporated in vacuum. The residue was washed with a small amount of dry ether and recrystallized from an acid-free 1:5 hexane-dichloromethane mixture to give blue-green crystals (78% yield). The compound was not analyzed since impurities could not be removed by recrystallization. Formation of the complex was confirmed by spectral data, hydrolysis, and benzoylating cleavage. The IR spectrum of the complex in KBr shows bands at 3440 and 1650 to 1620 cm^{-1} assigned to vibration modes of the water molecule. Bands at 1440 and 995 cm^{-1} were assigned to $(C_6H_5)_3P$ and at 1130 cm^{-1} to P=N vibration modes. The electronic spectrum of the complex dissolved in toluene shows absorption maxima at 620, 579, 536, 474, 446, 423, and 380 nm. The mass spectrum reveals peaks of $Mn(C_{18}H_{16}NP)(C_{48}H_{36}N_4)^+$ at 1000, $Mn(C_{48}H_{36}N_4)^+$ at 723, and $(C_{18}H_{15}NP)^+$ at 276 mass units. The complex is easily hydrolyzed in toluene: $[Mn(C_{18}H_{15}NP)(C_{48}H_{36}N_4)] + 3H_2O \rightarrow MnOH(C_{48}H_{36}N_4) \cdot H_2O + NH_3 + C_{18}H_{15}PO$. The benzoylation in toluene, $[Mn(C_{18}H_{15}NP)(C_{48}H_{36}N_4)] + C_6H_5COCl \rightarrow MnCl(C_{48}H_{36}N_4) + C_{18}H_{15}PNCOC_6H_5$, confirms the migration of nitrogen to phosphorus during the formation reaction of the complex.

Reference:

Buchler, J. W.; Dreher, C.; Lay, K.-L. (Chem. Ber. **117** [1984] 2261/74).

39.15 Complexes with Diphenylphosphinic Amide or Phenylphosphonic Diamide

$(C_6H_5)_2P(O)N(CH_3)_2$	$C_6H_5P(O)(N(CH_3)_2)_2$
ligand 1 $(=C_{14}H_{16}NOP)$	ligand 2 $(=C_{10}H_{17}N_2OP)$

$[Mn^{II}L_4]X_2$ ($X=ClO_4$, BF_4) and **$[Mn^{II}L_2Cl_2]$** complexes were prepared by dissolving the hydrated manganese(II) salt in methanol (or acetone) and adding an appropriate amount of trimethyl orthoformate as dehydrating agent followed by a slight excess of the ligand. The mixture was evaporated and upon standing the complexes appeared, in a few cases only after addition of dry ether. They were washed with ether and dried in vacuum at room temperature. All complexes are hygroscopic. Color, melting points, and band maxima observed in the reflectance spectra of the solid compounds are listed below:

complex	color	m.p. in °C	$\tilde{\nu}_{max}$ in cm^{-1}	Ref.
$[Mn(C_{14}H_{16}NOP)_4](ClO_4)_2$	white	144 to 146	23500, 24300	[1]
$[Mn(C_{10}H_{17}N_2OP)_4](ClO_4)_2$	white	210 to 212	23600, 24200	[2]
$[Mn(C_{10}H_{17}N_2OP)_4](BF_4)_2$	white	82 to 84	23600, 24200, 25200, 27700	[2]
$[Mn(C_{14}H_{16}NOP)_2Cl_2]$	light green	132 to 133	21600, 22900, 23800, 27700	[1]
$[Mn(C_{10}H_{17}N_2OP)_2Cl_2]$	light green	102 to 103	21600, 22800, 23900, 26800, 28600	[2]

An off-white tetrafluoroborate obtained with ligand 1 decomposes rapidly [1].

The IR spectra of the complexes (in Nujol) show shifts of the PO stretching and bending vibrations to lower and higher energy, respectively, in comparison to the free ligands. This indicates coordination of the amides through the phosphoryl oxygen atom. The P–N stretching frequency is shifted towards higher energy [1, 2]. The observed ν_3 and ν_4 bands of the perchlorate and tetrafluoroborate groups indicate their ionic nature in both complexes. A tetrahedral geometry is therefore assumed at Mn^{II} supported by the absorption maxima of the electronic reflectance spectra and by the isomorphism with the corresponding $[CoL_4](ClO_4)_2$ and $[CoL_4](BF_4)_2$ complexes, for which the tetrahedral coordination has been confirmed by their ligand field data. A pseudotetrahedral geometry is proposed for the chloro complexes $[Mn(C_{14}H_{16}NOP)_2Cl_2]$ and $[Mn(C_{10}H_{17}N_2OP)_2Cl_2]$. IR bands observed at 306 and 282 [1] or at 311 and 280 cm^{-1} [2], respectively, were assigned to $\nu(Mn–Cl)$ vibration modes.

References:

[1] de Bolster, M. W. G.; Groeneveld, W. L. (Z. Naturforsch. **27b** [1972] 759/63).
[2] de Bolster, M. W. G.; Groeneveld, W. L. (Recl. Trav. Chim. **90** [1971] 1153/65).

39.16 Complexes with Amides of Phosphoric Acid or Its Esters

39.16.1 With Diesters of Phosphoramidic Acid

$CH_3NHP(O)(OCH_3)_2$	ligand 1 $(=C_3H_{10}NO_3P)$
$C_6H_5SO_2NHP(O)(OC_6H_{13})_2$	ligand 2 $(=C_{18}H_{32}NO_5PS)$

Complexes in Solution. Stability constants of $Mn^{II}(C_{18}H_{31}NO_5PS)_n^{(2-n)+}$ complexes with ligand 2, log $K_1=4.39$ and log $K_2=3.82$, were calculated from distribution constants in the system water–benzene where the imidol –P(OH)=N– tautomer predominates over the amide –P(O)–NH– tautomer. The distribution ratio of the complex between benzene and water is 24 [1].

[$Mn^{II}(C_{18}H_{31}NO_5PS)(H_2O)$] was prepared by mixing the aqueous solution of manganese(II) sulfate and $NaC_{18}H_{31}NO_5PS$, the sodium salt of the imidol tautomer of ligand 2. The precipitate was extracted with ether, the organic layer distilled, and the remaining complex dried in a desiccator over concentrated sulfuric acid. The IR spectrum of the complex shows a band at 1340 cm^{-1}, which is not present for the free ligand, and spectral changes in the region 1250 to 950 cm^{-1}, probably caused by the chelating nature of the sulfonamide and phosphoryl groups. Coordination of the water molecule is indicated by a broad band between 3600 and 3400 cm^{-1} and a band at 1640 cm^{-1} assigned to ν(OH) and δ(HOH) vibrations, respectively. The solubility of the complex in water at 20°C is 7×10^{-4} mol/L, the solubility product is 1.2×10^{-9} [1].

[$Mn^{III}(C_3H_9NO_3P)(C_{48}H_{36}N_4)(H_2O)$]. The mixed ligand complex with the 5,10,15,20-tetrakis-(4-methylphenyl)porphyrin anion, $C_{48}H_{36}N_4^{2-}$, was prepared like the analogous triphenylphosphane imide complex (see p. 169), by refluxing the solution of 1.5 g of [$Mn^V(N)C_{48}H_{36}N_4$] in 400 mL of dry toluene with 0.75 mL of freshly distilled $P(OCH_3)_3$ for 3 h under nitrogen. The solvent and excess of $P(OCH_3)_3$ were evaporated in vacuum and the residue was recrystallized two times from a 1:1 cyclohexane-dichloromethane mixture (yield 87%). The IR spectrum of the green crystals (in KBr) shows bands at 2840, 1230(sh), and 1050 cm^{-1} assigned to ν(NCH_3), ν(P=O), and ν(C–O–P) vibrations, respectively. The electronic spectrum of the toluene solution reveals absorption maxima at 672, 615, 579, 530, 469, 444, 399, and 376 nm; the extinction values range from log ε = 3.76 to 5.09. The mass spectrum shows peaks of $Mn(C_3H_9NO_3P)$-$(C_{48}H_{36}N_4)^+$ at 848 mass units, $MnN(C_{48}H_{36}N_4)^+$ at 737, and $Mn(C_{48}H_{36}N_4)^+$ at 723. The complex dissolved in dichloromethane upon reaction with acetic acid forms $Mn(C_{48}H_{36}N_4)CH_3COO$ isolated as green crystals [2].

References:

[1] Shevchenko, F. D.; Voevutskaya, R. P.; Kuzina, L. O. (Visn. Kiiv'sk. Univ. Khim. No. 15 [1974] 47/51; C.A. **83** [1975] No. 153408).

[2] Buchler, J. W.; Dreher, C.; Lay, K.-L. (Chem. Ber. **117** [1984] 2261/74).

39.16.2 With N-Diethoxyphosphinoyldithiocarbamic Acid

$$(C_2H_5O)_2P(=O)-NH-C(=S)-SH \quad (=C_5H_{12}NO_3PS_2)$$

[$Mn^{III}(C_5H_{11}NO_3PS_2)_3$] was obtained on stirring aqueous solutions of manganese(II) salt and $KC_5H_{11}NO_3PS_2$ for 15 min at 0 to 5°C. Because of the air-sensitive nature of the corresponding Mn^{II} compound, only the Mn^{III} compound was obtained. The brown precipitate was washed with ice-cold water followed by ethanol and dried in vacuum over P_4O_{10}. The complex was recrystallized from an acetone solution by the addition of petroleum ether. The magnetic moment, $\mu_{eff} = 4.80\ \mu_B$, results from susceptibility measurements using a 10% solution in chloroform. The IR spectrum of the complex shows bands at ~3075, ~1350, and ~1260 cm^{-1}, which are assigned to ν(NH), ν(CN), and ν(PO) vibrations. A band at ~1000 cm^{-1} suggests a bidentate nature of the dithiocarbamato group. A band between 365 and 345 cm^{-1} was assigned to the ν(Mn–S) vibration. The complex is stable in inert atmosphere but decomposes on standing in air. It is soluble in common organic solvents. The complex dissolved in nitromethane behaves as a nonelectrolyte.

Reference:

Sodhi, G. S.; Kaushik, N. K. (Indian J. Chem. Soc. A **20** [1981] 922/4).

39.16.3 With Phosphoric Triamide $P(O)(NH_2)_3$ (= H_6N_3OP)

$[Mn^{II}(H_6N_3OP)_2]Cl_2$ was prepared in methanol from anhydrous $MnCl_2$ and phosphoric triamide. The mixture was stirred at room temperature for 3 h and then concentrated at 50 to 60°C. The white complex precipitated on addition of dry ether and was washed with ether and stored after drying [1]. The reaction of the components in aqueous solution at room temperature followed by the addition of acetone yields the solid complex containing 6 to 11% manganese(II) phosphoric diamidate [2]. The X-ray powder diffraction pattern of $[Mn(H_6N_3OP)_2]Cl_2$ is given in [1]. The IR spectrum of the complex (in KBr) shows characteristic bands assigned as follows (free ligand bands are given in parentheses): ν(NH) at 3420, 3400, 3310, 3300, 3250, 3200 (3370, 3280, 3140 (sh)), $\delta(NH_2)$ at 1570, 1535 (1610, 1570), ν(PO) at 1190 (1190), $\varrho(NH_2)$ at 1065, 1000 (1040), $\omega(NH_2)$ at 950, 905 (935), ν(PN) at 795 (700) cm^{-1}. The spectra suggest that the ligand is coordinated tridentately to Mn through the nitrogen atoms [1]. Thermal decomposition of $[Mn(H_6N_3OP)_2]Cl_2$ upon heating under reduced pressure (5×10^{-2} mm Hg) reveals oligomeric condensed phosphoric amide at 140 to 200°C and condensed phosphates with P=N–P and P–NH–P bonds at 300 to 700°C. By heating in air, highly cross-linked condensed phosphates at 300 to 400°C are formed, at 600°C linkages of the condensed phosphates are broken, and at 1000°C $Mn_2P_4O_{12}$ is formed [3].

The hygroscopic complex is insoluble in organic solvents, but soluble in methanol or water. The electrical conductivity, $\Lambda=119.8\ cm^2\cdot\Omega^{-1}\cdot mol^{-1}$, measured in 10^{-3} M methanol solution at 25°C corresponds to that of a 1:2 electrolyte [1]. The hydrolytic cleavage of $[Mn(H_6N_3OP)_2]Cl_2$ depending on time (up to 72 h) and temperature (up to 60°C) was examined by chemical analysis, IR spectroscopy, paper and ion exchange chromatography. Mixtures of phosphoric amidate, phosphoric diamidate, diphosphate, and triphosphate were demonstrated chromatographically [2].

References:

[1] Kumagai, N.; Mase, H. (Nippon Kagaku Kaishi **1975** 814/9; C.A. **83** [1975] No. 52612).
[2] Kumagai, N.; Mase, H. (Nippon Kagaku Kaishi **1978** 46/52; C.A. **88** [1978] No. 130139).
[3] Kumagai, N.; Mase, H. (Nippon Kagaku Kaishi **1976** 1831/7; C.A. **86** [1977] No. 64852).

39.16.4 With Hexamethylphosphoric Triamide $P(O)(N(CH_3)_2)_3$ (= $C_6H_{18}N_3OP$)

$[Mn^{II}(C_6H_{18}N_3OP)_4]X_2$ ($X=ClO_4$, BF_4). The complexes were prepared by dissolving the manganese(II) salts in methanol or acetone. If necessary an appropriate amount of trimethyl orthoformate was added for dehydration followed by a slight excess of ligand. Upon standing, or after concentration of the solution, crystals of white $[Mn(C_6H_{18}N_3OP)_4](ClO_4)_2$ [1] or light green $[Mn(C_6H_{18}N_3OP)_4](BF_4)_2$ [2] appeared; these were washed with dry ether or hexane and dried in vacuum. The perchlorate melts at 267 [1], the fluoroborate at 260°C [2], in both cases with decomposition. The perchlorate was also prepared as follows: The hydrated Mn^{II} perchlorate was stirred with 100% excess of 2,2-dimethoxypropane for 2 h. A slight excess of ligand was added and the solution stirred for 2 to 4 h at 0°C, then excess dry ether was added to complete the precipitation. The compound was washed with ether and dried in vacuum over P_4O_{10} [3].

Isomorphism of $[Mn(C_6H_{18}N_3OP)_4](ClO_4)_2$ with the corresponding complexes of Zn^{II}, Ni^{II}, Co^{II}, and Fe^{II} was concluded due to identical X-ray powder diffraction patterns [3]. A magnetic moment, $\mu_{eff}=5.24\ \mu_B$, for the perchlorate complex was determined by susceptibility measurements at 30°C [3]. The ESR spectrum of $[Mn(C_6H_{18}N_3OP)_4](BF_4)_2$, doped into the corresponding

Cd^{II} complex, was investigated by [4]. The very small zero-field parameter, $D \leqq 0.001\ cm^{-1}$, and the high value of the hyperfine coupling constant, $A = 84 \times 10^{-4}\ cm^{-1}$, are in accordance with a tetrahedral symmetry and indicate an essentially ionic bonding between Mn^{II} and the ligand. A value of $g = 2.00$ was assumed [4], whereas a value of $g = 1.80$ was estimated from susceptibility data at 28°C [5]. The IR spectra of both complexes (in Nujol) indicate a tetrahedral coordination at Mn^{II}. Negative shifts of the ν(PO) band from 1208 to 1190 cm^{-1} [1] or from 1212 to 1194 cm^{-1} [6] were observed in the spectrum of $[Mn(C_6H_{18}N_3OP)_4](ClO_4)_2$, compared to the free ligand. A ν(Mn–O) vibration is observed at 369 and 359 cm^{-1} [6] or at 365 cm^{-1} in the case of the fluoroborate complex [2]. Band positions of the anions are characteristic for ionic perchlorate [1, 3, 6] or ionic fluoroborate [2]. The electronic reflectance spectrum of $[Mn(C_6H_{18}N_3OP)_4]$-$(ClO_4)_2$ shows maxima at 21900, 23800, and 24500 cm^{-1} [1, 3]; that of $[Mn(C_6H_{18}N_3OP)_4](BF_4)_2$ at 21700, 23600, 24400, 25300, 27400, and 29400 cm^{-1} [2].

The electronic spectrum of solvated Mn^{2+} ions in neat hexamethylphosphoric triamide shows a maximum at 412 nm [7]. The extinction coefficient ($\log \varepsilon = -0.002$) agrees with the value observed in a nitromethane solution of $[Mn(C_6H_{18}N_3OP)_4](ClO_4)_2$, which reveals bands at ~412 ($\log \varepsilon = -0.022$), ~422 ($\varepsilon = 0.50$), and 444 (sh) nm [3]. Due to the spectral data and the calculated Stokes radius, r = 5.47 Å [7], a tetrahedral coordination of Mn^{II} in solution was assumed. The electrical conductivity of $[Mn(C_6H_{18}N_3OP)_4](ClO_4)_2$ determined in 5×10^{-3} M nitromethane solution at 29°C, $\Lambda = 185\ cm^2 \cdot \Omega^{-1} \cdot mol^{-1}$, agrees with a 1:2 electrolyte nature [3]. The limiting molar conductivity of $[Mn(C_6H_{18}N_3OP)_4](ClO_4)_2$ and its complex cation were determined in 10^{-3} M hexamethylphosphoric triamide at 25°C: $\Lambda_0 = 24.80$ and $\lambda_0 = 9.30\ cm^2 \cdot \Omega^{-1} \cdot mol^{-1}$, respectively [7]. The conductivity of the complex was also measured by [8] and values of $\Lambda_0 = 24.80$ and $\lambda_0 = 9.30\ cm^2 \cdot \Omega^{-1} \cdot mol^{-1}$ were found; values of 24.72 and 9.22 $cm^2 \cdot \Omega^{-1} \cdot mol^{-1}$ were calculated by use of the Fuoss-Edelson method [9], which revealed an association constant, log K = 0.95, for the complex formed according to $[Mn(C_6H_{18}N_3OP)_4]^{2+} + ClO_4^- \rightarrow [Mn(C_6H_{18}N_3OP)_4](ClO_4)^+$ [8].

The ^1H NMR measurements of $[Mn(C_6H_{18}N_3OP)_4](ClO_4)_2$ in nitromethane solution at 33°C show methyl proton contact shifts of $\Delta\nu = 370 \pm 50$ or 204 Hz relative to tetramethylsilane or $[Zn(C_6H_{18}N_3OP)_4](ClO_4)_2$, respectively. An electron spin–nuclear spin coupling constant, $A = 4.6 \times 10^{-3}$ G, was calculated. By comparison with NMR contact shifts of other tetrahedral hexamethylphosphoric triamide complexes an increase in covalent bonding ability was postulated for the series: $Mn^{II} < Fe^{II} < Co^{II} < Ni^{II} < Cu^{II}$ [5].

The ligand exchange reaction of $[Mn(C_6H_{18}N_3OP)_4](ClO_4)_2$ in neat ligand solution was also studied by ^1H NMR measurements. The calculated phosphorus and proton relaxation rates at 25°C are $1/T_1 = 2.8 \times 10^4\ s^{-1}$ and $1/T_2 = 2 \times 10^4\ s^{-1}$, respectively. The corresponding activation energies, $E_a = -3.4$ and -3 to -4 kcal, are comparable to those of octahedral systems. The small value indicates that the exchange rates are greater than the relaxation rates. A plot showing the relaxation times, T_1 and T_2, as functions of temperature in the range −10 to 127°C is given in the paper [10]. Another ^1H NMR study with $[Mn(C_6H_{18}N_3OP)_4](ClO_4)_2$ in neat ligand solution shows that the "excess" proton-spin relaxation is governed by the first solvent sphere relaxation at higher temperature and by chemical exchange near frozen temperature. The data indicate that dipolar interaction dominates the first solvation sphere relaxation. Free rotation of CH_3 groups about the P–N bond and ligand exchange were postulated to account for high-temperature broadening. The calculated correlation times at 25°C are: $\tau_h = 1.0 \times 10^{-7}$ s (lifetime of a solvent molecule in the first coordination sphere), $\tau_v = 2.97 \times 10^{-12}$ s (lifetime for impact of solvent molecule upon the solvated Mn^{II}), $\tau_s = 7.20 \times 10^{-9}$ s (correlation time of electron-spin relaxation); the corresponding activation energies are $E_h = 12.20$ and $E_v = 0.32$ kcal/mol. Entropy and enthalpy of activation for the exchange of solvent molecules are $\Delta H^* = 12.20$ kcal/mol and $\Delta S^* = 10.30\ cal \cdot mol^{-1} \cdot K^{-1}$. For comparison, a value of $\tau_h = 1.6 \times 10^{-7}$ s was estimated from

ESR relaxation data. The paper shows plots of ESR line width versus temperature (−17 to 97°C) and of NMR proton relaxation rate versus temperature (−40 to 85°C) [11].

$[Mn^{II}(C_6H_{18}N_3OP)_2X_2]$ (X = NO_2, NO_3, Cl, Br, I, and NCS). The nitrito complex, $[Mn(C_6H_{18}N_3OP)_2(NO_2)_2]$, was prepared by adding the stoichiometric amount of finely ground $AgNO_2$ to the solution of $[Mn(C_6H_{18}N_3OP)_2Cl_2]$ in acetone. The mixture was stirred in the dark until the chloride test on the solution was negative. After filtration, part of the solvent was evaporated and petroleum ether was added to yield yellow crystals that became brown after drying. The brown complex was recrystallized from nitromethane [12]. By another procedure the filtered reaction mixture was evaporated to dryness under reduced pressure and the remaining solid dissolved in dry chloroform. The resulting solution was concentrated and the deposited crystals were washed with dry ether and dried in vacuum [13].

The complexes with X = NO_3, Cl, Br, I, and NCS were prepared by dissolving the hydrated or anhydrous manganese(II) salts in methanol or acetone and, when necessary, an appropriate amount of trimethyl orthoformate was added for dehydration, followed by a slight excess of ligand. The mixture was evaporated to yield crystals, in some cases after the addition of dry ether. The precipitates were washed with dry ether or hexane and dried in vacuum [2, 3]. The chloro complex was also prepared by refluxing a suspension of finely divided $MnCl_2 \cdot 4H_2O$ (0.1 mol) and the ligand (0.2 mol) in 250 mL of xylene for 10 h. Then the water of crystallization was removed azeotropically and the solvent distilled off. The residue was recrystallized from benzene-petroleum ether [14]. To prepare the isothiocyanato or isoselenocyanato complexes acetone solutions of the chloro complex containing the appropriate amount of KSCN or KSeCN were cooled and the deposited KCl filtered off. Some of the acetone was evaporated, then the complexes were precipitated with petroleum ether [15]. They were also obtained from nonaqueous acetone solutions of $Mn(NO_3)_2$ by adding the ligand (~2.5 mol/mol Mn^{II} salt), then the stoichiometric amount of KSCN [16] or KSeCN [17]. After the potassium nitrate had been removed, the solutions were evaporated at room temperature in vacuum until crystallization began. Color, melting points, and magnetic moments of the $[Mn(C_6H_{18}N_3OP)_2X_2]$ complexes are listed below:

X	NO_2	NO_3	Cl	Br	I	NCS		NCSe	
color	brown	pink	white or greenish	light yellow	yellow	yellowish		yellowish	
m.p. in °C	76 to 77	92 to 93	111 to 113	94 to 95	96 to 97	—		—	
μ_{eff} in μ_B	5.92[a)]	—	—	—	—	5.96[a)]	5.8[b)]	6.0[a)]	5.6[b)]
Ref.	[12]	[2]	[2, 14]	[2]	[2]	[15]	[16]	[15]	[17]

[a)] Calculated from susceptibility measurements (Gouy method) at 297 K. – [b)] At room temperature (Faraday method).

The X-ray powder diffraction diagram of the complex with X = NO_2 is similar to that of the analogous Co^{II} compound for which a distorted octahedral structure was assumed from the diffuse reflectance spectrum [12]. The X-ray powder diffraction pattern of the complexes with X = NCS and NCSe are similar to those of the analogous Zn^{II} and Co^{II} compounds [15]; the interplanar spacings and their intensities are given in [16, 17].

The IR spectra of the complexes reveal coordination of the hexamethylphosphoric triamide molecules and the anionic groups to Mn^{II}. The ν(PO) band of the ligand at 1218 cm^{-1} is shifted to lower wavenumbers, those of the vibration modes of ν(PN) at 982 and 744 cm^{-1} and δ(PO) at 481 cm^{-1} to higher wavenumbers [2, 15 to 17]. Vibration modes of $\nu(Mn{-}O)_{ligand}$ and $\nu(Mn{-}X)_{anion}$ for X = O, Cl, Br, and N are tabulated on p. 175 together with band maxima observed in the electronic reflection spectra of the $[Mn(C_6H_{18}N_3OP)_2X_2]$ complexes:

X	ν(Mn–O) in cm^{-1}	ν(Mn–X) in cm^{-1}	$\tilde{\nu}_{max}$ in cm^{-1}	Ref.
NO_3	385	225	22000, 23100, 24000, 27400, 28300	[2]
Cl	390	303, 270	18200, 20200, 24200	[2]
Br	388	265, 230	21500, 22800, 23600, 26600, 27800	[2]
I	384	—	—	[2]
NCS	386	290	22700, 23500, 26300	[2]
NCSe	388	—	—	[15]

The monodentate bonding of NCS and NCSe via the nitrogen atom is confirmed by the ν(CN) band position shown in the range of 2070 to 2065 cm^{-1} [15 to 17].

The spectra of $[Mn(C_6H_{18}N_3OP)_2(NO_2)_2]$ recorded as KBr pellet [12] or Nujol mull [13] were interpreted differently. A monodentate coordination of the NO_2 group via the nitrogen atom was proposed due to bands assigned to $\nu_{as}(NO)$ at 1382, $\nu_s(NO)$ at 1270, and $\delta(NO_2)$ at 845 cm^{-1} [12]. Chelating O,O' bidentate coordination was assumed by [2] due to $\delta(NO_2)$ bands observed at 849 and 817 cm^{-1} [13]. This means hexacoordinated environment at Mn^{II} which is more probable than a tetrahedral one proposed by [12]. A broad band at 25000 cm^{-1} in the electronic reflectance spectrum was assigned to charge-transfer transitions covering the d-d transitions [13]. The bands in the electronic spectra (see the table above) indicate that Mn^{II} is tetrahedrally surrounded in the complexes, with the exception of $[Mn(C_6H_{18}N_3OP)_2(NO_3)_2]$ where an octahedral environment with bidentately coordinated nitrato groups was assumed. No ligand field bands could be detected in the spectrum of the iodo complex, owing to strong charge-transfer bands [2].

A pseudotetrahedral environment at Mn^{II} in the complexes with X = Cl, Br, I, and NCS was concluded from ESR data. All compounds have g values of 2.00 [4], g = 1.96 was found by [15] for the isothiocyanato complex. The values for zero-field parameters, λ, D and A, were determined from ESR X- and Q-band data. The λ values, λ = 0.22 to 0.25, calculated for the complexes with X = Cl, Br, and I, and for $[Mn(C_6H_{18}N_3OP)_2(NCS)_2]$ (doped into the corresponding Zn^{II} complex) indicate a pseudotetrahedral symmetry, C_{2v}, which is indicated also by the hyperfine coupling constant of $A = (74 \pm 2) \times 10^{-4}$ cm^{-1} (calculated for the iodo complex). The anion sequence with respect to increasing D values is $NCS^- < Cl^- < Br^- < I^-$. This sequence parallels the spectrochemical series for these anions. A low value, λ = 0.0(1), and a hyperfine coupling constant, $A = (86 \pm 3) \times 10^{-4}$ cm^{-1}, for $[Mn(C_6H_{18}N_3OP)_2(NO_3)_2]$ doped into the Zn^{II} complex is indicative of a pseudooctahedral D_{2h} environment at Mn^{II} with bidentate nitrato groups [4].

The complexes are hygroscopic and require handling in a drybox [2]. $[Mn(C_6H_{18}N_3OP)_2(NCS)_2]$ is unstable, forming a red powder, but is stable in acetone solution kept in a refrigerator [15]. The complexes with X = NCS and NCSe are soluble in water, alcohol, or acetone [16, 17]. They are nonelectrolytes in nitrobenzene [15]. The nitrito complex, soluble in many organic solvents, behaves as a nonelectrolyte in nitromethane [12].

$Mn^{II}(C_6H_{18}N_3OP)_2(C(CN)_3)_2$. The tricyanomethanide complex was obtained by dissolving $Mn(C(CN)_3)_2$, see "Manganese" D 7, 1990, pp. 15/6, in warm neat ligand. The complex was precipitated by adding carbon tetrachloride and, if oily, treated with acetone. The colorless complex, dried in vacuum, melts at 56°C. Interatomic distances and relative intensities from an X-ray powder diffraction diagram are listed in the paper. These data and IR spectral data of Nujol mulls correspond with neither a monodentate anion group in a tetrahedral environment (shown for the Zn^{II} complex) nor a bidentate anion group in a polymeric octahedral geometry, proposed for the analogous Ni^{II} and Co^{II} complexes [18].

$Mn^{II}(C_6H_{18}N_3OP)_{0.5}SO_4$ and **$Mn^{II}(C_6H_{18}N_3OP)_{0.5}(CH_3COO)_2$** complexes were prepared by the general procedure given for $[Mn(C_6H_{18}N_3OP)_2X_2]$ compounds on p. 174. The white sulfato complex does not melt below 300°C, the white acetato complex melts at 129 to 130°C. The IR spectra of the complexes in Nujol show bands at 380 and 375 cm^{-1}, respectively, which were assigned to the ligand ν(Mn–O) vibration modes. IR bands assigned to the anion group vibrations gave no definite information about their kind of coordination. The electronic reflectance spectra show maxima at 18200, 20200, and 24200 cm^{-1} for the sulfato complex and a maximum at 20100 cm^{-1} for the acetato complex, both of which indicate an octahedral environment at Mn^{II} [2].

References:

[1] de Bolster, M. W. G.; Groeneveld, W. L. (Recl. Trav. Chim. **91** [1972] 171/84).
[2] de Bolster, M. W. G.; Groeneveld, W. L. (Recl. Trav. Chim. **90** [1971] 477/507).
[3] Donoghue, J. T.; Drago, R. S. (Inorg. Chem. **2** [1963] 1158/61).
[4] de Bolster, M. W. G.; Nieuwenhuijse, B.; Reedijk, J. (Z. Naturforsch. **28b** [1973] 104/6).
[5] Wayland, B. B.; Drago, R. S. (J. Am. Chem. Soc. **87** [1965] 2372/8).
[6] Durney, M. T.; Marianelli, R. S. (Inorg. Nucl. Chem. Letters **6** [1970] 895/902).
[7] Abe, Y.; Wada, G. (Bull. Chem. Soc. Japan **53** [1980] 3547/51).
[8] Grzybkowski, W.; Pilarczyk, M.; Klinszporn, L. (J. Chem. Soc. Faraday Trans. I **84** [1988] 1551/61).
[9] Fuoss, R. M.; Edelson, D. (J. Am. Chem. Soc. **73** [1951] 269/73).
[10] Frankel, L. S. (Inorg. Chem. **8** [1969] 1784/7).

[11] Chen, Tzeng-ming; Morgan, L. O. (J. Phys. Chem. **76** [1972] 1973/81).
[12] LeCoz, E.; Guerchais, J. E. (Bull. Soc. Chim. France **1971** 409/15).
[13] de Bolster, M. W. G.; Wiegerink, F. J.; Groeneveld, W. L. (J. Inorg. Nucl. Chem. **35** [1973] 89/93).
[14] Gunther, P.; Herlinger, H. (Brit. 1209812 [1968]; C.A. **71** [1969] No. 49231).
[15] LeCoz, E.; Guerchais, J. E.; Goodgame, D. M. L. (Bull. Soc. Chim. France **1969** 3855/61).
[16] Skopenko, V. V.; Tryashin, A. S.; Zub, V. Ya. (Ukr. Khim. Zh. **40** [1974] 792/5; Soviet Progr. Chem. **40** No. 8 [1974] 5/7).
[17] Tryashin, A. S.; Skopenko, V. V.; Stakhov, D. A. (Zh. Neorgan. Khim. **18** [1973] 2658/61; Russ. J. Inorg. Chem. **18** [1973] 1412/4).
[18] Köhler, H.; Skirl, R.; Skopenko, V. W. (Z. Anorg. Allgem. Chem. **428** [1977] 115/9).

39.16.5 With 4,4′,4″-Phosphoryltris(morpholine)

$(O(CH_2CH_2)_2N-)_3PO$ $(=C_{12}H_{24}N_3O_4P)$

$[Mn^{II}(C_{12}H_{24}N_3O_4P)_4](BF_4)_2$ was prepared by adding the stoichiometric amount of ligand to the solution of the Mn^{II} salt in excess of triethyl orthoformate. Dry nitromethane was added and the mixture stirred for 3 h. The filtrate was then kept at 0°C and the deposited crystals were washed with dry chloroform and dried in vacuum. The white compound melts at 244 to 246°C. A tetrahedral environment at Mn^{II} is clearly indicated by a band at 1045 cm^{-1}, tentatively assigned to the vibration of ionic BF_4 groups [1].

$[Mn^{II}(C_{12}H_{24}N_3O_4P)_2X_2]$ complexes with X = NO_3, Br, I, and NCS were prepared by adding, if necessary, the appropriate amount of triethyl orthoformate for dehydration to the solution of the Mn^{II} salt in ethanol. Then a stoichiometric amount of the ligand dissolved in ethanol was added. The crystals, which separated on standing or after concentrating the solution by evaporation, were washed with dry ether and dried in vacuum. A yellow iodo complex was obtained by use of acetone as solvent instead of ethanol. $[Mn(C_{12}H_{24}N_3O_4P)_2Cl_2]$ was prepared by dissolving manganese(II) chloride in an excess of triethyl orthoformate and some chloroform. The ligand, dissolved in chloroform, was added (1:2 mole ratio) and the crystals deposited on standing were washed with chloroform and dried in vacuum. Colors and melting points of the $[Mn(C_{12}H_{14}N_3O_4P)_2X_2]$ complexes are as follows [1]:

X	NO_3	Cl	Br	I	I	NCS
color	light yellow	white	cream	cream	yellow	light yellow
m.p. in °C	230 to 232	242*)	261 to 264	270*)	287*)	208 to 212

*) Point of decomposition.

The cream and yellow iodo complexes reveal an X-ray type different from any other. The X-ray powder diagrams of $[Mn(C_{12}H_{24}N_3O_4P)_2(NO_3)_2]$ and $[Mn(C_{12}H_{24}N_3O_4P)_2Cl_2]$ are similar to those of the analogous Co^{II} complexes; these are proposed to have an octahedral and tetrahedral geometry, respectively, due to their electronic reflectance spectra [1].

The IR spectra of the complexes in Nujol show the ν(PO) vibration shifted from 1222 cm^{-1} to lower wavenumbers. A ν(CN) band at 2075 cm^{-1} of the nitrogen-bonded thiocyanate group and a ν(Mn–Cl) band at 308 cm^{-1} were observed for the complexes with X = NCS and Cl, respectively. Band positions observed for the NO_3 anion are attributed to coordinated nitrato groups. Most of the complexes are hygroscopic and must be handled in a drybox [1].

$[Mn^{II}(C_{12}H_{24}N_3O_4P)_3(H_2O)](PF_6)_2$ and **$[Mn^{II}(C_{12}H_{24}N_3O_4P)_2(H_2O)_2](ClO_4)_2$.** The hexafluorophosphate complex was prepared by mixing solutions of the Mn^{II} salt (1 mmol) and the ligand (4 mmol), each dissolved in 20 mL of anhydrous methanol containing 2 mL of triethyl orthoformate. The mixture was stirred under nitrogen at low temperature and then the solvent evaporated at reduced pressure. The solid obtained was washed with anhydrous methanol and dried in vacuum. The perchlorate was obtained by an analogous procedure: The reaction mixture was prepared in 30 mL of anhydrous ethanol (instead of methanol). The complex precipitated on standing in a desiccator over $CaCl_2$. The crystals were washed with small portions of a mixture of anhydrous ethanol and triethyl orthoformate and were dried in vacuum. The hexafluorophosphate melts at 182.2 to 184°C, the perchlorate at 283.6 to 285.6°C, respectively [2].

The IR spectra of the complexes in Nujol show the ν(PO) vibration shifted by about 20 cm^{-1} to lower, the ν(PN) by about 4 cm^{-1} to higher frequencies upon complexation. The bands at 1082 and 623 cm^{-1} were assigned to ionic perchlorate groups, those at 842 and 539 cm^{-1} to ionic fluorophosphate groups. The water molecules are assumed to be coordinated to Mn^{II} due to bands observed near 3375 and 1635 cm^{-1} in the spectrum of both complexes. The electrical conductivity at 25°C of $[Mn(C_{12}H_{24}N_3O_4P)_3(H_2O)](PF_6)_2$ in acetonitrile, nitromethane, or nitrobenzene is $\Lambda = 227.6$, 108.5, or 49.3 $cm^2 \cdot \Omega^{-1} \cdot mol^{-1}$, respectively. Corresponding conductivities of $[Mn(C_{12}H_{24}N_3O_4P)_2(H_2O)_2](ClO_4)_2$ are $\Lambda = 259.2$, 165.8, or 3.6 $cm^2 \cdot \Omega^{-1} \cdot mol^{-1}$. The complexes behave as 1:2 electrolytes, and the Mn^{II} is proposed to be tetrahedrally surrounded. The nonelectrolyte character observed with the perchlorate complex dissolved in nitrobenzene is in contrast to the IR data assigned to the perchlorate group. The complexes are soluble in acetonitrile, nitromethane, or nitrobenzene; slightly soluble in butanol, ether, ethyl acetate, dioxane, or chloroform; insoluble in benzene, petroleum ether, or carbon tetrachloride [2].

References:

[1] de Bolster, M. W. G.; Kortram, I. E.; Groenefeld, W. L. (J. Inorg. Nucl. Chem. **35** [1973] 1843/53).
[2] Dunstan, P. O.; Matos, F. A. P. (Anais Acad. Brasil. Cienc. **51** [1979] 211/6; C.A. **91** [1979] No. 203541).

35.17 Complex with Octamethyldiphosphoramide

$$\mathrm{[(CH_3)_2N]_2P(=O){-}O{-}P(=O)[N(CH_3)_2]_2} \quad (=C_8H_{24}N_4O_3P_2)$$

$[Mn^{II}(C_8H_{24}N_4O_3P_2)_3](ClO_4)_2$ was prepared by a procedure given in [1] for similar complexes with Ni^{II}, Co^{II}, or Zn: Hydrated manganese(II) perchlorate was first stirred with an excess of 2,2'-dimethoxypropane for 2 h. An excess of the ligand was added and stirring continued for 10 min. The white complex, precipitating on addition of dry ether, was washed with ether and dried in vacuum [1]. Single crystals were grown from acetone by slow evaporation under anhydrous conditions [2].

The compound is isomorphous with the corresponding complexes of Co^{II}, Cu^{II}, or Mg, which crystallize in the trigonal space group $P\overline{3}$–C_{3i}^{1} (No. 147) with the metal atom at a site of D_3 symmetry and the threefold axis of symmetry parallel to the c crystal axis [2]. Coordination of the bidentate ligand through both phosphoryl oxygen atoms is suggested, due to the negative shifts of ν(PO), observed in the IR spectra of the Co^{II}, Ni^{II}, or Zn complexes [1].

Single crystal absorption, emission, and excitation spectra of the complex are reported in [2]. The absorption spectrum was measured at 300, 80, and 5 K, while excitation and emission spectra were measured only at the two higher temperatures. Corrected phosphorescence excitation spectra of the crystalline powder were measured at 300 K only. Orthoaxial absorption and excitation spectra were recorded with light polarized parallel (π) and perpendicular (σ) to the crystal axis. The absorption spectrum (σ polarization, 300 K) shows bands at ~18200, 22500, 24800, 27900, and 29400 cm^{-1} with intensities of $\varepsilon = 0.011$, 0.021, 0.015, 0.042, and 0.028 $L \cdot mol^{-1} \cdot cm^{-1}$, respectively. Sharp bands that appear at ~27730 and 27790 cm^{-1} in the 5 K spectrum are not resolved at 300 K. Both corrected and uncorrected excitation spectra at room temperature, using unpolarized light and monitoring the emission at ~17100 cm^{-1}, are reported in the paper. (At 300 K the complex shows a broad phosphorescence band centered at ~17100 cm^{-1} which is shifted to 16100 cm^{-1} at 80 K.) The bands observed in the excitation spectra are shifted slightly to higher wavenumbers in comparison to the absorption bands. Additional bands were observed at ~26000, 29500, 30700, 31600, 34500, and 39700 cm^{-1}. When the temperature is lowered from 300 to 80 K, bands centered at ~19000 [$^4T_1(G)$] and 22700 cm^{-1} [$^4T_2(G)$] are shifted to lower energy. In addition fine structure appears in bands centered at ~24800 [$^4A_1$, $^4E(G)$], 27900 [$^4T_2(D)$], and 31600 cm^{-1} [$^2T_1(I)$]. All bands observed at a given temperature are observed for both π and σ polarizations of the exciting light with no detectable shift in maxima between polarizations [2].

The spectral behavior of $[Mn(C_8H_{24}N_4O_3P_2)_3](ClO_4)_2$ is independent of the Mn^{II} concentration, checked by doping the complex into the corresponding Zn complex. The crystal field calculations for D_3 symmetry support the assignments of the observed bands. The presence of D_3 instead of O_h symmetry is indicated by the rather strong polarizations of some of the bands in the orthoaxial absorption spectrum and by the light splitting of the ^4E, $^4A_1(G)$ pair of bands in both absorption and emission at low temperatures. Interelectronic repulsion, normalized spherical harmonic (NSH) trigonal potential, and spin-orbit interaction parameters are fitted, with B = 670, C = 3620, D(Q) = 26291, D(SIG) = − 2000, D(TAU) = 2000, and $\lambda = 60\ cm^{-1}$. The D(Q), D(SIG), and D(TAU) values correspond to the conventional crystal field parameter values Dq = 836, $D\sigma = -239$, and $D\tau = -93\ cm^{-1}$. The results are correlated with the previously reported [3, 4] zero-field parameters. The sign and magnitude of the zero-field parameter calculated are consistent with the experiment and with a trigonally compressed geometry about the Mn^{2+} ion [2].

An ESR study of $[Mn(C_8H_{24}N_4O_3P_2)_3](ClO_4)_2$, doped at 0.1% into single crystals of the corresponding Zn complex, shows five sets of six nuclear hyperfine lines in the spectra reported for $\Theta = 0°$ and 90°. The g values determined were $g_{\parallel} = 2.011 \pm 0.005$ and $g_{\perp} = 2.0100 \pm 0.0005$, i.e., the g value is isotropic. The magnitude of the hyperfine splitting parameter, $A = 99.7 \times 10^{-4}\ cm^{-1}$, indicates a metal–ligand bond almost devoid of covalent character [3]. Similar results obtained from the ESR spectrum of the complex doped at 0.1% into the corresponding Mg host lattice are reported. The electron spin–nuclear spin hyperfine coupling constant (106 G) is indicative of the highly ionic character of the Mn–O bond. The spectra were interpreted by using an axially symmetric spin-Hamiltonian for an $S = 5/2$ spin state. The zero-field splitting is predominantly due to interaction of Mn^{II} with the trigonal crystal field. The magnitude and sign of the zero-field splitting parameters, D and $(\alpha - F)$, were determined: D = 157 G, (a − F) = 9 G. The g value is 2.011. A nuclear quadrupole coupling constant of 0.154 G results from an analysis of the forbidden ESR lines [4].

References:

[1] Joesten, M. D.; Nykerk, K. M. (Inorg. Chem. **3** [1964] 548/50).
[2] Hempel, J. C.; Palmer, R. A.; Yang, M. Chin-Lan (J. Chem. Phys. **64** [1976] 4314/20).
[3] Woltermann, G. M.; Wasson, J. R. (Chem. Phys. Letters **16** [1972] 92/7).
[4] Woltermann, G. M.; Wasson, J. R. (J. Magn. Resonance **9** [1973] 486/94).

35.18 Complexes with Nonamethylimidodiphosphoramide

$$[(CH_3)_2N]_2P(=O){-}N(CH_3){-}P(=O)[N(CH_3)_2]_2 \quad (= C_9H_{27}N_5O_2P_2)$$

$[Mn(C_9H_{27}N_5O_2P_2)_3]X_2$ ($X = ClO_4$, BF_4). The perchlorate was prepared by stirring manganese(II) perchlorate with an excess of 2,2′-dimethoxypropane and adding a slight excess of ligand dissolved in acetone. Precipitation of the complex was completed by addition of dry ether. The white crystals were washed with ether and dried in vacuum [1]. A light pink perchlorate [2] and a white tetrafluoroborate [3] were prepared by reaction of the manganese salt (dehydrated with trimethyl orthoformate) with excess ligand in methanol or acetone. Both complexes must be handled in a drybox. $[Mn(C_9H_{27}N_5O_2P_2)_3](ClO_4)_2$ decomposes at 265°C [2], $[Mn(C_9H_{27}N_5O_2P_2)_3](BF_4)_2$ melts at 277 to 279°C [3]. Both compounds show X-ray powder diffraction patterns identical with the corresponding Zn, Ni^{II}, Co^{II}, and Fe^{II} complexes [2, 3].

The IR spectra of the complexes in Nujol indicate the existence of a six-membered chelate ring formed by the phosphoryl oxygen atoms of the ligand and the Mn^{II} ion [1 to 3]. The PO stretching and bending frequencies are shifted to lower and higher energies, respectively. Vibrations observed for both anions indicate their uncoordinated ionic nature [2, 3]. The electronic reflectance spectra show bands at 21900, 23800, and 24500 cm^{-1} for $[Mn(C_9H_{27}N_5O_2P_2)_3](ClO_4)_2$ [2] and at 19600 and 23200 cm^{-1} for $[Mn(C_9H_{27}N_5O_2P_2)_3](BF_4)_2$ [3]; these were assigned to an octahedral geometry at Mn^{II} [2, 3]. The ESR X- and Q-band spectra of $[Mn(C_9H_{27}N_5O_2P_2)_3](BF_4)_2$, doped into the corresponding solid Zn complex, reveal a g value of 2.00 and a hyperfine coupling constant, $A=(90\pm3)\times10^{-4}$ cm^{-1}, which is assigned to a (most probably) pseudooctahedral local symmetry at Mn^{II} [4]. $[Mn(C_9H_{27}N_5O_2P_2)_3](ClO_4)_2$ is soluble in methanol, acetone, nitromethane, or acetonitrile and insoluble in carbon tetrachloride, benzene, or ether. The conductivity in a 10^{-3}M nitromethane solution, $\Lambda=181\ cm^2\cdot\Omega^{-1}\cdot mol^{-1}$, is that of a 1:2 electrolyte [2]. The compound is soluble in water but decomposes to the hydrated manganese(II) perchlorate [1].

$[Mn^{II}(C_9H_{27}N_5O_2P_2)_2(NO_2)_2]$ and **$[Mn^{II}(C_9H_{27}N_5O_2P_2)_2I_2]$**. The nitrito complex was obtained from $[Mn(C_9H_{27}N_5O_2P_2)Cl_2]$ and $AgNO_2$ by a procedure given for the nitrito complex of hexamethylphosphoric triamide, $[Mn(C_6H_{18}N_3OP)_2(NO_2)_2]$, p. 174. The light ocher complex is sensitive to light and was washed with dry ether and dried in vacuum [5]. The iodo complex was prepared by reaction of anhydrous MnI_2 and ligand in acetone or methanol. The light yellow complex melts at 248 to 252°C [3]. The IR spectra of the complexes in Nujol show the ν(PO) vibration shifted to lower wavenumbers, which indicate that the ligand is coordinated to Mn via the phosphoryl oxygen atom [3, 5]. The $\delta(NO_2)$ bands observed at 832, 820, and 789 cm^{-1} do not elucidate the mode of anion bonding in $[Mn(C_9H_{27}N_5O_2P_2)_2(NO_2)_2]$ [5]. The bands in the electronic reflectance spectrum of $[Mn(C_9H_{27}N_5O_2P_2)_2I_2]$ at 20200, 23500, 26400, and 28200 cm^{-1} indicate a pseudooctahedral environment at Mn^{II} [3]. The ESR Q-band spectrum of the solid iodo complex reveals a g value of 2.00. A pseudooctahedral symmetry, D_{2h}, was proposed due to the zero-field parameters $D=0.70\pm0.02$ cm^{-1} and $\lambda=0.025$. The rather large D value and the small λ value are in accord with a *trans*-octahedral molecular unit [4]. The complexes were handled in a drybox [3, 5].

$[Mn^{II}(C_9H_{27}N_5O_2P_2)X_2]$ complexes with $X=NO_3$, Cl, Br, and NCS were prepared by reaction of the MnX_2 salt (if necessary, dehydrated with trimethyl orthoformate) and the ligand in methanol or acetone. The compounds were washed with dry ether and dried in vacuum. As shown by X-ray powder diffraction patterns, the nitrato complex is isomorphic with the analogous complex of Mg, and the chloro complex corresponds to the Co^{II} and Zn complexes. Colors, melting points, IR bands (in cm^{-1}) of the complexes in Nujol, and the bands observed in the electronic reflectance spectra are tabulated below [3]:

X	color	m.p. in °C	ν(PO)	ν(Mn–X)*)	$\tilde{\nu}_{max}$ in kK
NO_3	pink	82 to 83	1183	200	19.1, 22.5, 23.2, 24.3
Cl	light green	166 to 168	1186	318, 280	21.8, 22.8, 23.6, 26.3, 28.3
Br	white	203 to 204	1184	215	21.5, 22.4, 23.5, 23.8, 26.3, 27.9
NCS	white	150 to 151	—	295, 273	19.5, 21.9, 22.7, 25.2, 27.0

*) Stretching vibrations $Mn–O_{nitrate}$, Mn–Cl, Mn–Br, Mn–N, respectively.

The IR band positions support the anion groups being coordinated to Mn^{II}, the thiocyanato group via the nitrogen atom and the ligand molecule through the phosphoryl oxygen atom. Based on the electronic spectra the environment at Mn is proposed to be pseudotetrahedral with the exception of the nitrato complex having a geometry somewhat different from pseudooctahedral [3]. The ESR spectral data of the solid compound confirm the proposed structures. The zero-field parameters D and λ for the complexes were calculated: D = 0.075, 0.41, 0.44, and 0.02 cm^{-1} for $X=NO_3$, Cl, Br, and NCS, respectively, and $\lambda=0.033$, 0.21, and 0.20 for $X=NO_3$, Cl, and Br. All complexes have g values of 2.00 [4]. The complexes were handled in a drybox [3].

References:

[1] Lannert, K. P.; Joesten, M. D. (Inorg. Chem. **7** [1968] 2048/51).
[2] de Bolster, M. W. G.; Groeneveld, W. L. (Recl. Trav. Chim. **91** [1972] 171/84).
[3] de Bolster, M. W. G.; Groeneveld, W. L. (Recl. Trav. Chim. **90** [1971] 687/99).
[4] de Bolster, M. W. G.; Nieuwenhuijse, B.; Reedijk, J. (Z. Naturforsch. **28b** [1973] 104/6).
[5] de Bolster, M. W. G.; Wiegerink, F. J.; Groenefeld, W. L. (J. Inorg. Nucl. Chem. **35** [1973] 89/93).

39.19 Complexes with Dodecylmethyl-P′-amidodiimidotriphosphoramide

$((CH_3)_2N)_2P(O)-N(CH_3)-P(O)(N(CH_3)_2)-N(CH_3)-P(O)(N(CH_3)_2)_2$ $(= C_{12}H_{36}N_7O_3P_3)$

$[Mn^{II}(C_{12}H_{36}N_7O_3P_3)_2](BF_4)_2$ and **$[Mn^{II}(C_{12}H_{36}N_7O_3P_3)(NO_3)_2]$** complexes were prepared as follows: The hydrated metal salt was dissolved in acetone and trimethyl orthoformate was added (for dehydration). Then the ligand dissolved in acetone was added. Upon standing crystals of the complexes appeared, sometimes after concentration of the solution or after addition of dry ether. The crystals were washed with dry ether and dried in vacuum at room temperature. Changing the solvent to methanol or acetonitrile or changing the metal salt to ligand ratio (from 1:1 to 1:2 or vice versa) did not influence the results. The white fluoroborate melts at 161 to 163°C. It is isomorphous with the corresponding Ca or Cd compounds as indicated by the X-ray powder diffraction patterns. The white nitrato complex melts at 174 to 176°C.

The IR spectra of the complexes in Nujol show ν(PO) and δ(PO) vibrations shifted from 1226 and 502 cm^{-1} to lower and higher wavenumbers, respectively, indicating that coordination of the tridentate ligand occurs via the phosphoryl oxygen atoms. Bands observed for the BF_4 group are indicative of the ionic nature of the anion. The nitrato complex shows bands of $\nu_3(NO_3)$ at 1450, 1320, $\nu_1(NO_3)$ at 1033, and $\nu_2(NO_3)$ at 822 cm^{-1}; these were assigned to monodentate nitrato groups in a five-coordinate square-pyramidal geometry at Mn^{II}.

The ESR X-band spectrum of powdered solid $[Mn(C_{12}H_{36}N_7O_3P_3)_3](BF_4)_2$ consists of a single line near g = 2. Doped into the corresponding Zn or Cd compounds the complex reveals a hyperfine structure with the coupling constant A = 95 G. The electrical conductivity of $[Mn(C_{12}H_{36}N_7O_3P_3)(NO_3)_2]$ in 10^{-3} M acetonitrile solution at room temperature, $\Lambda = 160\ cm^2 \cdot \Omega^{-1} \cdot mol^{-1}$, corresponds to a 1:1 electrolyte, which is probably formed by solvation. The complexes were handled in a drybox.

Reference:

de Bolster, M. W. G.; den Heijter, J.; Groeneveld, W. L. (Z. Naturforsch. **27b** [1972] 1324/9).

39.20 Complexes with Heterocycles Containing Phosphorus

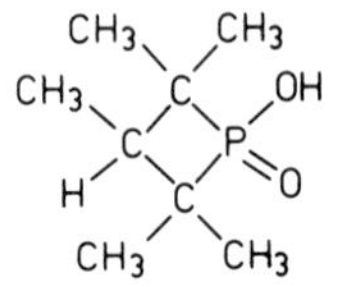

ligand 1 ($=C_8H_{17}O_2P$)

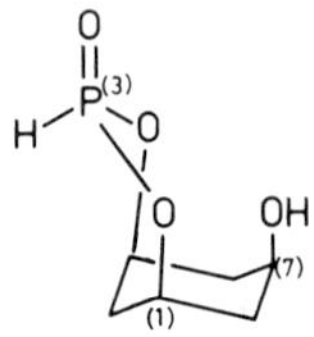

ligand 2 ($=C_6H_{11}O_4P$)

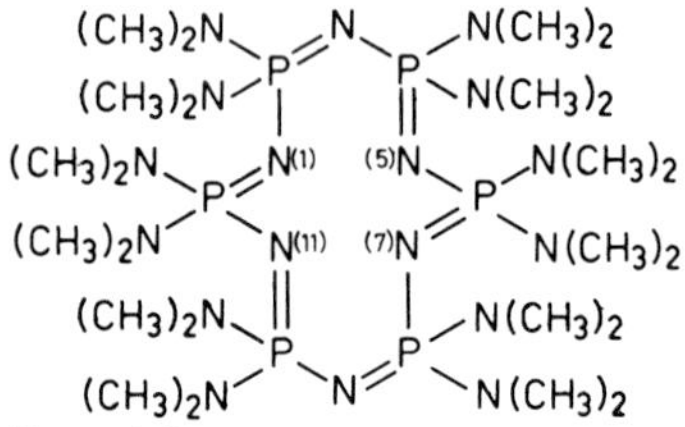

ligand 3 ($=C_{24}H_{72}N_{18}P_6$)

$Mn^{II}(C_8H_{16}O_2P)_2$ was prepared by refluxing Mn^{II} nitrate (10 mmol) and 28 mmol 1-hydroxy-2,2,3,4,4-pentamethylphosphetane 1-oxide (ligand 1) in 20 mL of acetone for 2 h. White microcrystalline needles separated on cooling. The complex does not melt below 360°C. The IR spectrum of the compound in Nujol shows bands at 1089 and 1056 cm^{-1}, assigned to $\nu_{as}(PO_2)$ and $\nu_s(PO_2)$ vibrations. Bands of the phosphetane ring vibrations were observed at 1239, 1147, 934, 758, 672, 641, 560, and 540 cm^{-1}. A band at 521 cm^{-1} was assigned to ν(Mn–O). A bidentate O,O-coordination of the ligand to Mn in a polymeric structure is proposed. The compound is insoluble in all solvents [1].

$Mn^{II}(C_6H_{11}O_4P)_4(ClO_4)_2$. The complex, in which the 3-hydroxy tautomer of ligand 2 is assumed to be bonded to manganese through the phosphorus atom, was prepared as follows: $Mn(ClO_4)_2 \cdot 6H_2O$ (12 mmol) dissolved in 5 mL acetone was added to a suspension of 4.5 mmol of ligand 2 (obtained by acid-catalyzed hydrolysis of the adamantane-like 2,8,9-trioxa-1-phosphatricyclodecane). The mixture was stirred for 1 h. The complex can also be prepared directly by heating a mixture of 1 mmol $Mn(ClO_4)_2 \cdot 6H_2O$ dissolved in 10 mL acetone and 4 mmol of 2,8,9-trioxa-1-phosphatricyclodecane ($=C_6H_9O_3P$) in 15 mL of acetone for 7 h. One ring of the phosphatricyclodecane derivative opens under reaction to yield the phosphabicyclononane derivative (ligand 2). The white complex was washed with ether or acetone and dried in vacuum. IR studies reveal that the PH stretching mode of ligand 2 disappears on complexation; a band at 1248 cm^{-1}, assigned to ν(PO), also disappears, whereas a broad ν(OH) band in the 3400 to 3300 cm^{-1} range appears. The complex is insoluble in most organic solvents. It dissociates in water, methanol, or acetonitrile to yield solvated Mn^{2+} ions and ligand 2 [2].

$[Mn^{II}(C_{24}H_{72}N_{18}P_6)NO_3]NO_3$. For preparation 1 mmol of $AgNO_3$ was added with stirring to a solution of $MnCl_2$ (0.49 mmol) in 20 mL acetonitrile. A slight excess (0.5 mmol) of ligand 3 was added to the filtered reaction mixture and the solution refluxed for 18 h. The solution was then filtered and the solvent removed to give a buff powder. The compound has a magnetic moment of $\mu_{eff}=5.74\ \mu_B$ at room temperature. Band positions observed in the IR spectrum of the complex in Nujol are tabulated in the paper. The stretching and deformation frequencies of the methyl groups are similar to those in the free ligand; thus, the exocyclic nitrogen atoms are not involved in coordination to Mn. The IR data show that the molecular structure is essentially the same as those established crystallographically for the $[Co(C_{24}H_{72}N_{18}P_6)Cl]^+$ and $[Cu(C_{24}H_{72}N_{18}P_6)Cl]^+$ cations, in which, as a consequence of steric interactions, the metal has a distorted trigonal-bipyramidal environment. The Mn^{2+} ion is coordinated by the tetradentate ligand through the N(1), N(5), N(7), and N(11) atoms. The molecule is folded about one N–Mn–N axis by repulsive interactions between the dimethylamino groups and the coordinated nitrate ion. The N(1) and N(7) atoms hold the apical, the N(5) and N(11) atoms and one NO_3 group the equatorial positions. The molar conductivity of $[Mn(C_{24}H_{72}N_{18}P_6)(NO_3)]NO_3$ in 10^{-2} M acetonitrile solution, $\Lambda=124\ cm^2 \cdot \Omega^{-1} \cdot mol^{-1}$, at room temperature is in the range of a 1:1 electrolyte [3].

A chloro complex, $\mathbf{Mn_2(C_{24}H_{72}N_{18}P_6)Cl_4}$, with variable and irreproducible incorporation of solvent molecules was formed by reaction of ligand 3 with dehydrated $MnCl_2$ in acetonitrile [3].

References:

[1] Emsley, J.; Dunning, A. P.; Parker, R. J.; Williams, J. K.; Brown, S.; Earnshaw, S.; Moore, D. S. (Polyhedron **3** [1984] 325/9).
[2] Jenkins, J. M.; Huttemann, T. J.; Verkade, J. G. (Advan. Chem. Ser. No. 62 [1967] 604/15).
[3] Calhoun, H. P.; Paddock, N. L.; Wingfield, J. N. (Can. J. Chem. **53** [1975] 1765/74).

39.21 Complexes with Phosphorous Trifluoride

[Mn(H)(F$_3$P)$_5$] was formed as a colorless liquid by treating K[Mn(CO)$_2$(F$_3$P)$_3$] with phosphorous trifluoride at 450 atm [1]. The compound was also obtained on reaction of [Mn(H)(CO)$_5$] with phosphorous trifluoride (1:7.5 mole ratio): The reagent at 0.75 atm and room temperature in a quartz vessel were irradiated by UV for 23 h and CO was removed regularly. The excess ligand gas was separated by vacuum distillation at −70°C and the reaction product was separated from a slightly volatile yellow residue by vacuum distillation at room temperature in a limited time of ~10 min. Separation from accompanying [Mn(H)(CO)$_n$(F$_3$P)$_{5-n}$] complexes (n=1 to 5) was accomplished by gas-liquid chromatography. [Mn(H)(F$_3$P)$_5$] melts at 18.5°C. A molecular weight of 493 (calc. 496) was determined by the vapor density method [2].

The IR spectrum of the complex dissolved in acetone shows ν(PF) vibrations at 971, 918, 877, 869, and 860 cm^{-1} and a band at 1882 cm^{-1}, assigned to ν(Mn–H). All band positions are nearly the same as found for the corresponding rhenium compound [1]. The IR spectra observed in the gas phase or in hexane solution show the ν(Mn–H) vibration at 1845 cm^{-1}. Bands of the ν(PF) vibrations were observed at 962, 914, 906, 901(sh), 880, 868(sh), and 849 cm^{-1}. It is proposed that the ligand molecules occupy five of six corners of an octahedron [2]. The He(I) photoelectron spectrum was investigated and was assigned by analogy to those of PF_3 and [Mn(H)(CO)$_5$]. The data indicate a minor distortion from O_h. The complex is assumed to have C_{4v} symmetry with the hydrogen atom occupying an apical position [3]. The complex is stable at ambient conditions when it is sealed in evacuated ampules [2].

[Mn(NO)$_3$(F$_3$P)] was prepared by condensation of manganese vapor (4 g) over 60 min into 4.8 L of a gas mixture at −196°C containing BF_3, NO, and the ligand (2:2:1 mole ratio). The resulting green liquid was separated from excess reagents by condensation in a trap at −100°C. The compound melts at −120±2°C. When manganese vapor was condensed with NO and PF_3 in the absence of BF_3, the condensate exploded violently on warming to a little above −196°C, and nitrogen was liberated [4]. A simple conventional synthesis, a two-step photolysis procedure, is given in [5]: [Mn(NO)(CO)$_4$] and a 20-fold excess of NO were photolyzed in the gas phase for 1 min in a quartz UV cell using a medium pressure Hg arc. The gases were frozen with liquid N_2. On warming to −80°C the displaced CO and unreacted NO were pumped off. The remaining pale green [Mn(NO)$_3$(CO)] was condensed at −196°C into a UV cell, together with a 20-fold excess of phosphorous trifluoride. Then the mixture was warmed to room temperature and photolyzed once more for 2 min. The color changed from pale green to dark green. The cell was cooled to −196°C and the unreacted ligand was pumped off at −80°C. The stability of the complex in the gas phase suggests that this preparative method could be used for larger quantities of the compound. Unreacted [Mn(NO)$_3$(CO)] may be removed by gas-liquid partition chromatography or, alternatively, by photolyzing the second stage gas mixture longer with a greater excess of phosphorous trifluoride [5]. The IR spectrum (70 mm gas cell) shows bands at 913, 899(sh), 894, 667, and 665 cm^{-1} of the ν(PF) vibrations [4]. Mixtures with N_2 or

CH_4 matrix gas condensed on a CsBr window at 20 K show bands at ≈1835 and 1745 cm^{-1}, which may be assigned to terminal ν(NO) vibrations. Terminal ν(PF) vibrations were observed at 910 cm^{-1} [5]. Peaks and relative intensities observed in the mass spectrum are listed in [4].

$[Mn(NO)_3(F_3P)]$ is stable in an inert atmosphere, but is readily oxidized by air. If exposed to carbon monoxide at 1 atm pressure at 20°C the compound forms $[Mn(NO)_3(CO)]$ almost quantitatively [4].

References:

[1] Kruck, T.; Engelmann, A. (Angew. Chem. **78** [1966] 820; Angew. Chem. Intern. Ed. Engl. **5** [1966] 836).
[2] Miles, W. J.; Clark, R. J. (Inorg. Chem. **7** [1968] 1801/6).
[3] Head, R. A.; Nixon, J. F.; Sharp, J.; Clark, R. J. (J. Chem. Soc. Dalton Trans. **1975** 2054/9).
[4] Middleton, R.; Hull, J. R.; Simpson, S. R.; Tomlinson, C. H.; Timms, P. L. (J. Chem. Soc. Dalton Trans. **1973** 120/4).
[5] Crichton, O.; Rest, A. J. (Inorg. Nucl. Chem. Letters **9** [1973] 391/2).

39.22 Complexes with Phosphoryl Halides or Chloromethylphosphonic Dichloride

$P(O)X_3$	$ClCH_2P(O)Cl_2$
ligand 1 with X = F (= F_3OP)	ligand 3 (= CH_2Cl_3OP)
ligand 2 with X = Cl (= Cl_3OP)	

$[Mn^{II}(F_3OP)_4](AsF_6)_2$ was formed in liquid sulfur dioxide by reaction of $[Mn(SO_2)_2(AsF_6)_2]$, see "Manganese" D 7, 1990, p. 112, and an excess of ligand 1. The complex was characterized by elemental analysis and IR spectroscopy. The band observed at 1360 cm^{-1} was assigned to the ν(PO) vibration of the ligand coordinated through the phosphoryl oxygen atom, bands at 1022 and 910 cm^{-1} to the ν(PF) vibrations. Corresponding bands of the free liquid ligand were observed at 1395, 982, and 875 cm^{-1}.

The complexes **$[Mn^{II}(Cl_3OP)_6](SbCl_6)_2 \cdot 2Cl_3OP$** and **$[Mn(Cl_3OP)_6](FeCl_4)_2$** as well as **$[Mn^{II}(Cl_3OP)_6][SnCl_5 \cdot Cl_3OP]_2$** were prepared by stirring $MnCl_2$ in phosphoryl chloride with $SbCl_5$, $FeCl_3$, or $SnCl_4$, respectively. The precipitated chloroantimonate and chlorostannate are white, the chloroferrate yellow-green. The complexes are proposed to have an octahedral environment at Mn^{II}, based on a broad ν(PO) band at 1270 cm^{-1} in their Nujol IR spectra. Phosphoryl chloride has a ν(PO) band at 1304 cm^{-1}, which was also found in the spectra of the complexes containing additional noncoordinated ligand molecules. The coordinated molecules are bonded through their phosphoryl oxygen atom. Band positions of the anions are shown at 340 cm^{-1} for $SbCl_6$, at 379 cm^{-1} for $FeCl_4$, and at 320 cm^{-1} for $SnCl_5 \cdot Cl_3OP$. The electronic reflection spectrum of $[Mn(Cl_3OP)_6](FeCl_4)_2 \cdot Cl_3OP$ shows bands at 450, 500, 535, 620, 690, and 725(sh) nm; these are characteristic of the $FeCl_4^-$ anion. The complexes are soluble in phosphoryl chloride. They are not stable toward water, as indicated by IR spectroscopy: The band at 1270 cm^{-1}, assigned to coordinated ligand molecules, moves to the 1304 cm^{-1} position of the free ligand after exposing the Nujol mull to the air [2].

$Mn^{II}O \cdot 3Cl_3OP$ was assumed to be formed on heating a mixture of MnO and excess phosphoryl chloride to 110°C in one end of an L-shaped sealed glass tube. The pale pink compound precipitated from the saturated liquid, which had been decanted carefully to the cool limb. With repeated decanting, heating, and cooling steps a sufficient yield was obtained. The complex is hygroscopic. It is sparingly soluble in phosphoryl chloride. If the $POCl_3$ is dissolved in acetone or ethyl acetate before the oxide is added, mixed ligand $[Mn(Cl_2O_2P)_2L_2]$ complexes are formed, see p. 186 [3].

$[Mn^{II}(CH_2Cl_3OP)_6](SbCl_6)_2 \cdot CH_2Cl_3OP$ precipitated on addition of a slight excess of ligand 3 to a nitromethane solution of $[Mn(CH_3NO_2)_6](SbCl_6)_2$, see "Manganese" D 7, 1990, p. 21. The white complex is extremely hygroscopic. The IR spectrum in Nujol shows the band of ν(PO) to be negatively shifted. There are additional bands indicating the presence of the uncoordinated ligand molecule. A broad band at 335 cm^{-1} was assigned to the octahedral $SbCl_6$ anion. Octahedral symmetry at Mn^{II} with ligands coordinated through the phosphoryl oxygen atom is proposed. The presence of an uncoordinated ligand molecule in the complex is confirmed by the fact that the X-ray powder diagram is very similar in d values and intensities to those of the corresponding Co^{II} and Ni^{II} complexes. The ligand field spectra of these compounds are typical for octahedral symmetry of the cation. $[Mn(CH_2Cl_3OP)_6](SbCl_6)_2 \cdot CH_2Cl_3OP$ was prepared and handled in a drybox [4].

References:

[1] Mews, R. (J. Chem. Soc. Chem. Commun. **6** [1979] 278/9).
[2] Drussen, W. L.; Groeneveld, W. L. (Recl. Trav. Chim. **87** [1968] 786/94).
[3] Bassett, H.; Taylor, H. S. (J. Chem. Soc. **99** [1911] 1402/11).
[4] de Bolster, M. W. G.; den Heijer, M.; Groenefeld, W. L. (J. Inorg. Nucl. Chem. **39** [1977] 1709/10).

39.23 Complexes with Diphenyl Phosphorochloridate or Phosphorodichloridic Acid

$(C_6H_5O)_2P(O)Cl$ ligand 1 ($=C_{12}H_{10}ClO_3P$) $Cl_2P(O)OH$ ligand 2 ($=Cl_2HO_2P$)

$[Mn^{II}(C_{12}H_{10}ClO_3P)_6](SbCl_6)_2 \cdot C_{12}H_{10}ClO_3P$. Preparation and properties of the white hygroscopic compound with ligand 1 are similar to those of the corresponding complex with chloromethylphosphonic dichloride, see above [1].

$[Mn^{II}(Cl_2O_2P)_2(Cl_3OP)_2]$ was prepared by reaction of dehydrated Mn^{II} acetate (3 mmol) with 40 mL (54 mmol) of phosphoryl chloride. The pale pink precipitate was separated after 4 h, dried, and once more stirred with 40 mL $POCl_3$ for 4 h. The compound was then washed and dried in high vacuum for 3 h. Moisture must be excluded during the preparation. The complex decomposes at 180°C [2].

The IR and Raman spectra have been investigated. According to the vibrational spectra the manganese atoms are linked to polymeric 8-ring structures by O–P–O bridges. Bands of $\nu(PO)_{POCl_3}$ were observed at 1296 (R) and 1278 (IR, R) cm^{-1}, of $\nu_{as}(PO_2)$ at 1141 (R) and 1134 (IR) cm^{-1}, and of $\nu_s(PO_2)$ at 1098 (IR) cm^{-1} [2]. The electronic reflection spectrum of $[Mn(Cl_2O_2P)_2(Cl_3OP)_2]$ observed at 77 K shows maxima at ~19000 ($^6A_{1g} \rightarrow {}^4T_{1g}$), ~22000 ($\rightarrow {}^4T_{2g}$), 23850 ($\rightarrow {}^4A_{1g}$, ^4E_g), 27000 ($\rightarrow {}^4T_{2g}$), and 28100 cm^{-1} ($\rightarrow {}^4E_g$), which agree with the values calculated for octahedral coordination. The ligand field parameters are $10Dq = 7200\ cm^{-1}$, $B = 600\ cm^{-1}$, and $\beta = 0.925$, indicating a stability lower than that of the $[Mn(H_2O)_6]^{2+}$ complex [3]. The compound is insoluble in benzene, dichloromethane, carbon tetrachloride, acetonitrile, and phosphorylchloride [2].

[$Mn^{II}(Cl_2O_2P)_2L_2$]. Mixed ligand complexes with ligand 2 and $L=CH_3COCH_3$ ($=C_3H_6O$) or $CH_3COOC_2H_5$ ($=C_4H_8O_2$), were obtained on reaction of MnO with a solution of $POCl_3$ in acetone or in ethyl acetate, if traces of moisture were present. Pale pink crystals separated on standing in a desiccator containing solid potassium hydroxide and concentrated sulfuric acid. The compounds were washed with small amounts of pure solvent and dried [4]. The compounds were obtained also on mixing MnO with $Cl_2P(O)OP(O)Cl_2$ in acetone or ethyl acetate containing some drops of water, according to $Cl_2P(O)OP(O)Cl_2 + H_2O \rightarrow 2\,P(O)(Cl_2)OH$ and $2\,P(O)(Cl_2)OH + MnO + 2\,L \rightarrow [Mn(Cl_2O_2P)_2L_2] + H_2O$. The complex $[Mn(Cl_2O_2P)_2(C_4H_8O_2)_2]$ was also prepared by the direct reaction of phosphorodichloridic acid with MnO, $MnCl_2$, $MnBr_2$, or $MnCO_3$ in ethyl acetate [5].

X-ray data reveal a monoclinic structure for $[Mn(Cl_2O_2P)_2(C_4H_8O_2)_2]$; space group $P2_1/a\text{–}C^5_{2h}$ (No. 14), with the lattice parameters a=14.27, b=13.87, c=10.04 Å, and $\beta=95.56°$; Z=4. Calculated density: D=1.671 g/cm^3 [6]. The structure was solved to R=0.15. The Mn atom is octahedrally surrounded by the four oxygen atoms of two bridging PO_2Cl_2 groups and by two carbonyl oxygen atoms of the *cis*-arranged ethyl acetate molecules, see **Fig. 39**. Eight-membered centrosymmetric rings are formed by manganese, oxygen, and phosphorus. Adjacent rings are almost perpendicular to each other in a polymer structure. The octahedra with connected dichlorophosphate ions and ethyl acetate molecules form infinite strings along the z axis and along 1/2, 1/2, z. The strings are connected by van der Waals forces. Atomic coordinates, bond lengths and angles, and van der Waals distances are listed. The four $Mn\text{–}O_{PO_2Cl_2}$ bond distances are from 2.11 to 2.14 Å, $Mn\text{–}O_{C_4H_8O_2}$ distances are 2.17 and 2.24 Å, $O\text{–}Mn\text{–}O_{vic}$ bond angles from 86.1° to 93.3°, $O\text{–}Mn\text{–}O_{ap}$ from 172.6° to 178.9° [7]. $[Mn(Cl_2O_2P)_2(C_4H_8O_2)_2]$ is isomorphous with the corresponding Mg and Ca complexes [5].

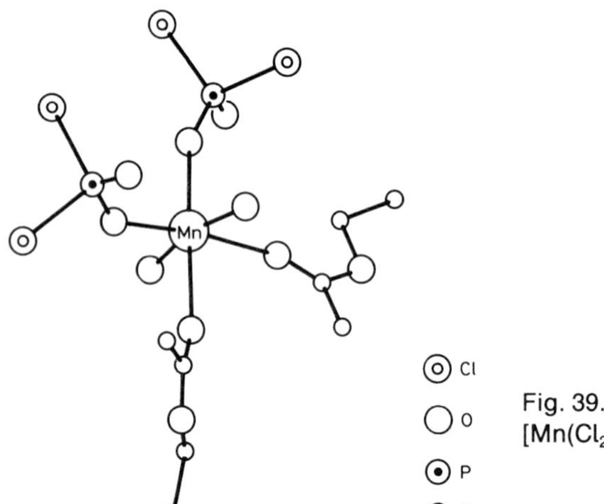

Fig. 39. Molecular structure of the $[Mn(Cl_2O_2P)_2(C_4H_8O_2)_2]$ complex [7].

The IR spectrum of $[Mn(Cl_2O_2P)_2(C_4H_8O_2)_2]$ in Nujol was investigated by [2]. The vibrational spectrum shows that the manganese atoms are linked to polymeric 8-ring structures by O–P–O bridges. The $\nu(CO)$ bands observed at 1700 and 1690 cm^{-1} are indicative of *cis*-arranged solvent molecules coordinated to Mn through their carbonyl oxygen atoms. Uncoordinated ethyl acetate has a $\nu(CO)$ band at 1742 cm^{-1}. The small separation of the $\nu_{as}(PO_2)$ and $\nu_s(PO_2)$ bands indicates that the bond between Mn^{II} and the dichlorophosphato groups is mainly ionic [2].

The ^{35}Cl nuclear quadrupole resonance spectrum of $[Mn(Cl_2O_2P)_2(C_4H_8O_2)_2]$ shows four signals from 26.112 to 26.660 MHz at 77 K and from 25.078 to 25.856 MHz at 273 K. The spectrum is only slightly changed in going from 273 to 77 K, due to similar thermal vibrations of four (crystallographically independent) different chlorine atoms. The results agree with the crystal structure (see p. 186) [8].

References:

[1] de Bolster, M. W. G.; den Heijer, M.; Groeneveld, W. L. (J. Inorg. Nucl. Chem. **39** [1977] 1709/10).
[2] Saavedra, A.; Dehnicke, K. (Z. Anorg. Allgem. Chem. **435** [1977] 82/90).
[3] Saavedra, A.; Reinen, D. (Z. Anorg. Allgem. Chem. **435** [1977] 91/7).
[4] Bassett, H.; Taylor, H. S. (J. Chem. Soc. **99** [1911] 1402/11).
[5] Grunze, H. (Z. Chem. [Leipzig] **6** [1966] 266/7).
[6] Danielsen, J.; Rasmussen, S. E. (Acta Chem. Scand. **15** [1961] 1398).
[7] Danielsen, J.; Rasmussen, S. E. (Acta Chem. Scand. **17** [1963] 1971/9).
[8] Fichtner, W.; Weiss, A. (J. Mol. Struct. **66** [1980] 289/99).

39.24 Complexes with Derivatives of Phosphoramidic or Phosphorodiamidic Chlorides

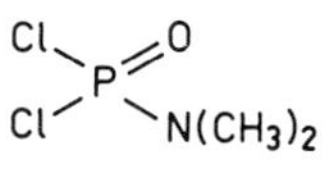

ligand 1 (= $C_2H_6Cl_2NOP$) 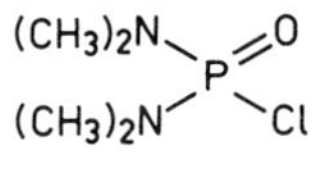 ligand 2 (= $C_4H_{12}ClN_2OP$) 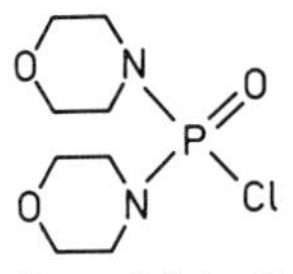 ligand 3 (= $C_8H_{16}ClN_2O_3P$)

$[Mn^{II}(C_2H_6Cl_2NOP)_6](SbCl_6)_2 \cdot C_2H_6Cl_2NOP$ and **$[Mn^{II}(C_4H_{12}ClN_2OP)_5](SbCl_6)_2$**. The compounds were precipitated by addition of ligand 1 or 2, respectively, to a nitromethane solution of $[Mn(CH_3NO_2)_6](SbCl_6)_2$, as described above for the complex with chloromethylphosphonic dichloride, $[Mn(CH_2Cl_3OP)_6](SbCl_6)_2 \cdot CH_2Cl_3OP$. According to the IR spectra the ν(PO) vibration is shifted in the complexes to lower energy, indicating that the ligands are coordinated via the phosphoryl oxygen atom. The spectrum of the complex with ligand 1 also shows the presence of an uncoordinated ligand molecule. Analysis of the X-ray powder diagrams show that the complexes are isostructural with the corresponding Ni^{II} compounds. The ligand-field spectrum of $[Ni(C_2H_6Cl_2NOP)_6](SbCl_6)_2 \cdot C_2H_6Cl_2NOP$ is typical for octahedral symmetry of the cation. A high-spin trigonal-bipyramidal structure might be possible for the complex with ligand 2 [1].

$[Mn^{II}(C_8H_{16}ClN_2O_3P)_2]X_2$ ($X = NO_3$, ClO_4). The complexes were prepared by mixing the solutions of the hydrated Mn^{II} salt (0.8 mmol) and of ligand 3 (1.6 mmol), each dissolved in 10 mL of anhydrous methanol containing 2 mL of 2,2-dimethoxypropane. The mixture was placed in a desiccator over $CaCl_2$ for 12 h and then evaporated completely in a vacuum line. The solid formed was washed with small portions of a mixture of chloroform and 2,2-dimethoxypropane then dried in vacuum over $CaCl_2$. The hygroscopic, pale rose complexes melt between 96 and 101.8°C ($X = NO_3$) or 98 and 101°C ($X = ClO_4$) [2].

The IR spectra of the complexes in Nujol show a ν(PO) and a ν(PN) band, both shifted to lower wavenumbers, and three bands each for both the NO_3 and ClO_4 anion groups, indicating their ionic uncoordinated nature. Coordination of the ligand molecule via the phosphoryl oxygen and the morpholine nitrogen atoms is proposed, probably in a polymeric structure containing bridging ligand molecules. $[Mn(C_8H_{16}ClN_2O_3P)_2](NO_3)_2$ has an X-ray powder diffraction pattern similar to that of the corresponding Zn^{II} and Co^{II} compounds. From its electronic spectrum Co^{II} is proposed to be pseudo-tetrahedrally surrounded. $[Mn(C_8H_{16}ClN_2O_3P)_2](NO_3)_2$ and $[Mn(C_8H_{16}ClN_2O_3P)_2](ClO_4)_2$ are slightly hygroscopic; they are insoluble in organic solvents but soluble in acetonitrile. The electrical conductivities in 10^{-3} M acetonitrile solution, $\Lambda=187$ and 287 $cm^2 \cdot \Omega^{-1} \cdot mol^{-1}$ for the nitrate and the perchlorate complexes, were interpreted in terms of an electrolyte intermediate between 1:1 and 1:2 and a 1:2 electrolyte, respectively [2].

References:

[1] de Bolster, M. W. G.; den Heijer, M.; Groeneveld, W. L. (J. Inorg. Nucl. Chem. **39** [1977] 1709/10).
[2] Dunstan, P. O.; Maieru, C. (Anais Acad. Brasil. Cienc. **50** [1978] 479/86).

39.25 Complexes with Phosphinothioic or Phosphinodithioic Acids

39.25.1 With Dialkyl- or Diphenylphosphinothioic Acids

$R_2P(S)OH \rightleftharpoons R_2P(O)SH$ (=HL)

ligands 1 to 3 1) $R=CH_3$ ($=C_2H_7OPS$)
2) $R=C_2H_5$ ($=C_4H_{11}OPS$)
3) $R=C_6H_5$ ($=C_{12}H_{11}OPS$)

$Mn^{II}(C_2H_6OPS)_2$ was prepared by reaction of manganese(II) chloride with the sodium salt of dimethylphosphinothioic acid in aqueous solution (details not given). The compound melts at 218 to 219°C. The IR spectrum of the complex (KBr pellets) has been investigated in the 2000 to 400 cm^{-1} region. The absorption bands of the PO vibrations at 1132, 1042, and 1024 cm^{-1} are much smaller than the ν(PO) band at 1184 cm^{-1} of the thiolic standard $(C_2H_5)_2P(O)SCH_3$. A strong vibration at 536 cm^{-1}, assigned to ν(PS), is shifted 45 cm^{-1} to lower frequencies with respect to the thionic standard $(CH_3)_2P(S)OCH_3$. A chelate structure is assumed, in which the thionic form of the ligand is predominant [1].

The polymeric complexes **$[Mn^{II}(C_4H_{10}OPS)_2]_n$** and **$[Mn^{II}(C_{12}H_{10}OPS)_2]_n$** were prepared by heating a mixture of solid manganese(II) acetate and the corresponding acid (mole ratio 1:2) to the melting points of the reactants or a little higher. The reaction mixture was dissolved in ethanol or benzene. Insoluble polymers were separated by filtration, then the soluble fraction by evaporation of the filtrate. Cryoscopic measurements in benzene reveal molecular weights of 3150 for $[Mn(C_4H_{10}OPS)_2]_n$ and 5290 for $[Mn(C_{12}H_{10}OPS)_2]_n$, indicating a degree of polymerization of 9 and 10, respectively. As shown by thermogravimetric analysis, the compounds undergo thermal decomposition in two stages. $[Mn(C_4H_{10}OPS)_2]_n$ decomposes between 250 and 545°C, and $[Mn(C_{12}H_{10}OPS)_2]_n$ between 385 and 700°C. The IR spectrum of the last complex is given in the paper [2]. A band at 780 cm^{-1} was assigned to ν(PS), and the doublet around 1050 and 1100 cm^{-1} to ν(PO). The ESR spectra give parameters of g=2.0174 for $[Mn(C_4H_{10}OPS)_2]_n$ and g=2.0215 for $[Mn(C_{12}H_{10}OPS)_2]_n$, consistent with a paramagnetic tetrahedral structure of the polymeric unit.

The coordination polymers are very slightly soluble in common organic solvents [2].

References:

[1] Kabachnik, M. I.; Mastryukova, T. A.; Matrosov, E. I.; Fisher, B. (Zh. Strukt. Khim. **6** [1965] 691/8; J. Struct. Chem. [USSR] **6** [1965] 657/63).
[2] Roşca, I. (Bul. Inst. Politeh. Iaşi [2] **17** [1971] 13/25).

39.25.2 With Dialkyl- or Diarylphosphinodithioic Acids $R_2P(S)SH$ (=HL)

ligand	R	formula	ligand	R	formula
1	CH_3	$C_2H_7PS_2$	6	4-$CH_3C_6H_4$	$C_{14}H_{15}PS_2$
2	C_2H_5	$C_4H_{11}PS_2$	7	4-$C_2H_5C_6H_4$	$C_{16}H_{19}PS_2$
3	C_3H_7	$C_6H_{15}PS_2$	8	3,4-$Cl_2C_6H_3$	$C_{12}H_7Cl_4PS_2$
4	CF_3	$C_2HF_6PS_2$	9	2-CH_3, 5-ClC_6H_3	$C_{14}H_{13}Cl_2PS_2$
5	C_6H_5	$C_{12}H_{11}PS_2$			

$Mn^{II}(R_2PS_2)_2$ compounds with ligand 1 or 5, $Mn(C_2H_6PS_2)_2$ or $Mn(C_{12}H_{10}PS_2)_2$, were prepared by mixing a concentrated aqueous solution of manganese chloride with the sodium salt of the ligand. The precipitates were washed with ethanol or 2-propanol, then with ether, and finally purified by fractional sublimation under vacuum [1]. $Mn(C_{12}H_{10}PS_2)_2$ was also obtained by reaction of $MnCl_2 \cdot H_2O$ with the sodium salt of ligand 5 in ethanolic solution. The mixture was kept over silica gel for 2 d and the precipitated NaCl was filtered off. The compound, precipitating after evaporation of the solvent on a steam bath, was washed with a small amount of ethanol and dried [2]. The complexes with ligands 7 to 9, $Mn(C_{16}H_{18}PS_2)_2$, $Mn(C_{12}H_6Cl_4PS_2)_2$, and $Mn(C_{14}H_{12}Cl_2PS_2)_2$, respectively, were obtained from $MnCl_2 \cdot 6H_2O$ (0.9 g) and the corresponding ligands (3.0 g) in methanol. The resulting precipitates were washed with methanol and dried under vacuum [3]. The volatile $Mn(C_2F_6PS_2)_2$ complex was prepared by a redox reaction of excess ligand 4 with powdered manganese in a sublimation apparatus with a cold water condenser. After completion of the reaction, volatile components were removed, and the solid product sublimed under vacuum at 70°C. The complex is air-sensitive and was handled in a conventional vacuum system [1]. Dimeric complexes with ligand 2 or 3, $[Mn(C_4H_{10}PS_2)_2]_2$ or $[Mn(C_6H_{14}PS_2)_2]_2$, were obtained by reaction of $Mn_2(CO)_{10}$ with the appropriate amount of the corresponding disulfane, $R_2P(S)S_2P(S)R_2$ ($R=C_2H_5$ or C_3H_7), in 1,2,4-trimethylbenzene at 125°C for 1.5 h. After filtration at 80°C, nearly colorless crystals separated from the solution; these were recrystallized from toluene. The complexes were also obtained by reacting the disulfane with powdered manganese; however, this reaction required a much longer reaction time (15 h) and gave considerably lower yields [4]. The coordination polymers, $[Mn(C_4H_{10}PS_2)_2]_n$ and $[Mn(C_{12}H_{10}PS_2)_2]_n$, were reported to form on heating solid manganese(II) acetate with the corresponding ligand in the same way as the complexes of diethyl- and diphenylphosphinothioic acids (p. 188) [5].

X-ray crystal data are given for the needle-shaped colorless crystals of $[Mn(C_4H_{10}PS_2)_2]_2$. The complex crystallizes in the orthorhombic space group Pbca-D_{2h}^{15} (No. 61) with the lattice constants a=20.164(4), b=13.837(3), and c=11.960(2) Å; Z=8. The structure was solved to R=0.05. Atomic coordinates for nonhydrogen atoms and equivalent isotropic displacement parameters are given in the original. As shown in **Fig. 40**, p. 190, the complex forms discrete dimers via ligand bridges with manganese(II) ions surrounded by four sulfur atoms in a tetragonal arrangement. The point symmetry of the dimer is $\bar{1}$. The puckered light-membered ring formed by the dimerization reveals a distorted chair conformation. The four atoms of the "back" (Mn, S(1), S(2), and P(1)) are planar within 0.11 Å, while the four atoms of the "seat" are exactly planar. The angle between the "back" and "seat" is 104.2°. The six S–Mn–S angles range from 81.7° to 131.6°; selected bond distances (in Å) are illustrated in Fig. 40. The nonbonding Mn–S(1) distance of 2.854(1) Å is remarkably short. The four-membered ring, Mn–S(3)–P(2)–S(4), appears to be planar within 0.002 Å, in contrast to many other metal complexes with dialkylphosphinodithioic acids. No evidence by calculation was obtained for an assumed disorder at the atoms S(4), C(6), and C(8). The calculated density is D_{calc}=1.439 g/cm³ [4].

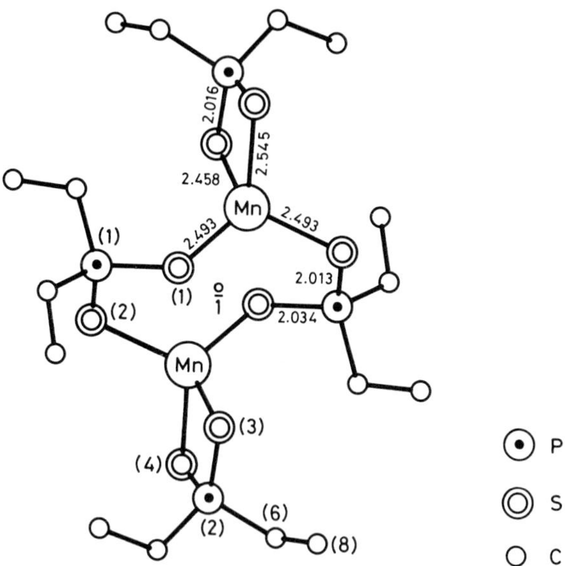

Fig. 40. Molecular structure of the $[Mn^{II}(C_4H_{10}PS_2)_2]_2$ complex with diethylphosphinodithioic acid [4]. H atoms are omitted for clarity.

The IR spectra (Nujol mulls) of the complexes with ligand 1, 4, or 5, investigated in the 1450 to 250 cm^{-1} range, exhibit the characteristic absorption bands of ν(PS) between 849 and 381 cm^{-1}, and ν(Mn–S) between 325 and 315 cm^{-1} [1]. The IR spectra of $[Mn(C_4H_{10}PS_2)_2]_n$ and $[Mn(C_{12}H_{10}PS_2)_2]_n$, taken in the 1350 to 400 cm^{-1} region, are given in the paper [5]. A band at 780 cm^{-1} was assigned to ν(P=S). Temperature-dependent magnetic susceptibility measurements in the 90 to 300 K range reveal Curie-Weiss behavior for $Mn(C_2H_6PS_2)_2$, $Mn(C_2F_6PS_2)_2$, and $Mn(C_{12}H_{10}PS_2)_2$ [1]. The Weiss constants (Θ in K) and magnetic moments, calculated by the Faraday method [1], are tabulated below, together with the magnetic moments (μ_{eff} in μ_B) at room temperature, obtained by the Gouy method [3, 4].

R	complex	μ_{eff}	Θ	Ref.
CH_3	$Mn(C_2H_6PS_2)_2$	5.86	−8	[1]
CF_3	$Mn(C_2F_6PS_2)_2$	5.86	−5	[1]
C_2H_5	$[Mn(C_4H_{10}PS_2)_2]_2$	5.85	—	[4]
C_3H_7	$[Mn(C_6H_{14}PS_2)_2]_2$	5.82	—	[4]
C_6H_5	$Mn(C_{12}H_{10}PS_2)_2$	5.92	−21	[1]

R	complex	μ_{eff}	Ref.
$4\text{-}C_2H_5C_6H_4$	$Mn(C_{16}H_{18}PS_2)_2$	5.81	[3]
$3,4\text{-}Cl_2C_6H_3$	$Mn(C_{12}H_6Cl_4PS_2)_2$	5.85	[3]
$2\text{-}CH_3, 5\text{-}ClC_6H_3$	$Mn(C_{14}H_{13}Cl_2PS_2)_2$	5.80	[3]

The ESR spectra of the powder samples for complexes with ligands 7 to 9 showed only a single line. When diluted with the analogous zinc matrix, six hyperfine lines (I = 5/2) were observed with g and A values in the ranges 2.02 to 2.03 and 80 to 90 Gauss, respectively [3].

The ESR spectra of $[Mn(C_4H_{10}PS_2)_2]_n$ and $[Mn(C_{12}H_{10}PS_2)_2]_n$ reveal g values of 2.0186 and 2.0233, respectively [5]. The comparison of the magnetic and spectral data of the complexes indicates that the substituent R does not markedly influence the properties of the paramagnetic chelate. Pseudotetrahedral coordination of manganese by four sulfur atoms, along with polymerization by ligand bridging in the solid state, is discussed [1, 5]. This is also indicated by sparing solubility in common solvents [1, 3, 5]. In the case of coordination polymers, cryoscopic measurements in benzene reveal molecular weights of 3670 for $[Mn(C_4H_{10}PS_2)_2]_n$ and 7790 for $[Mn(C_{12}H_{10}PS_2)_2]_n$, indicating degrees of polymerization of 10 and 14, respectively [5].

The electronic spectra in chloroform exhibit the following band maxima (in cm^{-1}): 21000 for $Mn(C_{16}H_{18}PS_2)_2$, 17540(sh) and 21200 for $Mn(C_{12}H_6Cl_4PS_2)_2$, and 21700 for $Mn(C_{14}H_{12}Cl_2PS_2)_2$ assigned to d–d transitions [3]. Mass spectral data were obtained for complexes with ligands 1, 4, and 5 at 200, 60, and 250°C, respectively [1]; electron impact-induced mass spectra for complexes with ligands 2 and 3 were reported in [4]. The highest intensity peak is given by the monomolecular parent ion of each complex, reflecting the thermal stability of the monomeric complexes and the breaking of the assumed polymeric structure under the conditions of the mass spectrometry [1, 3]. The complexes suffer substituent fragmentation, showing characteristic loss of CH_3, CF_2, or SC_6H_5 fragments for $Mn(C_2H_6PS_2)_2$, $Mn(C_2F_6PS_2)_2$, or $Mn(C_{12}H_{10}PS_2)_2$, respectively [1].

$[Mn(C_4H_{10}PS_2)_2]_2$ melts between 173 to 175°C, $[Mn(C_6H_{14}PS_2)_2]_2$ between 155 and 158°C. Both compounds decompose rapidly in contact with moist air, forming a brown precipitate. They are readily soluble in dichloromethane and less soluble in benzene or alcohols [4]. As shown by thermogravimetric analysis, $[Mn(C_4H_{10}PS_2)_2]_n$ decomposes between 210 and 535°C, $[Mn(C_{12}H_{10}PS_2)_2]_n$ between 320 and 695°C [5]. Only $Mn(C_{16}H_{18}PS_2)_2$ was found to be unstable at room temperature [3].

$Mn^{II}(R_2PS_2)_2\cdot$phen and **$Mn^{II}(R_2PS_2)_2\cdot$bpy** compounds with ligands 5 and 6, see p. 189, were prepared in methanolic solution by mixing manganese(II) chloride with a threefold excess of the corresponding ligand. The heterocyclic base 1,10-phenanthroline or 2,2′-bipyridine was added to the filtered reaction solution, the complexes (which precipitate slowly) were washed with methanol, then recrystallized from a chloroform-methanol mixture [6].

Magnetic moments (μ_{eff} in μ_B) at room temperature and absorption maxima in the electronic spectra of both the solid complexes and their chloroform solutions (ν_{max} in cm^{-1}) are tabulated below; molar extinction coefficients (in $L\cdot mol^{-1}\cdot cm^{-1}$) are given in parentheses:

R	complex	μ_{eff}	ν_{max}(solid)	ν_{max} in $CHCl_3$
C_6H_5	$Mn(C_{12}H_{10}PS_2)_2\cdot$phen	5.79	21300	37000(28780), 40500(25400)
4-$CH_3C_6H_4$	$Mn(C_{14}H_{14}PS_2)_2\cdot$phen	5.47	21300	36800(30790), 40800(27700)
C_6H_5	$Mn(C_{12}H_{10}PS_2)_2\cdot$bpy	5.88	21300	—
4-$CH_3C_6H_4$	$Mn(C_{14}H_{14}PS_2)_2\cdot$bpy	5.84	21300	33600(22220), 40800(50500)

A very weak band at 21300 cm^{-1} in the solid state electronic spectra is possibly associated with one of the spin-forbidden d–d transitions in an octahedral arrangement. Two well-defined solution absorption maxima in the range 34000 to 40000 cm^{-1} (bpy complexes reveal a number of shoulders on the main band) were assigned to intra-ligand π–π^* transitions. The magnetic moments in the 5.47 to 5.88 μ_B range are consistent with a high-spin octahedral configuration. The complexes are insoluble in water [6].

References:

[1] Cavell, R. G.; Day, E. D.; Byers, W.; Watkins, P. M. (Inorg. Chem. **11** [1972] 1759/72).
[2] Mukherjee, R. N.; Sonsale, A. Y.; Gupta, J. (Indian J. Chem. **4** [1966] 500/1).

[3] Mukherjee, R. N.; Venkateshan, M. S.; Vijaya, V. S.; Gogoi, P. K. (J. Indian Chem. Soc. **59** [1982] 170/2).
[4] Denger, C.; Keck, H.; Kuchen, W.; Mathow, J.; Wunderlich, H. (Inorg. Chim. Acta **132** [1987] 213/5).
[5] Roşca, I. (Bul. Inst. Politeh. Iaşi [2] **17** [1971] 13/25).
[6] Mukherjee, R. N.; Venkateshan, M. S.; Zingde, M. D. (Indian J. Chem. **13** [1975] 1341/4).

39.26 Complexes with Derivatives of Phosphonodithioic or Phosphorotrithious Acids

ligand 1 $C_2H_5(F)P(=S)SH$ $(=C_2H_6FPS_2=HL)$

ligand 2 $P(SC_6H_5)_3$ $(=C_{18}H_{15}PS_3)$

$Mn^{II}(C_2H_5FPS_2)_2$ was obtained by an exothermic reaction of ligand 1 with an excess of powdered metal under nitrogen at room temperature. The slightly pink, volatile complex, purified by sublimation at 110 to 120°C under reduced pressure (0.05 to 0.1 Torr), melts at 91°C [1]. The IR spectrum of the complex in Nujol is similar to the spectrum of the free phosphonodithioic acid, except for the strong ν(SH) band at 2485 cm^{-1} which was not observed [1, 2]. The absorption band at 312 cm^{-1} was assigned to ν(Mn–S) [1]. Susceptibility measurements yield a magnetic moment, $\mu_{eff}=5.44\ \mu_B$, indicating a high-spin octahedral complex [3]. The electronic reflectance spectrum shows strong bands at 4376, 5115, 5764, 6920, and 8584 cm^{-1} and a shoulder at 17860 cm^{-1}. A band at 17860 cm^{-1}, which may be assigned to the d–d transition $^6A_{1g} \rightarrow {}^4T_{1g}(G)$, is followed by one strong charge-transfer band [3]. The complex is soluble in acetone without decomposition [1].

$Mn(NO)_3(C_{18}H_{15}PS_3)$ was prepared by bubbling gaseous NO through a cyclohexane or benzene solution of $Mn(CO)_4P(SC_6H_5)_3I$ for 2 to 5 h at 80°C. The product was obtained as a green oil [4]. It was identified only by its IR spectrum. Two characteristic bands at 1792 and 1702 cm^{-1} were assigned to $\nu_s(NO)$ and $\nu_{as}(NO)$, respectively. The large difference in frequencies (90 cm^{-1}) and the intensity ratio of ~2:10 between symmetric and antisymmetric vibrations are consistent with an undisturbed C_{3v} tetrahedral symmetry. It is suggested that the complex contains coordinatively bonded NO^+, with significant metal-to-ligand back transfer (d–π*). A force constant of $f_{NO}=13.21$ mdyn/Å was determined in cyclohexane solution, and the influence of different ligands on the metal–nitrosyl bond discussed [5].

The compound is nonpolar and readily soluble in common organic solvents such as benzene and cyclohexane. It is a nonelectrolyte, rather thermically stable and slightly oxygen-sensitive [4].

References:

[1] Roesky, H. W. (Angew. Chem. **80** [1968] 844/5; Angew. Chem. Intern. Ed. Engl. **7** [1968] 815/6).
[2] Roesky, H. W. (Chem. Ber. **101** [1968] 3679/87).
[3] Roesky, H. W.; Dietl, M. (Z. Naturforsch. **28b** [1973] 707/10).
[4] Hieber, W.; Tengler, H. (Z. Anorg. Allgem. Chem. **318** [1962] 1236/54).
[5] Beck, W.; Lottes, K. (Chem. Ber. **98** [1965] 2657/73).

39.27 Complexes with Phosphorothioates, Phosphorodithioic Acid or Derivatives

39.27.1 With Phosphorothioates

ligand 1 $(C_2H_5O)_2P(S)OH \rightleftharpoons (C_2H_5O)_2P(O)SH = C_4H_{11}O_3PS$ (= HL)

ligand 2 $(CH_3O)_2P(O)S-CHC(O)OC_2H_5$ with $CH_2C(O)OC_2H_5$ substituent on CH (= Malaoxon = $C_{10}H_{19}O_7PS$)

$Mn^{II}(C_4H_{10}O_3PS)_2$ was obtained by reaction of manganese(II) chloride with the sodium salt of ligand 1 in aqueous solution. It melts at 135 to 137°C. The IR spectrum of the solid compound (KBr pellets) was studied in the 2000 to 400 cm^{-1} range. Absorption bands at 1166, 1145, and 1118 cm^{-1}, assigned to ν(PO) vibrations, are strongly shifted toward lower frequencies with respect to the P=O vibrational absorption at 1256 cm^{-1} of the thiolic standard $[(C_2H_5O)_2P(O)SC_2H_5]$. On the other hand, the absorption band at 618 cm^{-1}, assigned to ν(PS) vibrations, differs little in its position (−11 cm^{-1}) from the band of P=S vibration of the thionic standard $(C_2H_5O)_3P{=}S$. A polymeric chain structure with bridging oxygen and sulfur atoms in which the thionic character predominates is proposed [1].

$Mn^{II}Cl_2 \cdot C_{10}H_{19}O_7PS$. An ice-cold solution of ligand 2 (10 g) in 10 mL carbon tetrachloride was added slowly to an ice-cold solution of 3 g manganese(II) chloride in 12 mL carbon tetrachloride and the mixture was stored for 1 to 2 h at 0°C. The precipitate was filtered off in dry air and dried under vacuum. The IR spectra (KBr disks) of the complex and corresponding complexes of Cu^{II}, Zn^{II}, and Fe^{III} indicate coordination of malaoxon to manganese through the phosphorus oxygen: A shift of the ν(PO) ligand absorption at 1170 cm^{-1} to lower frequencies (−18 to −25 cm^{-1}) was observed, but only a slight shift of ν(P–O–C) (+2 to +5 cm^{-1}), and no shift of the strong ν(CO) band at 1720 cm^{-1} [2].

References:

[1] Kabachnik, M. I.; Mastryukova, T. A.; Matrosov, E. I.; Fisher, B. (Zh. Strukt. Khim. **6** [1965] 691/8; J. Struct. Chem. [USSR] **6** [1965] 657/63).
[2] Chopra, S. L.; Arora, C. L. (Pestic. Sci. **5** [1974] 271/4).

39.27.2 With Phosphorodithioic Acid or Its O,O′-Dialkyl Esters $(RO)_2P(S)SH$ (= HL)

ligand 1; R = H (= $H_3O_2PS_2$)
ligand 2; R = C_2H_5 (= $C_4H_{11}O_2PS_2$)
ligand 3; R = i-C_3H_7 (= $C_6H_{15}O_2PS_2$)
ligand 4; R = i-C_6H_{13} (= $C_{12}H_{27}O_2PS_2$)
ligand 5; R = $CH_3(CH_2)_3CH(C_2H_5)CH_2$ (= $C_{16}H_{35}O_2PS_2$)

General Remarks. Phosphorodithioic acid chelating through the sulfur donor atoms forms complexes that do not polymerize to such an extent as many phosphinodithioates. The MnL_2 complexes with ligands 1 to 5 are prepared by reaction of manganese halides or acetates with phosphorodithioic acids or their sodium salts. The complexes are generally purified by extraction with hydrocarbons or ether, and evaporation of the solvent. These compounds have found wide application in chemistry and chemical technology (pesticides, motor oil, war gases) and are used in a wide variety of analytical methods for metals (solvent extraction, flotation reagents for the recovery of metals from their solutions) [1, 2]. Manganese phosphorodithioates are also of importance from a theoretical point of view, since they may serve as models for some important biological systems [3].

MnL_2 Compounds. Quantum-chemical calculations of the bonding and the electronic structure have been made for the complex with ligand 1, $Mn(H_2O_2PS_2)_2$, and for a mixed ligand complex with dimethyldithiocarbamic acid, $Mn(H_2O_2PS_2)_2(C_3H_6NS_2)$. The calculations have been carried out by the MO-LCAO-SCR method with the CNDO/2 semiempirical approximation. Only a low-spin state and a planar configuration of the sulfur atoms in the coordination sphere of the metal were adopted. Numerical data demonstrate a high localization of the molecular orbitals in the coordination compound, $Mn(H_2O_2PS_2)_2$ [4]. The absolute values of the covalent components of the metal-donor bond energies constitute only about 10% of the total bond energy [5]. It has been shown that the covalent component is mainly due to the donor-acceptor interaction of the 4s,p orbitals of the metal with the orbitals of sulfur, and that the transfer of electron density occurs from the ligand to the metal. The role of the π dative ($M \rightarrow L$) bond is insignificant. It was also concluded that a metal-phosphorus bonding interaction makes a significant contribution to stabilization of the chelate ring [3]. It has been found that the formation of a mixed ligand complex is accompanied by polarization of the metal-ligand bonds which is manifested as a strengthening of the covalent component of the Mn-phosphorodithioates bond energy and a weakening of the ionic component. Reverse tendencies, i.e., the enhancement of the ionic component, were observed for the Mn-dithiocarbamate bonds [6].

For the complex with ligand 3, $Mn(C_6H_{14}O_2PS_2)_2$, melting points of 90.5 to 91.5°C [7] and 83 to 84°C [8, 9] are reported. The IR spectrum of the complex (KBr pellets, investigated in the 500 to 1200 cm^{-1} region) is depicted in the paper [7]. Absorption bands (in cm^{-1}) were assigned as follows: doublet at 517 and 543 to ν(P-S), singlet at 650 to ν(P=S), and doublet at ~985 and ~1010 to ν(P-O-R).

The polarographic behavior of $Mn(C_4H_{10}O_2PS_2)_2$ was investigated in ethanolic solution at 25°C and 0.12M ionic strength ($LiNO_3$ or $NaClO_4$), using a dropping mercury electrode. A single-step reduction of Mn^{2+}, giving an irreversible polarographic wave, was observed. The half-wave potential of −1.257 V (vs. SCE) shifts very little to more negative values upon adding an increasing excess of the complexing agent $(C_2H_5O)_2P(S)SNH_4$, indicating that the formation constant of the chelate is very small [10]. The polarographic reduction in 2:3 (v/v) benzene-ethanol solution at 25°C and 0.2 M ionic strength ($LiClO_4$) was reported in [11]. An attempt to obtain ESR spectra of frozen toluene solutions at liquid helium temperatures was unsuccessful, and the authors tried to explain it by the formation of an atypical low-spin tetrahedral complex [12]. Very large zero-field splittings may have been responsible for the failure of this experiment [1, 13]. $Mn(C_6H_{14}O_2PS_2)_2$ exhibits polyfunctional activity when reacting with peroxides: It is an effective peroxide decomposer, an inhibitor of peroxide radicals, and an initiator of oxidation [8, 9].

The complex with ligand 5, $Mn(C_{16}H_{34}O_2PS_2)_2$, was studied by reverse-phase paper chromatography. The R_f values in acid medium (0.016 to 7.8M HCl) are between 0.83 and 0.98. The method was also used for the separation of Co^{II}-Mn^{II} pairs of ions in 7.8M HCl [14].

Mixed Ligand Compounds. For quantum-chemical calculations on $Mn(H_2O_2PS_2)_2(C_3H_6NS_2)$, see above.

$Mn(C_4H_{10}O_2PS_2)_2 \cdot phen$ was prepared by anodic oxidation in acetone solution (50 mL) containing 1,10-phenanthroline (0.26 g), 0.2 mL of ligand 2, and $(C_2H_5)_4NClO_4$ (ca. 25 mg). The yellow adduct, which precipitated as electrolysis proceeded, was washed with acetone and then with n-pentane, and dried in vacuum. It melts at 224°C [15]. Solvent extraction and electrooxidation of the complex, $Mn(C_6H_{14}O_2PS_2)_2 \cdot 3\,phen$, was investigated in chloroform in the pH range 3 to 6 [16].

References:

[1] Wasson, J. R.; Woltermann, G. M.; Stoklosa, H. J. (Fortschr. Chem. Forsch. **35** [1973] 65/129, 68, 69, 113).
[2] Ertelt, H. R.; Wythe, S. L.; Furey, M. J.; Appeldoorn, J. K. (U.S. 2976122 [1961]).
[3] Pilipenko, A. T.; Savranskii, L. I.; Zubenko, A. I. (Koord. Khim. **7** [1981] 25/33; Soviet J. Coord. Chem. **7** [1981] 6/12).
[4] Pilipenko, A. T.; Savranskii, L. I.; Zubenko, A. I.; Kobylyashnyi, V. P.; Rosenfeld, A. L. (Dokl. Akad. Nauk SSSR **243** [1978] 378/80; Dokl. Chem. Proc. Acad. Sci. USSR **238/243** [1978] 541/3).
[5] Pilipenko, A. T.; Savranskii, L. I.; Zubenko, A. I. (Koord. Khim. **7** [1981] 1613/21; Soviet J. Coord. Chem. **7** [1981] 804/12).
[6] Pilipenko, A. T.; Savranskii, L. I.; Zubenko, A. I. (Koord. Khim. **8** [1982] 897/902; Soviet J. Coord. Chem. **8** [1982] 486/91).
[7] Shopov, D.; Ivanov, S. K.; Kateva, I. ; Karshalykov, K. (Izv. Otd. Khim. Nauki Bulg. Akad. Nauk. **3** [1970] 33/40).
[8] Kateva, I.; Ivanov, S. K. (Izv. Khim. [Sofia] **19** [1986] 78/88).
[9] Ivanov, S. K.; Kateva, I. (Izv. Khim. [Sofia] **19** [1986] 88/93).
[10] Shetty, P. S.; Fernando, Q. (J. Inorg. Nucl. Chem. **29** [1967] 1921/30).

[11] Budnikov, G. K.; Shakurova, N. K.; Ulakhovich, N. A.; Cherkasov, R. A.; Ovchinnikov, G. A.; Kutyrev, G. A.; Toropova, V. F. (Zh. Analit. Khim. **32** [1977] 1326/32; J. Anal. Chem. [USSR] **32** [1977] 1052/8).
[12] Garif'yanov, N. S.; Kamenev, S. E.; Kozyrev, B. M.; Ovchinnikov, I. V. (Dokl. Akad. Nauk SSSR **177** [1976] 880/2; Dokl. Phys. Chem. Proc. Acad. Sci. USSR **172/177** [1976] 850/1).
[13] Brackett, G. C.; Richards, P. L.; Wickmann, H. H. (Chem. Phys. Letters **6** [1970] 75/84).
[14] Lin, Chenke; Wu, Geng (Zhongnan Kuangye Xueyuan Xuebao **2** [1986] 83/6).
[15] Geloso, C.; Kumar, R.; Lopez-Grado, J. R.; Tuck, D. G. (Can. J. Chem. **65** [1987] 928/32).
[16] Ulakhovich, N. A.; Postnova, I. B.; Budnikov, G. K. (Zh. Analit. Khim. **40** [1985] 1502/7; J. Anal. Chem. [USSR] **40** [1985] 1182/6).

39.27.2 With Phosphorodifluoridodithioic Acid

$F_2P(S)SH$ ($=F_2HPS_2=HL$)

$Mn^{II}(F_2PS_2)_2$ was prepared under water- and oxygen-free conditions by addition of phosphorodifluoridodithioic acid (0.11 mol) to a suspension of manganese metal powder (0.27 g-atom) in 25 mL of toluene. After the initial exothermic reaction subsided, the stirred mixture was kept at 100°C for 1 h, then toluene was removed under vacuum. When larger quantities of reactants were employed, the materials were combined at −80°C and allowed to warm slowly to reaction temperature. The pale pink product sublimed at 120 to 140°C. Two further sublimations produced a sample which melted at 168 to 170°C with decomposition [1, 2]. Susceptibility measurements at temperatures in the 90 to 300 K range show that the complex obeys the Curie-Weiss law with the constant $\Theta=-38$ K and the magnetic moment $\mu_{eff}=6.03$ μ_B [3]. This solid-state result agrees well with the solution moment of 5.9 μ_B, which was obtained from NMR reference-shift measurements in 1,2-dichloroethane at 303 K [2]. The complex appears to possess pseudotetrahedral coordination of the metal by four sulfur atoms [2, 3]. The pink color of the solid complex (typical for octahedral environments) and the high Weiss constant suggest a polymeric structure in the solid state. A possible mode of polymerization through sulfur bridging is shown in the paper [2]. Osmometric investigations in toluene, 1,2-

dichloroethane, and n-heptane, as well as the yellow color of solutions of the complex, are consistent with a monomeric tetrahedral structure in the solution [2].

The IR spectrum (Nujol mulls), investigated in the 1450 to 250 cm^{-1} region, exhibits characteristic absorption bands of ν(PF) at 917 and 887 cm^{-1}; ν(PS) at 720, 699, 685, 550, and 401 cm^{-1}; ν(Mn–S) at ~320 cm^{-1} [2, 3]. The He(I) photoelectron spectrum of the solid complex was investigated. Three regions of ionization energy for the valence electrons were observed (in order of increasing energy): 1) d-orbital ionization with a band at 9.38 eV; 2) σ and π ionization of ligand-based orbitals of the coordination sphere with bands at 10.08 (π_S), 11.20, 12.44 (δMS), 13.00, and 15.14 eV (δPS); 3) σ and π ionization of fluorine-based orbitals with bands at 16.22, 17.55 (π_F), and 19.00 eV (δPF) [4]. The mass spectral data obtained at 40°C show that the parent ion is the strongest one, with the characteristic loss of F and S in the metastable ion patterns [3].

The complex is slightly volatile, insoluble in saturated hydrocarbons, and of limited solubility in 1,2-dichloroethane, chloroform, carbon tetrachloride, and aromatic hydrocarbons. It is inert to dry air in the solid state, but in solution reacts readily, and irreversibly, with oxygen undergoing a color change from yellow to deep red [1, 2].

References:

[1] Tebbe, F. N.; Roesky, H. W.; Rade, W. C.; Muetterties, E. L. (J. Am. Chem. Soc. **90** [1968] 3578/9).
[2] Tebbe, F. N.; Muetterties, E. L. (Inorg. Chem. **9** [1970] 629/37).
[3] Cavell, R. G.; Day, E. D.; Byers, W.; Watkins, P. M. (Inorg. Chem. **11** [1972] 1759/72).
[4] Andreocci, M. V.; Dragoni, P.; Flamini, A.; Furlani, C. (Inorg. Chem. **17** [1978] 291/4).

39.28 Complexes with Thiodiphosphoric Acid or Its Diethyl Ester

ligand 1 $HO-P(=O)(HO)-O-P(=O)(OH)-SH \rightleftharpoons HO-P(=O)(HO)-S-P(=O)(OH)-OH$ $(=H_4O_6P_2S=H_4L)$

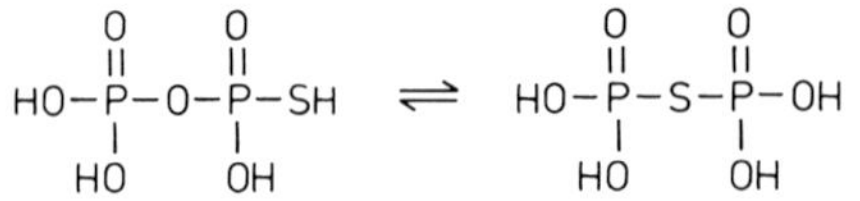

ligand 2 $C_2H_5O-P(=O)(HO)-S-P(=O)(OH)-OC_2H_5$ $(=C_4H_{12}O_6P_2S=H_2L)$

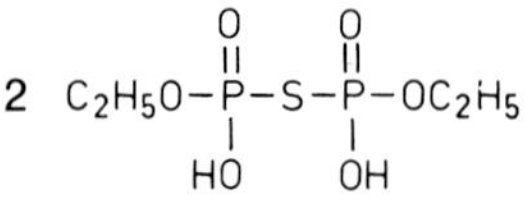

$[Mn^{II}_2(O_6P_2S)(H_2O)_6]$ and **$[Mn^{II}(C_4H_{10}O_6P_2S)(H_2O)]$**. The hexaaqua complex with ligand 1 was prepared by mixing a suspension of the anhydrous or hydrated manganese halide with an excess of the thiophosphate ester, $(C_4H_9O)_3PS$, at room temperature under N_2. An initially formed adduct was dissolved at ca. 100°C and the resulting solution was heated at 180 to 270°C for several minutes. Precipitation from the reaction mixture was effected by addition of an excess of ligroin (b.p. 63 to 75°C) after cooling to room temperature. The precipitate was washed with acetone and anhydrous diethyl ether and stored in a vacuum desiccator over $CaCl_2$. The aqua complex with ligand 2, $[Mn(C_4H_{10}O_6P_2S)(H_2O)]$, was prepared from anhydrous manganese halides with an excess of $(C_2H_5O)_3PS$ in the same way as $[Mn_2(O_6P_2S)(H_2O)_6]$, except the treatment with ligroin was not necessary for precipitation.

The manganese(II) complexes and the Zn^{II} analogs exhibit X-ray patterns similar to those of the corresponding Co^{II} and Ni^{II} compounds, for which a pentacoordinated configuration was suggested by their electronic spectra. Susceptibility measurements at room temperature yielded the magnetic moment μ_{eff} = 5.82 to 5.92 μ_B, indicative of high-spin $Mn^{II}(d^5)$ complexes. A dinuclear structure is assumed for $[Mn_2(O_6P_2S)(H_2O)_6]$ with pentacoordinated manganese atoms whereas $[Mn(C_4H_{10}O_6P_2S)(H_2O)]$ seems to be polymeric.

The IR spectra (Nujol mulls) in the 4000 to 700 cm^{-1} region show characteristic absorption bands in the same range as the bands of Ni^{II}, Co^{II} or Zn thiodiphosphate complexes. Absorptions found at 3360 and 1632 cm^{-1} were assigned to the ν(OH) and δ(HOH) vibration modes of the water molecules, respectively.The strong, sharp band between 1062 and 1026 cm^{-1} in the spectrum of $[Mn_2(O_6P_2S)(H_2O)_6]$ was assigned to $\nu_s(PO_3)$. For the complex with ligand 2, $[Mn(C_4H_{10}O_6PS_2)(H_2O)]$, the $\nu(PO_2)$ bands are observed at 1250 to 1200 and 1100 to 1050 cm^{-1}. Both complexes exhibit an absorption between 550 and 480 cm^{-1}, assigned to ν(P–S–P).

The complexes were not dehydrated, even after prolonged attempts at desiccation over a number of powerful drying agents. They are insoluble in most common organic solvents and of limited solubility in methanol and water.

Reference:

Mikulski, C. M.; Pytlewski, L. L.; Karayannis, N. M. (Inorg. Chem. **14** [1975] 1559/62).

39.29 Complexes with a Phosphinothioic Amide or Phosphoramidothioic Acids

ligand 1 $(C_6H_5)_2$–P(S)–NH–P(S)–$(C_6H_5)_2$ (= $C_{24}H_{21}NP_2S_2$ = HL)

ligand 2 O=P(NH$_2$)(OH)–SH (= H_4NO_2PS = H_2L)

ligand 3 O=P(NH$_2$)(NH$_2$)–SH (= H_5N_2OPS = HL)

$Mn^{II}(C_{24}H_{20}NP_2S_2)_2$ was prepared from manganese acetate and the sodium salt of ligand 1 in aqueous solution. Slow evaporation of a 1:10 hexane-dichloromethane solution produced single crystals suitable for X-ray study. Pink crystals in the form of rectangular prisms were formed. The air-stable compound melts at 288 to 289°C [1]. X-ray diffraction studies of single crystals reveal that the complex crystallizes in the triclinic space group P1–C_1^1 (No. 1) or P$\bar{1}$–C_i^1 (No. 2) with the lattice constants a = 13.550(4), b = 14.334(4), c = 13.824(3) Å, β = 110.50(2)° and Z = 2 for the Dirichlet-reduced cell. The calculated density, D_c = 1.377 g/cm^3, agrees well with the experimental value of 1.37(2) g/cm^3 measured by flotation in aqueous potassium iodide solutions at 23°C [1, 2]. Refinement of 8526 independent reflections produced a converged solution with R_f = 5.8%. Positional and thermal parameters of the nonhydrogen atoms with their standard deviations and positional parameters generated for hydrogen atoms are given in the paper [2]. The complex is monomeric. The four sulfur atoms are in an approximately tetrahedral arrangement with the manganese atom in the center, as shown in **Fig. 41**, p. 198. The two $Mn(C_{24}H_{20}NP_2S_2)$ chelate rings have twisted-boat configurations with sulfur and phosphorus atoms at the apices.

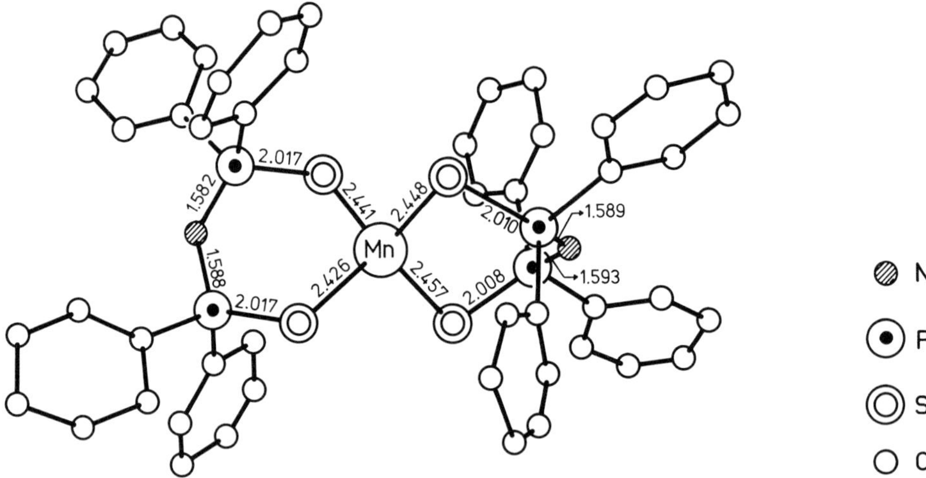

Fig. 41. Molecular structure of $Mn(C_{24}H_{20}NP_2S_2)_2$ [1, 2].

Average values of interatomic bond lengths and bond angles are tabulated below:

distances in Å		angles in °		angles in °	
Mn–S	2.443(12)	S–Mn–S	109.5(2.2)	S–P–C	107.9(0.9)
S–P	2.013(5)	Mn–S–P	99.9(3.4)	N–P–C	107.8(2.2)
P–N	1.588(6)	S–P–N	118.7(0.4)	C–P–C	106.1(1.3)
P–C	1.815(8)	P–N–P	133.5(1.8)		

The S–P and P–N bond lengths are consistent with a delocalized π-bond structure involving the five ligand atoms in the chelate rings. The S–P, P–N, and P–C distances, as well as the nonring average bond angles about the phosphorus atoms [S–P–C, N–P–C, and C–P–C], are in agreement with corresponding values for complexes of Ni^{II} and Fe^{II} with $[(CH_3)_2PS]_2NH$. The packing is largely determined by intermolecular contacts among the phenyl groups. The closest carbon–carbon contact of 3.555 Å compares well with the shortest C···C distance (3.5 Å) in the Ni^{II} and Fe^{II} complexes mentioned above [2]. Infrared (KBr pellets) and Raman spectra of the solid compound and the ligand 1 were investigated in the 1700 to 200 cm^{-1} range [3, 4]. Characteristic absorption bands (in cm^{-1}) with their assignments are given below:

compound	method	ν_{as}(PNP)	ν(PC)	ν_{as}(PS)	ν_s(PS)	ν_s(Mn–S)	ν_{as}(Mn–S)
$Mn(C_{24}H_{20}NP_2S_2)_2$	IR	1218	715, 700, 692	568	585	—	275, 285
	R	—	718, 705	567	580	245	275, 285
$C_{24}H_{21}NP_2S_2$ (=HL)	IR	926, 920, 935	720, 715, 688	648	610	—	—
	R	—	720, 713	646	616, 628	—	—

Compared with the free ligand, the broad strong ν_{as} (PNP) band is shifted by ca. 300 cm^{-1} to higher wavenumbers on complex formation, which indicates coordination of Mn^{II} through nitrogen. The lack of IR and Raman intensity in the symmetric P$\overset{...}{-}$N$\overset{...}{-}$P stretching band is explained in terms of opposing σ and π contributions to the dipole moment and polarizability changes during the symmetric vibrations. The strong low-frequency Raman band observed at ~245 cm^{-1} [2, 3] was assigned to ν_s(Mn–S) on the basis of vibrational fine structure in the low-temperature single crystal electronic absorption spectrum, reported in [2]. Normal coordinate analysis have been performed for a 1:1 metal–ligand model possessing C_2 symmetry. Results of the calculations support the assignments given above. Bond-stretching force constants, K(Mn–S) and K(P–S), are 0.45 and 3.70 mdyn/Å, respectively. The values are compared to those of a similar Co^{II} compound [4].

The solid-state magnetic moment, $\mu_{eff}=5.75\ \mu_B$, is consistent with a high-spin $d^5(^6A_1)$ electronic ground state [1]. The single crystal electronic absorption spectra of the tetrahedral complex have been measured between 13000 and 30000 cm^{-1} at 77 and 5 K [2]. Transitions from the $^6A_{1g}$ ground state to quartet-excited states are observed. Bands observed in the single crystal spectrum at 5 K, with their assignments, are shown below (in cm^{-1}):

$^6A_{1g}\rightarrow{}^4T_1(^4G)$	$\rightarrow{}^4T_2(^4G)$	$\rightarrow{}^4E(^4G)$	$\rightarrow{}^4A_1(^4G)$	$\rightarrow{}^4T_2(^4D)$	$\rightarrow{}^4E(^4D)$	$\rightarrow{}^4T_1(^4P)$
18520	20490	21139	21346	23700	25477	28011

The 77 K electronic excitation and emission spectra of a polycrystalline sample of the complex were reported in [1]. The excitation spectrum agrees well with the absorption spectra reported in [2]. Although the absorption maxima in dichloromethane solution and in a pressed pellet are shifted from their positions in the crystal spectra, the relative intensities of the bands do not change. The experimental transition energies are in excellent agreement with those calculated assuming 10 Dq = −4685.6, B = 559.2, and C = 3118.7 cm^{-1} [2].

$[Mn^{II}(H_2NO_2PS)(H_2O)_4]$ and **$[Mn^{II}(H_4N_2OPS)_2(H_2O)_2]$**. The ammonium salt of ligand 2 (2.2 mmol) or ligand 3 (4.4 mmol) in 20 mL water was added slowly to a solution of 2 mmol of hydrated manganese chloride in 20 mL of water, and stirred at room temperature for 2 h. The precipitate was filtered, successively washed with water, ethanol, and diethyl ether, and dried in vacuum above P_4O_{10} for 2 d. Significant absorption bands (in cm^{-1}) observed in the IR spectra of the complexes (KBr disks, 4000 to 250 cm^{-1} region) are shown below together with their assignments [5]:

complex	$\nu(H_2O)$	$\nu(NH_2)$	$\delta(H_2O)$	ν(PO)	ν(PS)	ν(Mn–O)	ν(Mn–S)
$[Mn(H_2NO_2PS)(H_2O)_4]$	3420 br	3180 s	1630 s	1130 m, 1050 s	520 m	432 m	350 s
$[Mn(H_4N_2OPS)_2(H_2O)_2]$	3410 br	3180 m	1650 m	1030 s	550 s	440 m	350 s

The assignments were based on the data reported for salts of phosphorodiamidic acid. The absorption band at 1650 to 1630, assigned to $\delta(H_2O)$, is absent in the spectra of anhydrous phosphorodiamidate complexes. Coordination to Mn^{II} through the sulfur and the oxygen atoms is indicated by the bands at 440 to 432 and 350 cm^{-1}, respectively. This is in agreement with quantum chemical calculations of the bonding capacities of the possible donor atoms sulfur, oxygen, or nitrogen.

Susceptibility measurements at room temperature yield the magnetic moments, $\mu_{eff}=5.89\ \mu_B$, for $[Mn(H_2NO_2PS)(H_2O)_4]$ and $\mu_{eff}=5.90\ \mu_B$ for $[Mn(H_4N_2OPS)_2(H_2O)_2]$. The high-spin character of the complexes as well as their IR and solid state electronic spectra (Nujol mulls) are consistent with an octahedral structure. The assignments of the electronic spectral bands (in cm^{-1}) are tabulated on p. 200 [5]:

assignment, ${}^6A_{1g}\rightarrow$	${}^4T_1(G)$	${}^4T_2(G)$	${}^4A_1(G)$, ${}^4E(G)$	${}^4T_2(D)$	${}^4E(D)$	other bands
$[Mn(H_2NO_2PS)(H_2O)_4]$	18500	23000	25800	28000	32000	44400
$[Mn(H_4N_2OPS)_2(H_2O)_2]$	18300	22600	25000	27500	31600	44000

The hydrate compounds are amorphous solids, air-stable, and nonconducting. They are insoluble in water and in many common organic solvents, such as chloroform, acetone, methylene chloride, ethanol, and diethyl ether [5].

References:

[1] Siiman, O.; Wrighton, M.; Gray, H. B. (J. Coord. Chem. **2** [1972] 159/61).
[2] Siiman, O.; Gray, H. B. (Inorg. Chem. **13** [1974] 1185/99).
[3] Siiman, O.; Vetuskey, J. (Inorg. Chem. **19** [1980] 1672/80).
[4] Siiman, O. (Inorg. Chem. **20** [1981] 2285/92).
[5] Syngollitou-Kourakou, A. S.; Tossidis, I. A.; Sigalas, M. P. (Inorg. Chim. Acta **118** [1986] 141/5).

39.30 Complexes with Esters of Phosphorodihydrazidothioic Acid or a Derivative of Phosphinothioic Amide

ligand 1 $(NH_2NH)_2P(S)OCH_3$ $(=CH_9N_4OPS)$
ligand 2 $(NH_2NH)_2P(S)OC_6H_5$ $(=C_6H_{11}N_4OPS)$
ligand 3 $(C_6H_5)_2P(S)$–NH–$C(S)N(CH_3)_2$ $(=C_{15}H_{17}N_2PS_2=HL)$

$[Mn^{II}(CH_9N_4OPS)_2Cl_2]$ and $[Mn^{II}(C_6H_{11}N_4OPS)_2Cl_2]$. The complex with ligand 1, $[Mn(CH_9N_4OPS)_2Cl_2]$, was prepared by refluxing the complex with ligand 2, $[Mn(C_6H_{11}N_4OPS)_2Cl_2]$, in absolute methanol for several hours. Colorless microcrystals formed on standing for several weeks, or by evaporation of the methanolic solution. $[Mn(C_6H_{11}N_4OPS)_2Cl_2]$ was prepared by adding a solution of $MnCl_2\cdot4H_2O$ (25 mmol) in 50 mL absolute methanol to a solution of ligand 2 (50 mmol) in 50 mL absolute methanol at 65°C. A colorless microcrystalline precipitate, which appeared gradually after slow cooling to room temperature, was filtered off, washed with methanol, and dried in a desiccator over P_4O_{10}. It melts at 185 to 187°C with decomposition [1]. As shown in [2] for the corresponding Ni^{II} complex, $[Ni(C_6H_{11}N_4OPS)_2Cl_2]$, the tridentate chelating ligand is coordinated through the sulfur atom and the nitrogen atoms of the NH_2 groups. Clear assignments of ν(PS) and ν(Mn–S) could not be documented. The IR spectra (KBr pellets) of complexes with ligands 1 and 2, taken in the 4000 to 400 cm^{-1} range, are given in the paper [1]. Bands at 1590 and 1490 cm^{-1} and those above 3000 cm^{-1}, corresponding to aromatic absorptions in the spectrum of the parent phenylester compound, are absent in the spectrum of $[Mn(CH_9N_4OPS)_2Cl_2]$. New bands appear between 2800 and 2960 cm^{-1}, which were assigned to ν(CH) vibration modes, and those between 1400 and 1450 cm^{-1} to δ(CH) modes of the methyl group, respectively [1, 2]. The Raman spectrum of $[Mn(C_6H_{11}N_4OPS_2)Cl_2]$ shows characteristic absorption bands (in cm^{-1}): ν(NH) and ν(NH_2) at 3282(m), 3241(s), 3155(s), 2944(m), and 2840(m); δ(NH_2) at 1594(vs); ν(PO) at 1217(s) and 1171(m); ν(ring) at 1045(m), 1031(s), and 1005(vs) [1].

Solvolysis of $[Mn(CH_9N_4OPS)_2Cl_2]$ in methanolic solution containing water results in slow formation of a complex, probably $[Mn\{(NH_2NH)P(O)S\}_2]$. However, this colorless microcrystalline product could not be obtained clean and homogeneous. The complex with ligand 2 is easily soluble in water, methanol, and ethanol. It is insoluble in dioxane, tetrahydrofuran, diethyl ether, and trichloromethane [1].

$Mn^{II}(C_{15}H_{16}N_2PS_2)_2$ was prepared by treating an aqueous solution (100 mL) of the potassium salt of ligand 3 (1 g) with the stoichiometric amount of $MnSO_4$. The flaky voluminous precipitate was dissolved by shaking the suspension with dichloromethane. After the organic phase was separated and dried, petroleum ether was added to induce crystallization. The pale cream-colored complex melts at 186 to 189°C. Osmometric determination of molecular weight in chloroform indicates a monomeric structure, somewhat high values may be due to an association equilibrium.

The IR spectra of the complex in Nujol or fluorinated hydrocarbon mulls and corresponding complexes of Ni^{II}, Zn^{II}, Cd^{II}, Hg^{II}, Fe^{II}, Pb^{II}, and Cu^{II} have been investigated. No band of a phosphazene vibration mode ν(P=N) was observed. Strong bands in the region 1524 to 1508 cm^{-1} were assigned to ν(C=N). Characteristic bands of ν(P–N) were observed in the region of 986 to 976 and 884 to 874 cm^{-1}. Bands in the range of 719 to 712 (vs) and 639 to 628 (m) cm^{-1} are very close to the characteristic absorption of phosphine sulfide and were therefore assigned to ν(P=S). The IR spectra suggest a chelate with the predominant contribution from a phosphine sulfide-isothiourea structure (a), and only a small participation of a phosphazene form (b):

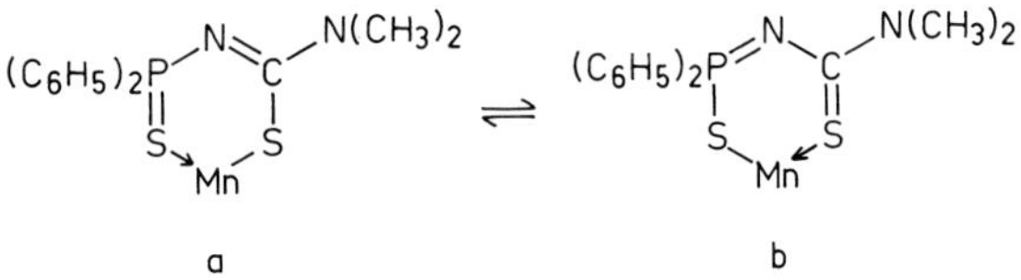

This assumption is in agreement with ^{31}P NMR spectra. A chemical shift of −41.6 ppm excludes coordination through the nitrogen atom and underscores that both the sulfur atoms are involved in coordinate bonding.

The complex is easily soluble in chloroform and dichloromethane and due to its hydrophobic character, stable in contact with cold dilute mineral acids [3].

References:

[1] Engelhardt, U.; Friedrich, B.; Kirner, I. (Z. Naturforsch. **36b** [1981] 791/6).
[2] Engelhardt, U.; Scherer, G. (Z. Naturforsch. **31b** [1976] 1553/61).
[3] Groeger, H.; Schmiedpeter, A. (Chem. Ber. **100** [1967] 3216/24).

40 Complexes with Ligands Containing Arsenic

Remark. For reasons of congruity complexes of manganese with Schiff Bases containing arsanyl groups have already been described in "Manganese" D 6, 1988, pp. 79/80 and 268/9. An azo compound containing arsonic acid is discussed in "Manganese" D 5, 1987, pp. 288/9. Arsonium tetrahalomanganates(II) were described in "Manganese C 5" 1978, pp. 209/10, 305, 326, and 342. Complexes of the $[MnL_n](AsF_6)_x$ or $[MnL_n(AsF_6)_2]$ type are described in the "Manganese" D series in the sections of the special ligand (L) type, e.g., the complex $[Mn(NSF_3)_4(AsF_6)_2]$ in "Manganese" D 7, 1990, p. 220.

40.1 Complexes with Arsanes or Related Compounds

40.1.1 With Arsane AsH_3

The formation of $\mathbf{[Mn^{VII}O_3(AsH_3)(OH)X]^-}$ species (X = Cl, Br, I) as intermediate complexes during the oxidation of AsH_3 by $KMnO_4$ in the presence of X^- ions in aqueous solution (in a flowing gas reactor under isothermal conditions) is indicated by kinetic and potentiometric studies. The kinetic activity of an inner-sphere redox decomposition of the complexes increases in the order: Cl < Br < I. The corresponding complexes with PH_3 are less active than those with AsH_3.

Reference:

Dorfman, Ya. A.; Polimbetova, G. S.; Mansurov, B. A.; Bikmukhametova, A. K.; Doroshkevich, D. M. (Koord. Khim. **14** [1988] 1219/23; C.A. **109** [1988] No. 238061).

40.1.2 With Triarylarsanes

$(C_6H_5)_3As$ ligand 1 (= $C_{18}H_{15}As$)

$(CH_3O-C_6H_4)_3As$ ligand 2 (= $C_{21}H_{21}AsO_3$)

Remark. Complexes of composition $[Mn(C_{18}H_{15}As)_4](I_3)_2$, $[Mn(C_{18}H_{15}As)_4]CdI_4$, and $[Mn(C_{18}H_{15}As)_2X_2]$ with X = Cl, Br, I, were reported to have been prepared by reaction of triphenylarsane with the corresponding manganese(II) salt in THF under "anhydrous" conditions [1]. However, it was shown that the corresponding complexes with arsane oxides were formed, owing to traces of water in the system [2, 3].

$\mathbf{Mn(NO)_2(C_{18}H_{15}As)_2I}$ was prepared by introducing dry NO gas through the solution of $Mn(CO)_3(C_{18}H_{15}As)_2I$ in benzene at 80°C for 8 h. The solution was filtered while hot and most of the solvent removed. A precipitate formed on addition of ethanol or pentane and was recrystallized from benzene-ethanol or benzene-petroleum ether. The reddish brown crystals were dried in high vacuum. Cryoscopic determinations of the molecular weight show that the complex is monomeric [4]. The IR spectrum of the complex in CCl_4 reveals two intense nitrosyl absorption bands at 1714 and 1670 cm^{-1}. A force constant of f_{NO} = 12.60 mdyn/Å was calculated [5]. The diamagnetic complex is soluble in benzene and behaves as a nonelectrolyte in acetone [4].

$Mn(NO)_3(C_{18}H_{15}As)$ and $Mn(NO)_3(C_{21}H_{21}AsO_3)$. The complex with ligand 1, $Mn(NO)_3(C_{18}H_{15}As)$, was prepared by passing dry NO gas through the cyclohexane solution of $Mn(CO)_4(C_{18}H_{15}As)I$ at 80°C for 2 h. The solution was filtered and the solvent removed to leave an oily green residue; this was precipitated by addition of n-hexane and the solid recrystallized from benzene-petroleum ether [4]. The dark green complex was also prepared by the reaction of $Mn(NO)_3 \cdot THF$ (by photo-induced nitrosylation of $Mn_2(CO)_{10}$ in THF) with the stoichiometric amount of ligand 1. After stirring overnight, the solvent was removed completely and the dark green residue was recrystallized from benzene or purified by chromatography with silica gel [6]. $Mn(NO)_3(C_{21}H_{21}AsO_3)$ was prepared by reacting $Mn(NO)_3CO$ (obtained from $Mn(CO)_5I$ and NO at 60°C) and ligand 2 in cyclohexane [5]. A kinetic study of the reaction $Mn(NO)_3CO + C_{18}H_{15}As \rightarrow Mn(NO)_3(C_{18}H_{15}As) + CO$ was made in p-xylene and it appears that the reaction follows a two-term rate law, depending on the concentration of ligand 1 [7]. The values for ν(NO) absorption bands (in cm^{-1}) and force constants (f_{NO} in mdyn/Å) for both complexes are given in the table:

complex	ν(NO)	f_{NO}	solvent	Ref.
$Mn(NO)_3(C_{18}H_{15}As)$	1787, 1694	13.10	c-hexane	[5]
	1781, 1686	—	THF	[6]
	1792, 1705	—	p-xylene	[7]
$Mn(NO)_3(C_{21}H_{21}As)$	1783, 1691	13.05	c-hexane	[5]
	1788, 1705	—	p-xylene	[7]

The IR spectra display two N–O absorption bands, indicating C_{3v} symmetry and therefore a tetrahedral environment around the manganese atom. Due to the position of the nitrosyl bands it is assumed that NO is coordinated as the NO^+ ion, with considerable metal-ligand back bonding (d-π*). The results show decreasing ν(NO) frequencies and force constants with increasing σ donor ability of the ligand [5].

$Mn(NO)_3(C_{18}H_{15}As)$ is soluble in common organic solvents like benzene or cyclohexane and almost insoluble in petroleum ether. It is air-stable and a nonelectrolyte [4].

References:

[1] Naldini, L. (Gazz. Chim. Ital. **90** [1960] 1337/42).
[2] Casey, S.; Levason, W.; McAuliffe, C. A. (J. Chem. Soc. Dalton Trans. **1974** 886/9).
[3] McAuliffe, C. A.; Levason, W. (Stud. Inorg. Chem. **1** [1979] 95).
[4] Hieber, W.; Tengler, H. (Z. Anorg. Allgem. Chem. **318** [1962] 136/54).
[5] Beck, W.; Lottes, K. (Chem. Ber. **98** [1965] 2657/73).
[6] Herberhold, M.; Razavi, A. (J. Organometal. Chem. **67** [1974] 81/6).
[7] Wawersik, H.; Basolo, F. (J. Am. Chem. Soc. **89** [1967] 4626/30).

40.1.3 With Diorganylarsanylbenzoic Acid

COOH
AsR2

(=HL)

ligand	R	formula
1	CH_3	$C_9H_{11}AsO_2$
2	C_2H_5	$C_{11}H_{15}AsO_2$
3	C_6H_5	$C_{19}H_{15}AsO_2$
4	$4\text{-}CH_3C_6H_4$	$C_{21}H_{19}AsO_2$

Light pink complexes, **$[Mn(C_{11}H_{14}AsO_2)_2(H_2O)_2]\cdot H_2O$** or **$[Mn(C_{19}H_{14}AsO_2)_2(H_2O)_2]$** with ligand 2 or 3, and a white complex with ligand 4, **$[Mn(C_{21}H_{18}AsO_2)_2(H_2O)_2]$**, were prepared by reaction of the appropriate ligand and $MnCl_2\cdot 4\,H_2O$ in ethanolic solutions (2:1 mole ratio). The compounds were filtered, washed with ethanol and ether, and dried in vacuum.

Melting points (in °C), magnetic moments (μ_{eff} in μ_B) at room temperature, and absorption bands in the IR spectra (in cm^{-1}) are given below:

complex	m.p.	μ_{eff}	$\nu(OH)$	$\delta(H_2O)$	$\nu_{as}(CO_2)$	$\nu_s(CO_2)$
$[Mn(C_{11}H_{14}AsO_2)_2(H_2O)_2]\cdot H_2O$	300	5.37	3500, 3410	1610	1587	1385
$[Mn(C_{19}H_{14}AsO_2)_2(H_2O)_2]$	144	5.98	3410	—	1590	1375
$[Mn(C_{21}H_{18}AsO_2)_2(H_2O)_2]$	175	5.82	3410	1620	1595	1395

The observed magnetic moments suggest the presence of a high-spin manganese(II) complex. An octahedral structure, where Mn^{II} is coordinated by the uninegative bidentate ligand via carboxylate oxygen and As^{III} atoms plus the coordinated water molecules, was suggested. The complexes with ligands 3 and 4 show an X-band EPR signal at room temperature that is a broad single line with $g\approx 2$, indicating nearly axial symmetry with small distortions from octahedral geometry.

$[Mn(C_{21}H_{18}AsO_2)_2(H_2O)_2]$ is soluble and a nonelectrolyte in nitrobenzene.

The white complex, **$Mn^{II}(C_9H_{10}AsO_2)Cl\cdot 0.5\,H_2O$**, was prepared analogously (?) to the foregoing compounds from $MnCl_2\cdot 4H_2O$ and ligand 1. It melts at 198°C. The low magnetic moment, $\mu_{eff}=5.40\ \mu_B$, is attributed to a polymeric structure, confirmed by a $\nu(Mn{-}Cl)$ absorption band at 280 cm^{-1} that is typical of bridged chloro complexes. Other bands in the IR spectrum are at 3420, 1615, 1590, and 1388 cm^{-1}, assigned to $\nu(OH)$, $\delta(H_2O)$, $\nu_{as}(CO_2)$, and $\nu_s(CO_2)$, respectively.

Reference:

Parmar, S. S.; Bharaj, H. K.; Saighal, M. L. (Polyhedron **6** [1987] 1699/701).

40.1.4 With Dimethylarsanylbenzenamine

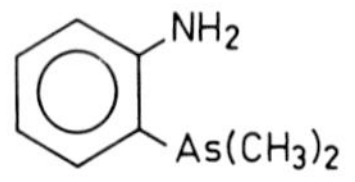

NH_2

$(=C_8H_{12}AsN)$

$As(CH_3)_2$

$[Mn^{II}(C_8H_{12}AsN)_2(ClO_4)_2]$ was obtained by addition of the ligand (1.4 mmol) in benzene to an ethanolic solution of hydrated manganese(II) perchlorate (1 mmol) under nitrogen. After removing about three quarters of the solvent by distillation, a precipitate of fibrous white crystals was formed. It was collected and washed with boiling benzene and boiling petroleum ether under nitrogen. A high-spin six-coordinate complex with coordinated perchlorate groups is proposed. The complex hydrolyzes quite rapidly in air.

$[Mn^{II}(C_8H_{12}AsN)Br_2]$ was prepared from anhydrous $MnBr_2$ and the ligand (1:2 mole ratio) in a procedure analogous to that described above for $Mn^{II}(C_8H_{12}AsN)_2(ClO_4)_2$. The magnetic moment of the complex at room temperature, $\mu_{eff}=5.78\ \mu_B$, is slightly less than the spin-only value (5.92 μ_B) of high-spin manganese(II) complexes. Thus a polymeric six-coordinate compound with a weak antiferromagnetic interaction of the bromide bridges was proposed.

Reference:

Chiswell, B.; Plowman, R. A.; Verrall, K. (Inorg. Chim. Acta **5** [1971] 579/89).

40.1.5 With Tris(2-pyridinyl)arsane

$(\text{C}_5\text{H}_4\text{N})_3\text{As}$ (= $C_{15}H_{12}AsN_3$)

[Mn^{II}($C_{15}H_{12}AsN_3$)$_2$](ClO_4)$_2$. Manganese(II) perchlorate hexahydrate was refluxed for several hours in absolute ethanol and 2,2-dimethoxypropane under nitrogen before adding the ligand (1:2 mole ratio) dissolved in 95% ethanol. Precipitation occurred within a few minutes. The reaction mixture was heated for 1 to 2 h and then cooled. The solid was filtered off and washed with ethanol. Heating in ethanol and 2,2-dimethoxypropane was repeated before the precipitate was collected and dried at 90 to 100°C in vacuum. The IR spectrum of the complex (in mineral oil or hexachlorobutadiene mulls) exhibits bands at 1575, 1550, 1455, and 1010 cm^{-1}, which were assigned to the pyridinyl group. The presence of these bands and the band shifts from 1550 and 990 cm^{-1} (in the free ligand) to 1575 and 1010 cm^{-1} indicate coordination of the symmetric tridentate ligand by the pyridine N atoms. The unsplit vibrations at 1100 to 1080 cm^{-1} and 615 cm^{-1} were ascribed to the ionic, noncoordinated perchlorate groups. The molar conductivity of a 10^{-3}M solution in acetonitrile at 25°C, $\Lambda = 311\ cm^2 \cdot \Omega^{-1} \cdot mol^{-1}$, indicates a 1:2 electrolyte.

Reference:

Boggess, R. K.; Zatko, D. A. (J. Coord. Chem. **4** [1975] 217/24).

40.1.6 With 1,2-Phenylenebis(dimethylarsane)

$1,2\text{-}C_6H_4[As(CH_3)_2]_2$ (= $C_{10}H_{16}As_2$)

[Mn^{II}($C_{10}H_{16}As_2$)$_2$X$_2$] (X = Cl, Br, I). A freshly prepared and filtered solution of anhydrous manganese(II) halide in THF was treated with the ligand (1:2 mole ratio). After stirring for 1 h, the solution was concentrated under reduced pressure and the white crystalline product was filtered, washed with THF, and dried in vacuum [1]. The complexes were also prepared from MnX_2 and the ligand in sodium-dried redistilled dioxane. The white precipitate obtained on cooling was separated by decantation and centrifuging, washed with dioxane and ether, and dried in vacuum [2]. The chloro complex, [Mn($C_{10}H_{16}As_2$)$_2$Cl$_2$], also formed in low yield from the reaction of $MnCl_3$ in ether solution with excess ligand. The crude product was recrystallized from dichloroethane-ether [3]. Magnetic moments (μ_{eff} in μ_B) from susceptibility measurements at room temperature, the band (in cm^{-1}) assigned to ν(Mn–X) in the far-IR spectrum (Nujol mulls), and the bands observed in the electronic reflectance spectra (in cm^{-1}; electronic transitions from $^6A_{1g}$) are tabulated below [1]:

complex	μ_{eff}	ν(Mn–X)	$\rightarrow {}^4T_{1g}$	$\rightarrow {}^4T_{2g}$	$\rightarrow {}^4A_{1g}, {}^4E_g$	$\rightarrow {}^4T_{2g}$
[Mn($C_{10}H_{16}As_2$)$_2$Cl$_2$]	6.00	305	19600	22800	23600	27400
[Mn($C_{10}H_{16}As_2$)$_2$Br$_2$]	5.96	210	19100	22500	~23800	27300
[Mn($C_{10}H_{16}As_2$)$_2$I$_2$]	5.99	—	18500	21500	23000	27000

Earlier data for magnetic moments of 6.1, 5.95, and 5.96 μ_B were reported for the complexes with X = Cl, Br, and I [2]. The IR spectra of the complexes reveal absorption bands in the 4000 to 500 cm^{-1} range, attributed to coordinated ligands and ruling out the presence of ligand oxides. The single ν(Mn–X) band in the far-IR is consistent with a *trans*-octahedral structure of the complexes. The observed bands in the electronic spectra were assigned on the basis of an essentially O_h geometry and the magnetic moments are indicative of a high-spin configuration. The EPR spectrum of the chloro complex recorded in frozen THF at ~115 K is characterized by a broad single line at $g_{eff} \approx 2$ with shoulders ($D < 0.05$ cm^{-1}; $\lambda \approx 0.033$). The EPR spectra of the bromo and iodo complex are characterized by lines at $g_{eff} \approx 6$ and 2, together with other lines. The lowest field lines were split further by nuclear hyperfine coupling to manganese with line spacing of ~107 G ($D \approx 0.21$ cm^{-1}; $\lambda \approx 0.033$). The low values of λ are consistent with a *trans*-octahedral structure of little departure from axial symmetry [1].

The complexes are very sensitive to hydrolysis by moisture and develop the repulsive odor of the free ligands within a few minutes. THF solutions of the compounds, open to air, subsequently show similar EPR spectra at room temperature, assuming the formation of a $[MnX_4]^{2-}$ species with X = Cl, Br, and I [1]. $[Mn(C_{10}H_{16}As_2)_2Br_2]$ and $[Mn(C_{10}H_{16}As_2)_2I_2]$ decompose on heating at 196 and 194°C, respectively [2]. Decomposition occurs also in alcohols, acetone, and nitromethane [1].

The complexes are slightly soluble in THF, dioxane, and nitrobenzene and almost insoluble in benzene, halocarbons, and diethyl ether [1]. They are nonelectrolytes in nitrobenzene [1, 2].

$[Mn^{III}(C_{10}H_{16}As_2)(H_2O)X_2]\ ClO_4$ (X = Cl, Br) complexes are claimed to result from the reaction of manganese(III) acetate, HX, $HClO_4$, ligand and a little more than the stoichiometric amount of water in ethanol-acetic acid-acetic anhydride solution at 0°C [2]. It seems probable that the trace of water converts the ligand to the corresponding bis(arsane oxide) to give $[Mn(C_{10}H_{16}As_2O_2)X_2]ClO_4$ having elemental analysis close to the postulated $[Mn(C_{10}H_{16}As_2)(H_2O)X_2]ClO_4$ complexes [1].

References:

[1] Jones, M. H.; Levason, W.; McAuliffe, C.A.; Parrott, M. J. (J. Chem. Soc. Dalton Trans. **1976** 1642/6).
[2] Nyholm, R. S.; Sutton, G. J. (J. Chem. Soc. **1958** 564/6).
[3] Levason, W.; McAuliffe, C.A. (J. Inorg. Nucl. Chem. **37** [1975] 340/2).

40.2 Complexes with Arsane Oxides

40.2.1 With Trimethylarsane Oxide $(CH_3)_3AsO$ (= C_3H_9AsO)

$[Mn^{II}(C_3H_9AsO)_5]\ (ClO_4)_2$ was obtained from a reaction of hydrated manganese(II) perchlorate with a slight excess of trimethylarsane oxide in hot acetone containing triethyl orthoformate as dehydrating agent. The white compound precipitated upon cooling the mixture, was washed with dry ether, and dried in high vacuum. The X-ray powder photographs show that the complex is isomorphous with the corresponding Ni^{II} and Co^{II} compounds, assumed to have a square-pyramidal structure.

The magnetic moment, $\mu_{eff} = 6.05$ μ_B, was determined by susceptibility measurements at room temperature, indicating a high-spin five-coordinated complex. The IR spectrum reveals absorption bands at 869 and 863 cm^{-1}, assigned to ν(AsO) and at 366 cm^{-1} to ν(Mn–O). An additional band was observed at 284 cm^{-1}. There is no evidence of coordinated anion groups. The complex is less sensitive to atmospheric moisture than its phosphane oxide analog, but decomposes in acetone solution [1].

$Mn^{II}(C_3H_9AsO)_4(ClO_4)_2$ separated from hot anhydrous ethanol, following the addition of ligand to manganese perchlorate (4:1 mole ratio) under anhydrous conditions. It is favored using a deficiency of ligand.

The X-ray powder photograph is similar to that of the corresponding Ni^{II} complex, for which a square-pyramidal geometry was proposed from its electronic spectrum. In the IR spectrum two absorption bands at 868 and 855 cm^{-1} are assigned to ν(AsO). The ν(Mn–O) frequency appears at 368 cm^{-1} and two other absorption bands at 306 and 279 cm^{-1}. Although there is no conspicuous splitting of the perchlorate bands, as usually observed for anion coordination, a pentacoordinate distorted square-pyramidal structure with one coordinated perchlorate group was suggested [1]. A binuclear oxygen-bridged structure is also discussed [2].

References:

[1] Brodie, A. M.; Hunter, S. H.; Rodley, G. A.; Wilkins, C. J. (Inorg. Chim. Acta **2** [1968] 195/8).
[2] Karayannis, N. M.; Mikulski, C. M.; Strocko, M. J.; Pytlewski, L. L.; Labes, M. M. (J. Inorg. Nucl. Chem. **33** [1971] 269/3).

40.2.2 With Triphenylarsane Oxide $(C_6H_5)_3AsO$ (= $C_{18}H_{15}AsO$)

40.2.2.1 Manganese(II) Compounds

$[Mn(C_{18}H_{15}AsO)_4X]X$ (X = I, ClO_4, NCS). The white complex $[Mn(C_{18}H_{15}AsO)_4NCS]NCS$ was obtained on mixing solutions of calculated quantities of manganese(II) salt and triphenylarsane oxide in ethanol [1]. The complex with X = ClO_4 was prepared from a reaction of Mn^{II} perchlorate and a slight excess of ligand in acetone or ethanol containing ethyl orthoformate [2]. Also, a reaction of $MnCl_2 \cdot 4\,H_2O$ and triphenylarsane oxide, dissolved in hot ethanol (containing a small amount of ethanol saturated with $NaClO_4$) yielded $[Mn(C_{18}H_{15}AsO)_4ClO_4]ClO_4$ as sparkling, colorless plates [3]. The iodo complex $[Mn(C_{18}H_{15}AsO)_4I]I$ was obtained as cream-colored crystals in an attempt to prepare $[Mn(C_{18}H_{15}AsO)_2I_2]$ from manganese(II) iodide and triphenylarsane oxide in hot ethanol [4], as were the corresponding $[Mn(C_{18}H_{15}AsO)_2X_2]$ complexes (see p. 208). The yellow $[Mn(C_{18}H_{15}AsO)_4I]I$, with a melting point of 230°C, was also obtained by decomposition of $[Mn(C_{18}H_{15}AsO)_4I][$*cis*-$Mn^{I}(CO)_4I_2]$, when exposed to light or refluxed in benzene, THF, acetone, or ethanol under nitrogen atmosphere [5].

The X-ray powder photographs show that $[Mn(C_{18}H_{15}AsO)_4ClO_4]ClO_4$ is isomorphous with the corresponding Co^{II} and Ni^{II} complexes; for the Ni^{II} compound a square-pyramidal geometry with an axial monodentate ClO_4 group was assumed [2]. Susceptibility measurements at room temperature yielded a magnetic moment of $\mu_{eff} = 6.02\ \mu_B$ for the perchlorato complex [3]. For the iodo complex magnetic moments of 5.99 μ_B [4] and 6.1 μ_B [5] were determined at room temperature. Characteristic vibration modes (in cm^{-1}) from the IR spectra in Nujol in the 4000 to 100 cm^{-1} region [1 to 4] and Raman spectra in the 450 to 100 cm^{-1} [1] region are shown below:

compound	anion bands	ν(AsO)[a]	ν(Mn–O) or metal-sensitive bands	Ref.
$[Mn(C_{18}H_{15}AsO)_4I]I$	—	870	380, 311, 304	[1]
$[Mn(C_{18}H_{15}AsO)_4ClO_4]ClO_4$	1104, 1036, 621, 614	872	380, 305, 302	[1, 3, 4]
		873	386	[2]
$[Mn(C_{18}H_{15}AsO)_4NCS]NCS$	2062	875	380, 306, 304	[1]

[a] ν(AsO) of the free ligand at 880 cm^{-1} [1, 2].

The perchlorato complex shows a splitting of the ν_3 and ν_4 anion bands at 1104 and 1036 and 621 and 614 cm^{-1}, respectively. According to deductions from the ESR spectrum, at least one perchlorate is bonded to manganese although no absorption band assignable with certainty to a metal-anion mode could be identified down to 100 cm^{-1} [1]. The reflectance spectrum of the solid $[Mn(C_{18}H_{15}AsO)_4I]I$ complex shows absorption bands at 27400, 28990, and 30300 cm^{-1} [4]. Band maxima in the electronic spectrum of the perchlorate complex in dichloromethane solution are at 23100, 23700, 28800, 29800, 30600, and 31500 (sh) cm^{-1} with extinction coefficients of $\varepsilon = 1.7$, 1.7, 7.0, 11.2, and 13.6 $L \cdot mol^{-1} \cdot cm^{-1}$, respectively [6]. The ESR X-band spectra of the complexes are very similar, showing a strong band near $g_{eff} = 6$ with many weaker bands extending to higher field. Distortion parameters (D) from 0.178 to 0.19 cm^{-1} and $\lambda = 0.100$ have been deduced from comparison with calculated spectra. These values are larger than that of the corresponding complexes with triphenylphosphane oxide, indicating a weaker metal–anion bonding in the $[Mn(C_{18}H_{15}AsO)_4X]X$ complexes. The complexes are formulated as five-coordinated compounds, analogous to the methyldiphenylarsane oxide complex, $[Mn(C_{13}H_{13}AsO)_4ClO_4]ClO_4$, see p. 212. Due to the close ESR parameters very similar ligand fields in the $[Mn(C_{18}H_{15}AsO)_4X]X$ complexes were suggested [1]. A binuclear, oxide-bridged structure is postulated for the perchlorato complex from IR data and stereochemical considerations [7]. $[Mn(C_{18}H_{15}AsO)_4ClO_4]ClO_4$ shows a molar electrical conductivity of $\Lambda = 43.8$ $cm^2 \cdot \Omega^{-1} \cdot mol^{-1}$ at 25°C in a 10^{-3} M nitrobenzene solution [3]. 10^{-3} M solutions of the corresponding iodo complex in nitromethane at 27.3°C and nitrobenzene at 26.4°C showed $\Lambda = 136$ and 32.3 $cm^2 \cdot \Omega^{-1} \cdot mol^{-1}$, respectively [4].

The perchlorato complex is soluble in ethanol, acetone, nitrobenzene, and nitromethane and insoluble in benzene [3]. $[Mn(C_{18}H_{15}AsO)_4I]I$ is soluble in cold methanol, nitromethane, and nitrobenzene and, on heating, in higher alcohols, acetone, and chlorobenzene. It is insoluble in benzene and cyclohexane and decomposes in hot dioxane [4].

The iodo complex reacts with $Mn(CO)_5I$ in THF according to: $[Mn(C_{18}H_{15}AsO)_4I]I + Mn(CO)_5I \rightleftharpoons [Mn(C_{18}H_{15}AsO)_4I][\textit{cis}\text{-}Mn^I(CO)_4I_2] + CO$. The reaction can be reversed by bubbling CO through the THF solution [5].

$[Mn(C_{18}H_{15}AsO)_4I][\textit{cis}\text{-}Mn^I(CO)_4I_2]$ was prepared from $Mn(CO)_5I$ and triphenylarsane oxide in THF at 60°C for 8 h. It was also obtained on reacting $Mn(CO)_5I$ and $[Mn(C_{18}H_{15}AsO)_4I]I$ in THF at 50°C for 6 h (see above). On cooling, yellow-orange crystals with a melting point of 182°C separated [5]. The complex will be treated in detail, together with other dihalotetracarbonyl manganates(I), in the volumes describing organometallic compounds.

$[Mn(C_{18}H_{15}AsO)_2X_2]$ (X = Cl, Br, I, NCS). The complexes with X = Cl, Br, and NCS were prepared from the appropriate manganese(II) salt and triphenylarsane oxide in a 1:2.2 mole ratio in hot absolute ethanol. On cooling, crystals were obtained, washed with ethanol, and dried in vacuum [4, 8]. For X = I this procedure yielded $[Mn(C_{18}H_{15}AsO)_4I]I$ (see p. 207) instead of $[Mn(C_{18}H_{15}AsO)_2I_2]$. The greenish yellow halo complexes were also obtained, like their corresponding phosphane oxide complexes (see pp. 90/4), by heating MnX_2 (dehydrated at 130°C in vacuum) and triphenylarsane in dry THF at 40°C for 3 d [9]. $[Mn(C_{18}H_{15}AsO)_2Cl_2]$ was also prepared by mixing a solution of $MnCl_2 \cdot 4\,H_2O$ and ligand in absolute ethanol and allowing the volume-reduced mixture to crystallize in a desiccator over $CaCl_2$ [3]. It was also formed by adding an excess of $MnCl_3$ in ether to a solution of triphenylarsane oxide in diethyl ether or THF [10].

X-ray powder data show that $[Mn(C_{18}H_{15}AsO)_2Cl_2]$ is isomorphous with the corresponding cobalt compound [11]. Melting points (in °C), magnetic moments at room temperature (in μ_B), and IR absorption bands of the ν(AsO) vibration mode (in cm^{-1}) for the $[Mn(C_{18}H_{15}AsO)_2X_2]$ complexes are given in the table:

X	m.p.	μ_{eff}	ν(AsO)[a]	Ref.	X	m.p.	μ_{eff}	ν(AsO)[a]	Ref.
Cl	—	5.90	—	[3]	Br	214	6.03	911, 883	[4]
	236	6.03	923, 892	[4]		212	—	910, 882	[9]
	230	—	920, 900	[9]		—	—	910, 880	[12]
	—	—	920, 894	[10]		—	—	912, 885	[14]
	—	—	894	[12]	I	193	—	898, 884	[9]

[a] ν(AsO) of the free ligand at 880 cm^{-1} [4, 12].

IR absorption bands at 370 cm^{-1} for $[Mn(C_{18}H_{15}AsO)_2Cl_2]$ and at 405 and 395 cm^{-1} for $[Mn(C_{18}H_{15}AsO)_2Br_2]$ were attributed to ν(Mn−O) [12]. Vibrations of the chloro complex, assigned to ν(Mn−Cl), are observed at 295 and 278 [9], at 311 and 287 [12], and at 310 cm^{-1} [13]. For the bromo complex ν(Mn−Br) was detected at 225 [9], at 230 [12], and at 243 and 237 cm^{-1} [13]. The ν(Mn−X) bands observed in this region are typical of complexes with tetrahedral structures [13]. The IR frequencies in the 1650 to 450 cm^{-1} range with detailed assignments for the bromo complex are given in [14].

The electronic reflectance spectra of the solid compounds reveal absorption maxima at ~20620, 22470, 23260, 26670, and 30300 cm^{-1} for $[Mn(C_{18}H_{15}AsO)_2Cl_2]$ and at ~21050, 22320, 23260, 26670, ~28820, and 30300 cm^{-1} for $[Mn(C_{18}H_{15}AsO)_2Br_2]$. The first three bands can be assigned to the $\rightarrow {}^4T_{1g}(G)$, $\rightarrow {}^4T_{2g}(G)$, and $\rightarrow {}^4A_{1g}$, ${}^4E_g(G)$ transitions arising from a ${}^6A_1(S)$ ground state in a tetragonal environment of the Mn^{II} ion. Due to the tetrahedral geometry the solid bromo complex shows fluorescence and triboluminescence on cooling to 80 K [4].

ESR X- and Q-band spectra of the $[Mn(C_{18}H_{15}AsO)_2X_2]$ complexes, powdered or diluted with the corresponding zinc compound (containing 0.5 to 1 mol% manganese), show band positions and hyperfine structure depending mainly on the X group. Good agreement between observed and calculated ESR transitions of both X-band and Q-band spectra was obtained. The band positions were not altered by diluting the complexes, and a similar environment for each manganese ion can be postulated. The g values and distortion parameters D (in cm^{-1}) and λ for $[Mn(C_{18}H_{15}AsO)_2X_2]$ complexes are listed below [8]:

complex	g_{eff}[a]	g_{eff}[b]	D	λ
$[Mn(C_{18}H_{15}AsO)_2Cl_2]$	4.3	—	0.290	0.244
$[Mn(C_{18}H_{15}AsO)_2Br_2]$	4.3	4.3	0.425	0.267
$[Mn(C_{18}H_{15}AsO)_2(NCS)_2]$	2	2	0.045	0.333

[a] From X-band spectrum at ~9 GHz. – [b] From Q-band spectrum at ~35 GHz.

A very strong signal near $g_{eff}=4.27$ for $[Mn(C_{18}H_{15}AsO)_2Br_2]$ was also reported in [15]. The λ values are close to 1/3, in accordance with calculated data for tetrahedral C_{2v} symmetry. From comparison with the analogous complexes of triphenylphosphane oxide and γ-picoline, it was assumed that steric factors (i.e., the sizes of the ligand and anion X) exert the most influence on the variation of the D values. The X-band spectra of the diluted $[Zn(Mn)(C_{18}H_{15}AsO)_2Cl_2]$ and $[Zn(Mn)(C_{18}H_{15}AsO)_2Br_2]$ complexes also show only the strong line near g = 4.3, but this is split into x, y, and z components [8].

$[Mn(C_{18}H_{15}AsO)_2Cl_2]$ and $[Mn(C_{18}H_{15}AsO)_2Br_2]$ are nonelectrolytes in nitrobenzene [3, 4]. They are slightly soluble in acetone, acetonitrile, nitrobenzene, and nitromethane [3, 4]. The chloro complex is also soluble in ethanol, insoluble in benzene, chloroform, and ether and is decomposed by cold water, turning colorless immediately [3].

[Mn($C_{18}H_{15}AsO$){($(CH_3)_3CCOO$)}$_2$]$_2$ was obtained by adding triphenylarsane oxide (1 mL) to a brown benzene solution of Mn{$(CH_3)_3CCOO$}$_2$ (9 mmol in 30 mL) and concentrating the mixture at 50 to 60°C under 10 Torr of argon until crystals appeared. They were washed with a benzene-hexane mixture and dried in vacuum. The IR spectrum (KBr pellets) reveals two vibrations at 1635 and 1440 cm^{-1}; these were assigned to ν_{as}(COO) and ν_s(COO) and indicate bridging carboxylato groups [16]. Susceptibility measurements at various temperatures (by the Faraday method) yielded a magnetic moment of $\mu_{eff} = 4.70\ \mu_B$ at 293 K, decreasing monotonically to 2.18 μ_B at 78 K [16, 17]. The data are characteristic of an antiferromagnetic dimeric compound. The dependence of μ_{eff} on temperature can be described by the Heisenberg-Dirac-van Vleck model for a system with two exchange-coupled ions of spin S = 5/2 [16]. The g factor of 2.005 (±3% error) and the exchange parameter $J = -16\ cm^{-1}$ were evaluated using the isotropic Hamiltonian for metal-metal interaction in dimeric molecules. The magnetic properties and the close coincidence of the IR spectrum with that of the corresponding cobalt(II) complex (the structure of which is known from X-ray data) indicate a dimeric structure of the so-called lantern type with four bridging carboxylic groups and two axial triphenylarsane oxide ligand molecules. It is easily oxidized to form a trinuclear oxo cluster, [Mn_3O{$(CH_3)_3CCOO$}$_7$] [16].

[Mn($C_{18}H_{15}AsO$)$_3$(SO_2)$_2$]I_2. An evacuated flask containing a solution of [Mn($C_{18}H_{15}AsO$)$_4$I]I in dry toluene was filled with gaseous SO_2, producing a color change from beige to orange. After stirring for 3 d the orange product was isolated by standard Schlenk techniques [18]. It was also prepared, analogously to the [Mn($C_{18}H_{15}OP$)$_4$(SO_2)$_2$]I_2 complex with triphenylphosphane oxide (see p. 95), by reacting SO_2 with a suspension of MnI_2 and ligand in propyl acetate [19]. The IR spectrum of the complex shows absorption bands of ν(AsO) at 875 cm^{-1} and of ν(SO) at 1248 and 1080 cm^{-1}. Heating [Mn($C_{18}H_{15}AsO$)$_3$(SO_2)$_2$]I_2 in vacuum to ca. 100°C yielded red-violet crystals of $C_{18}H_{15}As \cdot I_2$, determined by X-ray diffraction to be a rare complex with an As−I−I moiety. Thus, it is assumed that SO_2 is activated by the manganese bonded to it, causing reduction of triphenylarsane oxide and oxidation of iodide [19].

References:

[1] Goodgame, D. M. L.; Goodgame, M.; Hayward, P. J. (J. Chem. Soc. A **1970** 1352/6).
[2] Hunter, S. H.; Nyholm, R. S.; Rodley, G. A. (Inorg. Chim. Acta **3** [1969] 631/4).
[3] Phillips, D. J.; Tyree, S. Y. (J. Am. Chem. Soc. **83** [1961] 1806/10).
[4] Goodgame, D. M. L.; Cotton, F. A. (J. Chem. Soc. **1961** 3735/41).
[5] Sartorelli, U.; Canziani, F.; Garlaschelli, L. (Gazz. Chim. Ital. **104** [1974] 567/80).
[6] Ciampolini, M.; Mengozzi, C. (Gazz. Chim. Ital. **104** [1974] 1059/66).
[7] Karayannis, N. M.; Mikulski, C. M.; Strocko, M. J.; Pytlewski, L. L.; Labes, M. M. (J. Inorg. Nucl. Chem. **33** [1971] 2691/3).
[8] Dowsing, R. D.; Gibson, J. F.; Goodgame, D. M. L.; Goodgame, M.; Hayward, P. J. (J. Chem. Soc. A **1969** 1242/8).
[9] Casey, S., Levason, W.; McAuliffe, C.A. (J. Chem. Soc. Dalton Trans. **1974** 886/9).
[10] Levason, W.; McAuliffe, C.A. (J. Inorg. Nucl. Chem. **37** [1975] 340/2).

[11] Goodgame, D. M. L.; Goodgame, M.; Cotton, F. A. (Inorg. Chem. **1** [1962] 239/45).
[12] Rodley, G. A.; Goodgame, D. M. L.; Cotton, F. A. (J. Chem. Soc. **1965** 1499/505).
[13] Allan, J. R.; Brown, D. H.; Nuttall, R. H.; Sharp, D. W. A. (J. Inorg. Nucl. Chem. **27** [1965] 1305/9).
[14] Deacon, G. B.; Green, J. H. S. (Spectrochim. Acta A **25** [1969] 355/64).
[15] Dowsing, R. D.; Gibson, J. F.; Goodgame, D. M. L.; Goodgame, M.; Hayward, P. J. (Nature **219** [1968] 1037/8).

[16] Pasynskii, A. A.; Idrisov, T. Ch.; Suvorova, K. M.; Kalinnikov, V. T. (Koord. Khim. **2** [1976] 1060/8; Soviet J. Coord. Chem. **2** [1976] 813/9).
[17] Novotortsev, V. M.; Rakitin, Yu. V.; Pasynskii, A. A.; Kalinnikov, V. T. (Dokl. Akad. Nauk SSSR **240** [1978] 355/7; Dokl. Chem. Proc. Acad. Sci. USSR **238/243** [1978] 220/2).
[18] Beagley, B.; El-Sayrafi, O.; Gott, G. A.; Kelly, D. G.; McAuliffe, C.A.; Mackie, A. G.; MacRory, P. P.; Pritchard, R. G. (J. Chem. Soc. Dalton Trans. **1988** 1095/7).
[19] McAuliffe, C.A.; Beagley, B.; Gott, G. A.; Mackie, A. G.; MacRory, P. P.; Pritchard, R. G. (Angew. Chem. **99** [1987] 237/8).

40.2.2.2 Manganese(III) Compounds

[Mn($C_{18}H_{15}$AsO)$_3$Cl$_3$] was obtained on adding a cold solution of triphenylarsane oxide to a solution of $MnCl_3$ in ether (free from HCl and kept at −30°C) under dry conditions. The blue precipitate is sensitive to moisture but stable at room temperature, if kept in a closed vessel. It melts at 169°C [1, 2]. The magnetic susceptibilities, measured by the Faraday method at five different temperatures between 65 and 294 K, obey the Curie-Weiss law with $\Theta = 0.88$ K. A magnetic moment of $\mu_{eff} = 5.41$ μ_B has been calculated [2]. The IR spectrum (Nujol mulls) shows absorption bands at 830 cm^{-1} [1] or 870, 860, and 835 cm^{-1} [2], assigned to ν(AsO) vibrations shifted towards lower frequencies from the free ligand value (880 cm^{-1}). Further bands at 455 [1] or 498 and 455 cm^{-1} [2] were attributed to ν(Mn−O), those at 330 and 280 cm^{-1} [1, 2] to ν(Mn−Cl). The electronic spectrum of the complex dissolved in $CHCl_3$ shows an absorption maximum at 14710 cm^{-1} ($\varepsilon = 510$ $L \cdot mol^{-1} \cdot cm^{-1}$), assigned to the $^5B_{1g} \rightarrow {}^5B_{2g}$ transition in a distorted octahedral environment at Mn^{III} ion. The pyridine solution shows a single band at 23810 cm^{-1}, in accordance with its red color and probably due to the substitution of the ligand by the solvent molecules [1, 2]. The complex is a nonelectrolyte in nitromethane, acetone, or pyridine. It is soluble in common organic solvents with variations of color and hydrolyzes in water or alkaline solutions to form MnO_2. Solutions in acidic aqueous media change color from blue to pale pink [2].

[Mn($C_{18}H_{15}$AsO)$_2$Cl$_3$] was prepared from the dioxane solvate, $MnCl_3 \cdot 2\,C_4H_8O_2$, and triphenylarsane oxide in ethanol, analogously to the corresponding phosphane oxide compound (see p. 97). The blue precipitate is stable at room temperature and melts at 185 to 187°C with decomposition. Magnetic susceptibilities, measured by the Faraday method at temperatures between 295 and 65 K, obey the Curie-Weiss law with $\Theta = 0.84$ K. The resulting magnetic moments, $\mu_{eff} = 4.76$ to 4.82 μ_B, are independent of temperature and indicative of a high-spin $Mn^{III}(d^4)$ complex. The IR spectrum (Nujol mulls) reveals absorption bands at 860 and 840 cm^{-1}, assigned to ν(AsO). Bands at 440, 363, 343, 290, and 260 cm^{-1} are attributed to ν(Mn−O) and ν(Mn−Cl) stretching modes; however, they are not useful for structural assignments. Observed band maxima in the diffuse reflectance spectrum of the solid state are at 14300 and 29400 cm^{-1}; in dichloromethane solution at 14700 ($\varepsilon = 920$ $L \cdot mol^{-1} \cdot cm^{-1}$) and 30800 cm^{-1} ($\varepsilon = 2750$ $L \cdot mol^{-1} \cdot cm^{-1}$). The lower frequency band was ascribed to a d−d electron transition and the other band to a charge transfer. Conductivity measurements in nitromethane and nitrobenzene indicate that the pentacoordinated complex is a nonelectrolyte [3].

References:

[1] Contreras, E.; Riera, V.; Usón, R. (Inorg. Nucl. Chem. Letters **8** [1972] 287/91).
[2] Contreras, E.; Riera, V.; Usón, R. (Rev. Acad. Cienc. Exact. Fis. Quim. Nat. Zaragoza [2] **28** [1973] 43/66; C.A. **80** [1974] No. 43582).
[3] Usón, R.; Riera, V.; Ciriano, M. A.; Valderrama, M. (Transition Metal Chem. [Weinheim] **1** [1976] 122/6).

40.2.3 With Methyldiphenylarsane Oxide, $CH_3(C_6H_5)_2AsO$ ($=C_{13}H_{13}AsO$)

$[Mn^{II}(C_{13}H_{13}AsO)_4ClO_4]ClO_4$ was prepared by reaction of a slight excess of ligand with manganese perchlorate in anhydrous acetone or ethanol. The product was washed with solvent and ether and dried in vacuum [1, 2]. The X-ray powder photographs show that the complex is isomorphous with the corresponding Fe^{II}, Co^{II}, Ni^{II}, Zn^{II}, and Cu^{II} compounds [1].

A thermochemical study of the reaction $[M(H_2O)_6](ClO_4)_2+4\,C_{13}H_{13}AsO \rightarrow [M(C_{13}H_{13}AsO)_4ClO_4]ClO_4+6\,H_2O$ ($M=Mn^{II}$, Co^{II}, Ni^{II}, Cu^{II}, Zn^{II}) was done using a thermochemical cycle. The enthalpy changes for the reaction were determined by a calorimetric method in 50 vol% DMF. The stability of isomorphous complexes decreases in the order $Cu^{II}>Zn^{II}>Mn^{II}\approx Co^{II}>Ni^{II}$. These thermodynamic results are in agreement with those obtained by crystal field calculations based on spectroscopic data [3].

Susceptibility measurements yielded the magnetic moment, $\mu_{eff}=6.1\,\mu_B$ [2]. In the IR spectrum of the complex in Nujol, a marked splitting of the perchlorate bands enabled the assignment of two absorption bands at 1136 and 1046 cm^{-1} to coordinated perchlorate and a third band at 1092 cm^{-1} to ionic perchlorate [1, 2]. The complex shows a ν(AsO) band at 871 cm^{-1} (free ligand at 1171 cm^{-1}) and a $\nu(Mn\text{–}O_{ligand})$ band at 383 cm^{-1}. A pentacoordinate square-pyramidal structure with distortion towards a square-planar arrangement is suggested from IR data and comparison with the known crystal structure of the isomorphous Co^{II} complex [1]. The energy level for the five-coordinated Mn^{II} complex of C_{4v} symmetry was calculated and compared with its electronic spectrum. Absorption maxima (in cm^{-1}), recorded from the solid state and solution in dichloromethane, assigned to transitions from the $^6A_1(S)$ ground state, are given below [4]:

state	$\rightarrow {}^4E(G)$	$\rightarrow 2\,{}^4A_1(G)$	$\rightarrow {}^4E(P)$	$\rightarrow {}^4A_1(D)$	$\rightarrow {}^4A_2(P)$
solid	19600	23700	28200	29700	30400
CH_2Cl_2	20000(sh)	24000	28900	29900	30800(sh)

From these values the ligand field parameters $Dq=700\ cm^{-1}$ and $\beta=0.91$ were calculated [4].

$[Mn^{II}(C_{13}H_{13}AsO)_2X_2]$ (X = Cl, Br). The complexes were prepared from manganese(II) halide and ligand in absolute ethanol [5], as described previously for the synthesis of the corresponding complexes with triphenylarsane oxide, $[Mn^{II}(C_{18}H_{15}AsO)_2X_2]$, see pp. 208/9. The X-ray diffraction patterns suggest isomorphism of $Mn(C_{13}H_{13}AsO)_2Cl_2$ with the related complexes of Co^{II}, Ni^{II}, and Zn^{II} and of $Mn(C_{13}H_{13}AsO)_2Br_2$ with those of Fe^{II}, Co^{II}, and Zn^{II}. The essentially tetrahedral structure of these compounds is discussed. The IR spectra in Nujol reveal the following absorption bands (in cm^{-1}):

complex	$\varrho(As\text{–}CH_3)+\nu(As{=}O)$	ν(Mn–O)	ν(Mn–X)
$Mn(C_{13}H_{13}AsO)_2Cl_2$	896, 881, 874, 865, 847	395	298, 272
$Mn(C_{13}H_{13}AsO)_2Br_2$	880, 873, 859	397	236

The assigning of the ν(AsO) bands is complicated by the methyl rocking mode, which gives rise to bands in the 900 to 800 cm^{-1} region. The free ligand spectrum shows an absorption band at 875 cm^{-1} and a shoulder at 866 cm^{-1} which involve a mixture of (AsO) stretching and methyl rocking modes. As would be expected, the complexes show more than one band in this region. In the IR spectrum of the chloro complex both antisymmetric and symmetric ν(Mn–Cl) bands appear and are separated by 16 cm^{-1}. The ν(Mn–O) vibration modes were assigned to those bands which were neither ligand nor ν(Mn–X) absorptions [5].

References:

[1] Hunter, S. H.; Nyholm, R. S.; Rodley, G. A. (Inorg. Chim. Acta **3** [1969] 631/4).
[2] Lewis, J.; Nyholm, R. S.; Rodley, G. A. (Nature **207** [1965] 72/3).
[3] Dei, A.; Gatteschi, D.; Sacconi, L. (Gazz. Chim. Ital. **104** [1974] 1033/9).
[4] Ciampolini, M.; Mengozzi, C. (Gazz. Chim. Ital. **104** [1974] 1059/66).
[5] Rodley, G. A.; Goodgame, D. M. L.; Cotton, F. A. (J. Chem. Soc. **1965** 1499/505).

40.2.4 With Tribenzylarsane Oxide

$(C_6H_5-CH_2)_3AsO$ $(=C_{21}H_{21}AsO)$

$Mn^{II}(C_{21}H_{21}AsO)_nX_2$ ($n=4$ for $X=ClO_4$, $n=3$ for $X=NO_3$, and $n=2$ for $X=Br$. For preparation a solution of the required hydrated manganese(II) salt in absolute ethanol ($Mn(NO_3)_2$ in 50% aqueous solution) was added rapidly with stirring to the stoichiometric amount of the ligand (21% excess for $X=NO_3$, 5% excess for $X=Br$) in absolute ethanol. The oily phase which usually separated after several minutes could be crystallized by scratching with a glass rod. If the complex remained in solution, some of the solvent was evaporated slowly over $CaCl_2$. The crystalline product was washed with absolute ethanol and/or ether and dried in vacuum over $CaCl_2$. The composition of the complexes, the absorption bands (in cm^{-1}) of the IR spectra (Nujol mulls), and the magnetic moments (μ_{eff} in μ_B, resulting from susceptibility measurements by the Faraday method at 24.5 ± 0.5°C, corrected for diamagnetism) are listed below:

complex	ν(AsO)[a]	ν(Mn–O)	μ_{eff}
$Mn(C_{21}H_{21}AsO)_4(ClO_4)_2$	882, 870, 830	360 (sh)	6.04 ± 0.04
$Mn(C_{21}H_{21}AsO)_3(NO_3)_2$	882, 875, 849, 832	360 to 318	5.98 ± 0.13
$Mn(C_{21}H_{21}AsO)_2Br_2$	889, 873	385 to 370	6.38 ± 0.14

[a] Free ligand ν(AsO) band at 870 cm^{-1}.

The magnetic moments of the complexes are in accordance with a paramagnetic high-spin type. The apparent molecular weight of $Mn(C_{21}H_{21}AsO)_4(ClO_4)_2$, measured in acetonitrile by the isothermal molecular weight method, is approximately onethird (492) of the formula weight (1711). This indicates a 1:2 electrolyte, in agreement with $\Lambda_0=239.8\ cm^2\cdot\Omega^{-1}\cdot mol^{-1}$ in acetonitrile at 25 ± 0.1°C, extrapolated to zero concentration. The value of $\Lambda_0=123.2\ cm^2\cdot\Omega^{-1}\cdot mol^{-1}$ for the nitrate complex was discussed in terms of a 1:1 electrolyte; i.e., $[Mn(C_{21}H_{21}AsO)_3NO_3]NO_3$. Molar electrical conductance of the bromo compound in 9.35×10^{-4} M acetonitrile solution at 24.7°C, $\Lambda_M=34.0\ cm^2\cdot\Omega^{-1}\cdot mol^{-1}$, is indicative of a weak electrolyte.

Reference:

Parris, G. E.; Long, G. G. (J. Inorg. Nucl. Chem. **32** [1970] 1585/92).

40.2.5 With Bis(arsane Oxides)

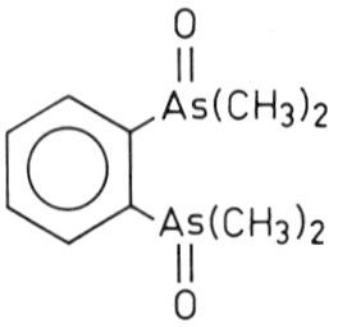

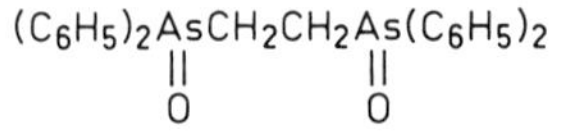

C_6H_4-1,2-$[As(O)(CH_3)_2]_2$ $(C_6H_5)_2As(O)CH_2CH_2As(O)(C_6H_5)_2$

ligand 1 $(= C_{10}H_{16}As_2O_2)$ ligand 2 $(= C_{26}H_{24}As_2O_2)$

$Mn^{II}(C_{10}H_{16}As_2O_2)Cl_2$ and $Mn^{II}(C_{26}H_{24}As_2O_2)Cl_2$. The complex with ligand 1, $Mn(C_{10}H_{16}As_2O_2)Cl_2$, was prepared by adding an excess of $MnCl_3$ in ether to a tetrahydrofuran solution of the corresponding diarsane. After standing overnight, the mixture was evaporated to dryness and the residue was extracted with hot dichloromethane. The insoluble solid (largely $MnCl_2$) was filtered off, the filtrate evaporated to a small volume and allowed to crystallize after standing for 2 to 3 d. The complex was recrystallized from dichloromethane [1]. The complex with ligand 2, $Mn(C_{26}H_{24}As_2O_2)Cl_2$,was obtained by reacting equimolar amounts of anhydrous $MnCl_2$ and corresponding diarsane in tetrahydrofuran. It melts at 225°C [2]. The IR spectra of the complexes show bands of ν(AsO) at 923 cm^{-1} for $Mn(C_{10}H_{16}As_2O_2)Cl_2$ [1] and at 886 and 870 cm^{-1} (free ligand at 885 and 865 cm^{-1}) for $Mn(C_{26}H_{24}As_2O_2)Cl_2$ [2]. The complex with ligand 1 does not hydrolyze in air [1].

References:

[1] Levason, W.; McAuliffe, C. A. (J. Inorg. Nucl. Chem. **37** [1975] 340/2).
[2] Casey, S.; Levason, W.; McAuliffe, C. A. (J. Chem. Soc. Dalton Trans. **1974** 886/9).

40.3 Complexes with Diorganylarsinic or Organylarsonic Acids

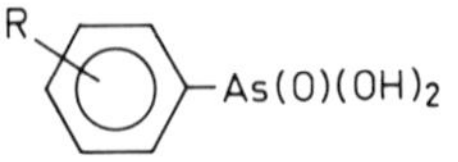

R-C_6H_4-$As(O)(OH)_2$ $(= H_2L)$

ligand	R	formula
1	H	$C_6H_7AsO_3$
2	2-NO_2	$C_6H_6AsNO_5$
3	3-NO_2	$C_6H_6AsNO_5$
4	4-NO_2	$C_6H_6AsNO_5$
5	4-Cl	$C_6H_6AsClO_3$
6	4-Br	$C_6H_6AsBrO_3$
7	2-OCH_3	$C_7H_9AsO_4$
8	4-OCH_3	$C_7H_9AsO_4$
9	4-CH_3	$C_7H_9AsO_3$
10	3-NO_2, 4-OH	$C_6H_6AsNO_6$

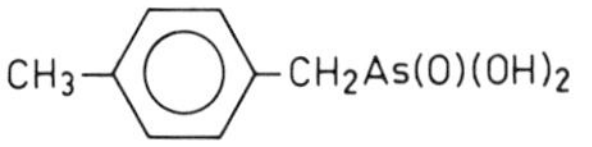

CH_3-C_6H_4-$CH_2As(O)(OH)_2$

ligand 11 $(C_8H_{11}AsO_3 = HL)$

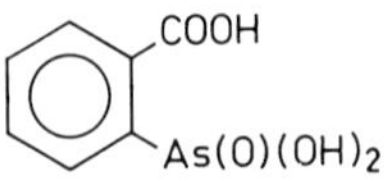

C_6H_4(COOH)-$As(O)(OH)_2$

ligand 12 $(C_7H_7AsO_5 = H_2L)$

Remark. The above ligand list contains only those substituted organylarsonic acids that have been described as forming manganese complexes of known composition. In addition to these the precipitation of manganese(II) complexes with **diphenylarsinic acid** and various mono-, di-, or trisubstituted derivatives of **phenylarsonic acids** in aqueous solution was investigated. The influence of steric effects on the pH value at the beginning of the precipitation is discussed, but the composition of the complexes is not given. They are therefore not treated in this volume. The references concerning these investigations are [1 to 9].

40.3.1 Manganese(II) Compounds

$Mn^{II}(C_8H_{10}AsO_3)_2$. The solubility product, $S = 1.259 \times 10^{-12}$ mol/L, of the complex with monodeprotonated ligand 11 is reported in [10].

$Mn^{II}L$ and $Mn^{II}L \cdot nH_2O$. Anhydrous complexes with ligands 1, 6, 8, and 9 and hydrated complexes with ligands 5 (n = 0.5), 2, 7, and 12 (n = 1) were prepared by refluxing alcoholic solutions of manganese(II) acetate and a slight excess of the appropriate ligand for 30 min. The resulting precipitate was washed with alcohol and dried in vacuum [11]. The complex with ligand 1, $Mn(C_6H_5AsO_3) \cdot H_2O$, precipitated immediately on mixing aqueous solutions of $MnCl_2$ and the ligand (or its mono- or disodium salt), regardless of the mole ratio of reactants. The complex was isolated by centrifuging, washing with methanol, and drying in vacuum [12]. The white $Mn(C_6H_4AsNO_6)$ complex was isolated from aqueous solutions of manganese(II) nitrate (50 mmol) and ligand 10 (60 mmol); aqueous KOH (1 molar) was added to reach a pH value of 3.5. The precipitate was washed free of nitrate and dried at 110°C [13]. X-ray powder patterns of various monohydrated metal(II) complexes with ligand 1, $M(C_6H_5AsO_3) \cdot H_2O$ with M^{II} = Mg, Mn, Co, Ni, Zn, and Cd, show that they form an isomorphous series [12]. The IR spectra of the complexes indicate bonding of the ligand to manganese through the oxygen of the arsonic acid moiety: The ν(AsO) frequencies for ligands 1, 2, 5 to 9, and 12 in the 935 to 845 cm^{-1} range are shifted downwards by 10 to 100 cm^{-1} in the spectra of the complexes [11]. The ligand bands at ~2800 cm^{-1} assigned to ν(OH) disappear on complexation, confirming the dibasic nature of the arsonic acids [11, 13]. From the comparison of the IR spectra of ligand 12 and its metal complexes it was concluded that the COOH group is still present in the complexes; i.e., that loss of both protons takes place from the arsonic group alone [11]. Reinvestigation of the IR data has shown that deprotonation of ligand 12 by metal ions takes place from both the COOH and $As(O)(OH)_2$ groups. Thus, the absorption bands in the spectrum of $Mn(C_7H_5AsO_5) \cdot H_2O$ at 1630 and 1620, 1300, 825, and 745 cm^{-1} were assigned to ν(C=O), ν(C–O), ν(AsO), and $\nu_s(AsO_2)$ vibration modes, respectively [14]. $Mn(C_7H_5AsO_5) \cdot H_2O$ exhibits two intense bands in the 3500 to 3000 cm^{-1} range and one band at ~1600 cm^{-1}; these are typical when a water molecule is present [12]. Transitions from the $^6A_{1g}$ ground state (in cm^{-1}), shown in the diffuse reflectance spectra of the complexes with ligand 1, are listed below [12]:

complex	$\to {}^4T_{1g}(G)$	$\to {}^4T_{2g}(G)$	$\to {}^4A_{1g}, {}^4E_g(G)$	$\to {}^4T_{2g}(D)$	$\to {}^4E_g(D)$	$\to {}^4T_{1g}(P)$
$Mn(C_6H_5AsO_3)$	18000	21980	24250	27150	29250	30250
$Mn(C_6H_5AsO_3) \cdot H_2O$	18691	—	23810	27397	29240	33220

The electronic spectra of complexes with ligands 1, 2, 5 to 9, and 12 show one band at 23800 cm^{-1} and one in the 18500 to 17300 cm^{-1} region, assigned to the $^6A_{1g} \to {}^4A_{1g}(G)$, $^4E_g(G)$ and $^6A_{1g} \to {}^4T_{1g}(G)$, $^4T_{2g}(G)$ transitions, respectively. The bands above 35000 cm^{-1} are due to the charge-transfer transitions [11]. For the complex with ligand 10, $Mn(C_6H_4AsNO_6)$, only a band at 17857 cm^{-1} was reported [13]. The electronic spectra indicate a distorted octahedral geometry with polymeric units [11 to 13]. It was suggested that layers consisting of octahedrally-coordinated manganese atoms are linked by multibridging acid groups and held together by weak van der Waals forces [12].

The magnetic moments (μ_{eff} in μ_B) of the complexes with ligands 1, 2, 5 to 10, and 12 measured at room temperature by the Gouy method, are given below [11]:

ligand	complex	μ_{eff}	ligand	complex	μ_{eff}
1	$Mn(C_6H_5AsO_3)$	4.31	8	$Mn(C_7H_7AsO_4)$	4.21
2	$Mn(C_6H_4AsNO_5) \cdot H_2O$	5.50	9	$Mn(C_7H_7AsO_3)$	4.10
5	$Mn(C_6H_4AsClO_3) \cdot 0.5\ H_2O$	4.19	10	$Mn(C_6H_4AsNO_6)$	4.2[a)]
6	$Mn(C_6H_4AsBrO_3)$	4.20	12	$Mn(C_7H_5AsO_5) \cdot H_2O$	5.87
7	$Mn(C_7H_7AsO_4) \cdot H_2O$	5.35			

[a)] From [13].

The magnetic moments are considerably lower than those expected for spin-only complexes with five unpaired electrons, due to metal-metal interaction in the polymeric structure. The higher values for the ortho-substituted ligands 2, 7, and 12 may be explained by steric hindrance [11]. $Mn(C_6H_5AsO_3)$ and $Mn(C_6H_5AsO_3) \cdot H_2O$ obey the Curie-Weiss law from 20 to −180°C. Magnetic moments from 5.49 to 4.93 μ_B and $\Theta = 54$ K for $Mn(C_6H_5AsO_3) \cdot H_2O$ also support antiferromagnetic interactions. Magnetic moments from 4.32 to 2.07 μ_B, the Weiss constant $\Theta = -624$ K, and a sharply defined Curie point at −123°C for $Mn(C_6H_5AsO_3)$ are characteristic of extensive antiferromagnetic interactions. It is probable that the anhydrous complex has a three-dimensional polymeric structure in which the manganese ions occupy octahedral sites [12].

Thermogravimetric analysis of $Mn(C_6H_5AsO_3) \cdot H_2O$ reveals that the water molecule is rapidly lost at 130°C. The next weight loss (at 460°C) probably indicates thermal decomposition [12]. The complex with ligand 10, $Mn(C_6H_4AsNO_6)$, decomposes at 250°C [13]. The anydrous complex with ligand 1, $Mn(C_6H_5AsO_3)$, is completely stable towards hydration [12] and all complexes with ligands 1, 2, 5 to 10, and 12 are insoluble in common organic solvents [11 to 13].

References:

[1] Pietsch, R. (Mikrochim. Acta **1957** 699/704).
[2] Pietsch, R. (Mikrochim. Acta **1957** 705/13).
[3] Pietsch, R. (Mikrochim. Acta **1955** 1019/25).
[4] Pietsch, R. (Mikrochim. Acta **1955** 954/61).
[5] Pietsch, R. (Mikrochim. Acta **1957** 161/6).
[6] Pietsch, R. (Mikrochim. Acta **1959** 861/9).
[7] Pietsch, R. (Mikrochim. Acta **1959** 854/60).
[8] Pietsch, R.; Ludwig, P. (Mikrochim. Acta **1963** 1007/13).
[9] Pietsch, R. (Mikrochim. Acta **1963** 1027/32).
[10] Zhu, Y. (Zhongnan Kuangye Xueyuan Xuebao **3** [1983] 29/38; C. A. **100** [1984] No. 124536).

[11] Sandhu, S. S.; Manhas, B. S.; Kohli, H. S. (J. Indian Chem. Soc. **53** [1976] 985/7).
[12] Cunningham, D.; Hennelly, P. J. D.; Decney, T. (Inorg. Chim. Acta **37** [1979] 95/102).
[13] Garcia, J.; Borras, J.; Molla, C.; Estevan, F. (Anales Quim. B **81** [1985] 143/6).
[14] Parmar, S. S.; Basra, T. S.; Bhasin, H. K.; Jain, A.; Sandhu, S. S. (Indian J. Chem. A **20** [1981] 669/72).

40.3.2 Manganese(III) Compounds

[MnL(OH)(H_2O)$_3$] and **[MnL(OH)(H_2O)$_3$]·nH_2O.** The complexes of the composition listed below were prepared by mixing methanolic solutions of $Mn(CH_3COO)_3 \cdot 2\,H_2O$ and the appropriate ligand in a 2:3 mole ratio and adding 0.5 to 1 mL of glacial acetic acid, except for ligand 1. In the case of ligand 8 the reaction was carried out in methanol-dioxane (2:1). The precipitate, whose formation was enhanced by the addition of a few drops of water, was washed with ethanol and dried in vacuum. Color, magnetic moments (μ_{eff} in μ_B at 28°C), and absorption maxima (in cm^{-1}) in the electronic spectra of the complexes are given in the table:

ligand	complex	color	μ_{eff}	ν_{max}
1	$[Mn(C_6H_5AsO_3)(OH)(H_2O)_3]$	gray	4.51	19680, 22020
2	$[Mn(C_6H_4AsNO_5)(OH)(H_2O)_3] \cdot 2H_2O$	gray	4.34	19490, 21800
3	$[Mn(C_6H_4AsNO_5)(OH)(H_2O)_3]$	gray	4.58	19530, 22320
4	$[Mn(C_6H_4AsNO_5)(OH)(H_2O)_3]$	gray	4.36	—
5	$[Mn(C_6H_4AsClO_3)(OH)(H_2O)_3] \cdot H_2O$	grayish brown	4.68	19420, 22120
6	$[Mn(C_6H_4AsBrO_3)(OH)(H_2O)_3]$	grayish brown	4.62	19340, 21930
7	$[Mn(C_7H_7AsO_4)(OH)(H_2O)_3]$	brown	4.38	—
8	$[Mn(C_7H_7AsO_4)(OH)(H_2O)_3]$	brown	4.40	19350, 22000

In the IR spectra of the complexes recorded as KBr disks or Nujol mulls the absorption bands at ~2700 and 2300 cm^{-1}, assigned to ν(OH) in the free ligand spectra disappeared, indicating both deprotonation of the OH group and its coordination to the manganese(III) ion via a bidentate bridge. This is supported by the negative shift (ca. −15 cm^{-1}) of the absorption bands at 885 and 775 cm^{-1}, attributed to the vibration of the AsO_3 group, and by the positive shift of the ν(As−ring) vibration mode. The appearance of a broad band at ~3380 cm^{-1} has been tentatively assigned to coordinated water molecules and confirmed by TGA (see below). The ν(Mn–O) band appears at ~480 cm^{-1} in all complexes. The observed absorption bands in the electronic spectra indicate the same environment for the complexes with ligands 1 to 8 and a band splitting due to the Jahn-Teller effect has been detected. A weak shoulder at 38000 cm^{-1} in the UV region has been assigned to π–π^* transitions of the aromatic ring and the nitro group (ligands 2 to 4). The complexes show low magnetic moments at room temperature. The compounds with ligands 1, 2, 3, and 5 have been studied down to liquid nitrogen temperature, showing decrease of the magnetic moments with lowering of the temperature. The plot of the reciprocal molar susceptibilities ($^1/\chi$) vs. the absolute temperature (T) is linear for complex with ligand 1 ($\Theta = -25$ K), 2 ($\Theta = -40.6$ K) and 3 ($\Theta = -26.5$ K). The plot of μ_{eff} vs. T is linear only for the complex with ligand 1. The magnetic behavior of the complexes underlines the existence of magnetic interactions due to oxygen-bridging arsonic acid groups, rather than oxo or hydroxo groups. Considering the tetrahedral geometry of the arsonic acid group and the stoichiometry of the complexes, a distorted octahedral structure is assumed for the Mn^{III} compounds. In each case the TGA in the 30 to 800°C temperature range shows an endothermic peak in the 105 to 165°C region (starting at 50°C for the complex with ligand 2), indicating the loss of all water and an exothermic peak due to complete decomposition into MnO_2 and Mn_2O_3.

The complexes are nonelectrolytes. They are stable towards dilute mineral acids, except hydrochloric acid, and are sparingly soluble in polar organic solvents.

Reference:

Sen, D.; Sen, S. S.; Panda, A. K.; Pramanik, P. (Indian J. Chem. A **23** [1984] 582/5).

40.4 Complex with Dimethylarsinodithioic Acid $(CH_3)_2As(S)SH$ (= $C_2H_7AsS_2$ = HL)

$Mn^{II}(C_2H_6AsS_2)_2$ was obtained as a pale pink-white precipitate by mixing hot ethanolic solutions of $MnCl_2 \cdot 4H_2O$ and the sodium salt of the ligand. The IR spectra recorded as KBr disks and Nujol or Vaseline mulls in the 1500 to 250 cm^{-1} region show absorption bands at 1395, 1267, 1254, 912, 873, 840, 820, 618, 596, 449, 424, 298, and 282 cm^{-1}. The frequencies between 1395 and 596 cm^{-1} are associated with the vibration and deformation modes of a $(CH_3)_2As$ fragment. The bands at 449 and 424 cm^{-1} are assigned to ν(AsS), and those at 298 and 282 cm^{-1} to ν(Mn–S) vibration modes in the four-atom chelate ring, AsS_2Mn [1]. The diffuse reflectance spectrum of the complex at 80 K exhibits bands at 25600, 23600, 20800, 21500, 18900, and 16500 cm^{-1}, assigned to electron transitions $^6A_{1g} \rightarrow {}^4E_g(D)$, $\rightarrow {}^4T_{2g}(D)$, $\rightarrow {}^4E_g(G) + {}^4A_{1g}(G)$, $\rightarrow {}^4T_{2g}(G)$, and $^4T_{1g}(G)$, respectively. The ligand field parameters 10 Dq = 6800 and 7000 cm^{-1} can be estimated from the energies of the $^4T_{1g}(G)$ and $^4T_{2g}(G)$ level, respectively, assuming octahedral coordination of manganese by six sulfur atoms, similar to MnS. A splitting of the band near 21300 cm^{-1} into two components (20800 and 21500 cm^{-1}) at 80 K indicates that the $^4E_g(G)$ and $^4A_{1g}(G)$ terms differ in energy due to some lowering of the symmetry below O_h [2]. Octahedral geometry is supported by the ESR spectrum of the powdered complex, which is a simple, symmetrical line centered at g = 2.007 [2]. Also, a singlet with g = 2.024 and ΔH = 200 Oe in the solid state or g = 2.003 and ΔH = 96 Oe from dichloromethane solution was found by [3]. The magnetic susceptibilities show Curie-Weiss behavior at liquid nitrogen temperatures with $\mu_{eff} = 5.92\ \mu_B$ and $\Theta = -12$ K. The low uncorrected value of the magnetic moments, $\mu_{eff} = 5.79$ and 5.55 μ_B at 291 and 92 K, respectively, together with $\Theta = -12$ K reflect the presence of weak antiferromagnetic exchange in the compound. It is assumed that adjoining Mn^{2+} ions are bridged by the uninegative ligand groups, providing a superexchange pathway for the spin interaction [2].

The complex is insoluble in chloroform, but soluble in polar solvents such as water and ethanol. A molar electrical conductance of a 1.15×10^{-2} M aqueous solution, $\Lambda = 150$ $cm^2 \cdot \Omega^{-1} \cdot mol^{-1}$, indicates a 1:2 electrolyte [2].

References:

[1] Casey, A. T.; Ham, N. S.; Mackey, D. J.; Martin, R. S. (Australian J. Chem. **23** [1970] 1117/23).
[2] Casey, A. T.; Mackey, D. J.; Martin, R. L. (Australian J. Chem. **24** [1971] 1587/98).
[3] Baratova, Z. R.; Solozhenkin, P. M.; Semenov, E. V. (Dokl. Akad. Nauk SSSR **236** [1977] 1137/9; Dokl. Phys. Chem. Proc. Acad. Sci. USSR **232/237** [1977] 977/9).

41 Complexes with Ligands Containing Antimony or Tin

Remark. Complexes with stannanylphosphanes are described on pp. 73/4.

41.1 With Organylstibanes

$(C_6H_5)_3Sb$ (= $C_{18}H_{15}Sb$)

ligand 1

C_6H_4-1-$As(CH_3)_2$-2-$Sb(CH_3)_2$ (= $C_{10}H_{16}AsSb$)

ligand 2

$Mn(NO)_2(C_{18}H_{15}Sb)_2I$. The brown complex was obtained analogously to the corresponding arsane compound, see p. 202, by introducing dry NO gas through the solution of $Mn(CO)_3$-$(C_{18}H_{15}As)_2I$ in benzene for 15 h [1].

$Mn(NO)_3(C_{18}H_{15}Sb)$ was prepared by passing dry NO gas through a solution of $Mn(CO)_4(C_{18}H_{15}Sb)I$ in benzene for 3 h at 80°C. After filtering and removing the solvent, the green, oily product was treated with pentane. The resulting dark green crystals were recrystallized from alcohol then washed with pentane and ether [1]. A cyclohexane solution of the complex shows IR absorption bands at 1788 and 1696 cm^{-1}, assigned to ν(NO). The comparison of the calculated force constant, f_{NO} = 13.13 mdyn/Å, with those of corresponding phosphane and arsane complexes shows an increase of f_{NO} in the order $C_{18}H_{15}P < C_{18}H_{15}As < C_{18}H_{15}Sb$ [2].

$[Mn(C_{10}H_{16}AsSb)_2X_2]$ complexes with X = Cl, Br, I were obtained as white or cream-colored crystals from tetrahydrofuran solution of anhydrous manganese(II) halides and ligand 2. The preparation and properties of the complexes are the same as for the corresponding complexes with 1,2-phenylenebis(dimethylarsane), $[Mn(C_{10}H_{16}As_2)_2X_2]$, described on pp. 205/6. Magnetic moments (in μ_B) measured at room temperature, ν(Mn–X) absorption bands (in cm^{-1}) in the IR spectra, and bands observed in the diffuse reflectance spectra (in cm^{-1}), assigned to the transitions from $^6A_{1g}$, are given in the following table:

complex	μ_{eff}	ν(Mn–X)	$\rightarrow ^4T_{1g}$	$\rightarrow ^4T_{2g}$	$\rightarrow ^4A_{1g}, ^4E_g$	$\rightarrow ^4T_{eg}$
$[Mn(C_{10}H_{16}AsSb)_2Cl_2]$	6.03	287	~19600	22000	23200	26400
$[Mn(C_{10}H_{16}AsSb)_2Br_2]$	6.00	215	19100	21800	~23400	26000
$[Mn(C_{10}H_{16}AsSb)_2I_2]$	6.12	—	~18700	21000	22400, 23800	—

The EPR spectra of the complexes measured at 115 K in frozen tetrahydrofuran solutions show a single line at g_{eff} ~2 for the chloro complex and lines at g_{eff} ~6 and ~2 for the bromo and iodo compound [3].

References:

[1] Hieber, W.; Tengler, H. (Z. Anorg. Allgem. Chem. **318** [1962] 136/54).
[2] Beck, W.; Lottes, K. (Chem. Ber. **98** [1965] 2657/73).
[3] Jones, M. H.; Levason, W.; McAuliffe, C. A.; Parrott, M. J. (J. Chem. Soc. Dalton Trans. **1976** 1642/6).

41.2 With Bis(tributyltin) Oxide, $\{(C_4H_9)_3Sn\}_2O$ (= $C_{24}H_{54}OSn_2$)

$Mn(C_{24}H_{54}OSn_2)_6(ClO_4)_2$ and **$Mn(C_{24}H_{54}OSn_2)_4X_2$** (X = NO_3, Cl, Br, I, NCS). The complexes were obtained on addition of the appropriate anhydrous manganese salt (dehydrated with 2,2-dimethoxypropane) in absolute ethanol to a methanolic solution of the ligand. After stirring the mixture for 3 h, the precipitate was filtered, washed with methanol, and dried in vacuum.

Magnetic moments (in μ_{eff}) and absorption bands in the IR spectrum (in cm^{-1}) are shown below:

No.	complex	μ_{eff}	ν(Sn–O–Sn)	ν(Mn–O)
I	$Mn(C_{24}H_{54}OSn_2)_6(ClO_4)_2$	6.0	730	370
II	$Mn(C_{24}H_{54}OSn_2)_4(NO_3)_2$	5.85	750	340
III	$Mn(C_{24}H_{54}OSn_2)_4Cl_2$	5.9	740	350
IV	$Mn(C_{24}H_{54}OSn_2)_4Br_2$	5.8	760	345
V	$Mn(C_{24}H_{54}OSn_2)_4I_2$	5.9	730	360
VI	$Mn(C_{24}H_{54}OSn_2)_4(NCS)_2$	5.9	750	340

The observed shifts of ν(Sn–O–Sn) in comparison to the ν(Sn–O–Sn) band of the free ligand (at 783 cm^{-1}) show that manganese is coordinated through the oxygen atom. Bands at 1080 and 620 cm^{-1} in the spectrum of the perchlorate complex indicate that the anion is not coordinated. A ν(Mn–Cl) vibration at 310 cm^{-1} was observed in the spectrum of the chloro complex. An intense ν(CN) band at 2050 cm^{-1} reveals bonding of the thiocyanate group through the nitrogen atom. Absorption bands (ν_{max} in cm^{-1}) in the electronic spectra of the complexes assigned to transitions from the $^6A_{1g}$ ground state, together with calculated ligand field parameters 10 Dq, B and C (in cm^{-1}) and β, are given in the table:

complex	λ_{max}	10 Dq	B	C	β
I	15150, 23200, 25300, 31000	694	631	3398	0.66
II	14928, 23000, 25250, 30900	684	622	3806	0.65
III	14925, 23250, 25250, 31250	684	622	3806	0.65
IV	14810, 23200, 25100, 31000	679	617	3786	0.74
V	14705, 23150, 25000, 30750	674	613	3774	0.64
VI	14705, 22850, 25200, 20700	674	613	3814	0.64

The compounds are stable at room temperature but decompose on heating beyond 200°C. They are insoluble in polar and nonpolar solvents, they are toxic and allergenic towards human skin.

Reference:

Srivasta, M.; Srivastava, A. K.; Agarwal, R. K. (Proc. Indian Natl. Sci. Acad. A **50** [1984] 249/55; C.A. **102** [1985] No. 6755).

Ligand Formula Index

The ligands treated in this volume are arranged in the index according to the system of Hill, A. (J. Am. Chem. Soc. **22** [1900] 478/94). In this system, the first criterion for the location of a ligand is the number of carbon atoms. The second is the number of hydrogen atoms, and finally the number of atoms of the other elements, in alphabetical order.

The first column contains the empirical formulas of the ligands. The second column shows their linearized sructural formulas. Additional ligands of mixed ligand complexes and of adducts are placed in subheadings. The last column lists the pertinent pages.

The formulas of ligands coordinated as deprotonated acids are given in their nondeprotonated form. Ligands occurring in tautomeric equilibria are presented only in the form commonly known. The numbering of locants for substituents corresponds to IUPAC rules.

List of abbreviations for ligands or solvents used in the volume:

bpy	2,2′-bipyridine	phen	1,10-phenanthroline
dmf = DMF	dimethylformamide	py	pyridine
DMSO	dimethyl sulfoxide	thf = THF	tetrahydrofuran

C_8

Physical Constants and Conversion Factors

Avogadro constant N_A (or L) = 6.02214×10^{23} mol^{-1}
Faraday constant F = 9.64853×10^{4} C/mol
molar gas constant R = 8.31451 $J \cdot mol^{-1} \cdot K^{-1}$
molar volume (ideal gas) $V_m = 2.24141 \times 10^{1}$ L/mol
(273.15 K, 101325 Pa)

Planck constant $h = 6.62608 \times 10^{-34}$ J·s
elementary charge $e = 1.60218 \times 10^{-19}$ C
electron mass $m_e = 9.10939 \times 10^{-31}$ kg
proton mass $m_p = 1.67262 \times 10^{-27}$ kg

1 kg = 2.205 pounds
1 m = 3.937×10^{1} inches = 3.281 feet
1 m^3 = 2.642×10^{2} gallons (U.S.)
1 m^3 = 2.200×10^{2} gallons (Imperial)

Force	N	dyn	kp
1 N	1	10^{5}	1.019716×10^{-1}
1 dyn	10^{-5}	1	1.019716×10^{-6}
1 kp	9.80665	9.80665×10^{5}	1

Pressure	Pa	bar	kp/m^2	at	atm	Torr	lb/in^2
1 Pa = 1 N/m^2	1	10^{-5}	1.019716×10^{-1}	1.019716×10^{-5}	9.86923×10^{-6}	7.50062×10^{-3}	1.450378×10^{-4}
1 bar = 10^6 dyn/cm^2	10^{5}	1	1.019716×10^{4}	1.019716	9.86923×10^{-1}	7.50062×10^{2}	1.450378×10^{1}
1 kp/m^2 = 1 mm H_2O	9.80665	9.80665×10^{-5}	1	10^{-4}	9.67841×10^{-5}	7.35559×10^{-2}	1.422335×10^{-3}
1 at (technical)	9.80665×10^{4}	9.80665×10^{-1}	10^{4}	1	9.67841×10^{-1}	7.35559×10^{2}	1.422335×10^{1}
1 atm = 760 Torr	1.01325×10^{5}	1.01325	1.033227×10^{4}	1.033227	1	7.60×10^{2}	1.469595×10^{1}
1 Torr = 1 mm Hg	1.333224×10^{2}	1.333224×10^{-3}	1.359510×10^{1}	1.359510×10^{-3}	1.315789×10^{-3}	1	1.933678×10^{-2}
1 lb/in^2 = 1 psi	6.89476×10^{3}	6.89476×10^{-2}	7.03069×10^{2}	7.03069×10^{-2}	6.80460×10^{-2}	5.17149×10^{1}	1

Work, Energy, Heat	J	kW·h	kcal	Btu	eV
1 J = 1 W·s = 1 N·m = 10^7 erg	1	2.778×10^{-7}	2.39006×10^{-4}	9.4781×10^{-4}	6.242×10^{18}
1 kW·h	3.6×10^6	1	8.604×10^2	3.41214×10^3	2.247×10^{25}
1 kcal	4.1840×10^3	1.1622×10^{-3}	1	3.96566	2.6117×10^{22}
1 Btu (British thermal unit)	1.05506×10^3	2.93071×10^{-4}	2.5164×10^{-1}	1	6.5858×10^{21}
1 eV	1.602×10^{-19}	4.450×10^{-26}	3.8289×10^{-23}	1.51840×10^{-22}	1

1 cm^{-1} = 1.239842×10^{-4} eV
1 hartree = 27.2114 eV

1 Hz = 4.135669×10^{-15} eV
1 eV $\triangleq$ 23.0578 kcal/mol

Power	kW	hp	$kp \cdot m \cdot s^{-1}$	kcal/s
1 kW = 10^3 J/s	1	1.35962	1.01972×10^2	2.39006×10^{-1}
1 hp (horsepower, metric)	7.3550×10^{-1}	1	7.5×10^1	1.7579×10^{-1}
1 $kp \cdot m \cdot s^{-1}$	9.80665×10^{-3}	1.333×10^{-2}	1	2.34384×10^{-3}
1 kcal/s	4.1840	5.6886	4.26650×10^2	1

References:

Mills, I. (Ed.), International Union of Pure and Applied Chemistry, Quantities, Units and Symbols in Physical Chemistry, Blackwell Scientific Publications, Oxford 1988.
The International System of Units (SI), National Bureau of Standards Spec. Publ. 330 [1972].
Landolt-Börnstein, 6th Ed., Vol. II, Pt. 1, 1971, pp. 1/14.
ISO Standards Handbook 2, Units of Measurement, 2nd Ed., Geneva 1982.
Cohen, E. R., Taylor, B. N., Codata Bulletin No. 63, Pergamon, Oxford 1986.